GENERALIZED MEASURE THEORY

International Federation for Systems Research
International Series on Systems Science and Engineering

Volume 20 *FUZZY RELATIONAL SYSTEMS: Foundations and Principles*
Radim Bělohlávek

Volume 21 *ARCHITECTURE OF SYSTEMS PROBLEMS SOLVING, Second Edition*
George J. Klir and Doug Elias

Volume 22 *ORGANIZATION STRUCTURE: Cybernetic Systems Foundation*
Yasuhiko Takahara and Mihajlo Mesarovic

Volume 23 *CONSTRAINT THEORY: Multidimensional Mathematical Model Management*
George J. Friedman

Volume 24 *FOUNDATIONS AND APPLICATIONS OF MIS: Model Theory Approach*
Yasuhiko Takahara and Yongmei Liu

Volume 25 *GENERALIZED MEASURE THEORY*
Zhenuyan Wang and George J. Klir

IFSR was established "to stimulate all activities associated with the scientific study of systems and to coordinate such activities at international level." The aim of this series is to stimulate publication of high-quality monographs and textbooks on various topics of systems science and engineering. This series complements the Federation's other publications.

A Continuation Order Plan is available for this series. A continuation order will bring delivery of each new volume immediately upon publication. Volumes are billed only upon actual shipment. For further information please contact the publisher.

Volumes 1–6 were published by Pergamon Press.

GENERALIZED MEASURE THEORY

Zhenyuan Wang
George J. Klir

Springer

Zhenyuan Wang
Department of Mathematics
University of Nebraska at Omaha
Omaha, NE 68182-0243
U.S.A
zhenyuanwang@mail.unomaha.edu

George J. Klir
Thomas J. Watson School
of Engineering and Applied Sciences
Department of Systems Science
and Industrial Engineering
Binghamton University
Binghamton, NY 13902
U.S.A
gklir@binghamton.edu

ISBN 978-1-4419-4576-1 e-ISBN 978-0-387-76852-6

DOI: 10.1007/978-0-387-76852-6

Mathematics Subject Classification (2000): 28-01, 28E-05, 28E-10, 28A-12, 28A-25, 28B-15

Printed on acid-free paper

springer.com

Preface

In 1992 we published a book entitled *Fuzzy Measure Theory* (Plenum Press, New York), in which the term "fuzzy measure" was used for set functions obtained by replacing the additivity requirement of classical measures with weaker requirements of monotonicity with respect to set inclusion and continuity. That is, the book dealt with nonnegative set functions that were monotone, vanished at the empty set, and possessed appropriate continuity properties when defined on infinite sets.

It seems that *Fuzzy Measure Theory* was the only book available on the market at that time devoted to this emerging new mathematical theory. Some ten years after its publication we began to see that the subject had expanded so much that a second edition of the book, or even a new book on the subject, was needed. We eventually decided to write a new book because the new material we wished to include was too extensive for—and far beyond the usual scope—of a second edition. More importantly, we felt that some fundamental changes regarding this topic's scope and terminology would be desirable and timely.

As far as the scope of the new book, *Generalized Measure Theory*, is concerned, we felt, on the basis of recent developments in the literature, that the material should not be restricted to set functions that had to be nonnegative and monotone. Rather, it needed to capture a broader class of set functions; a function in this class would have only one requirement to qualify as a "measure": it would vanish at the empty set. Then, various special requirements could be introduced as needed to restrict this broad class of set functions to specialized subclasses. One of these subclasses would consist of nonnegative, monotone, and continuous set functions that vanish at the empty set—or fuzzy measures—the subject of our previous book.

Regarding terminology, it was obvious that we needed to revise it completely in view of the expanded scope of the book. First, we had to introduce a name for the most general measures. We did so by referring to nonnegative set functions that vanish at the empty set as *general measures* and referring to those that are not required to be nonnegative as *signed general measures*. Second, we needed to introduce appropriate names of the various subclasses of general measures or signed general measures. This we did in Chapters 3 and 4, where we followed, by and large, the terminology established in the literature. However, it should be emphasized that we made a deliberate decision to abandon the central term of our previous book, the term "fuzzy measure." We judge this term to be highly misleading. Indeed, the so-called fuzzy measures do not involve any fuzziness. They are just special set functions that are defined on specified classes of classical sets, not on classes of fuzzy sets. Since the primary characteristic of such functions is monotonicity, we deemed it reasonable to call these set functions *monotone measures* rather than fuzzy measures.

However, contrary to the concept of fuzzy measures in our previous book, monotone measures as understood in *Generalized Measure Theory* need not be continuous. If, in fact, they are continuous then they are here specifically referred to as *continuous monotone measures*. Moreover, if they are only semicontinuous from below or from above, then they are called, respectively, *lower-semicontinuous* or *upper-semicontinuous monotone measures*. Clearly, any continuous monotone measure is both lower-semicontinuous and upper-semicontinuous.

There is another reason why abandoning the term "fuzzy measure" is justified: It is certainly meaningful to fuzzify any class of measures, as we show in Chapter 14. A given class of measures is "fuzzified" when it is defined on fuzzy sets rather than on classical sets. However, the resulting term—"fuzzified fuzzy measures" we find awkward, not properly descriptive, and quite confusing. For all these reasons, we decided to replace the term "fuzzy measure" with "continuous monotone measure" and to use the term "monotone measure" when continuity or even semicontinuity is not required. When they *are* fuzzified we refer to these measures as "fuzzified monotone measures." When measures of any other type are defined on classes of fuzzy sets we refer to them as *fuzzified measures* of the respective type. We thus use names such as *fuzzified general measures, fuzzified monotone measures, fuzzified continuous monotone measures*, and the like.

We realize it is not likely that the confusing term "fuzzy measures" for "measures defined on classes of crisp sets" will soon disappear in the literature. However, we are confident that the time is ripe to stop using it. In a sense we have joined some major contributors to generalized measure theory who have already abandoned this ill-descriptive term.

We have made in this book a few additional terminological changes with respect to our previous book. However, all these changes affect special concepts, so we explain our rationale for making these changes as we introduce each concept.

Our previous book contains, in addition to its original material, six of our reprinted papers. In this book, no reprinted papers are included. Instead the original material is substantially expanded. Major expansions are in the area of integration, methods for constructing generalized measures, fuzzification of generalized measures, and applications of generalized measure theory.

Much like our previous book, this book is primarily a text for a one-semester graduate or upper division course. Such a course is suitable not only for programs in mathematics, where it might be offered at the junior or senior level, but also for programs in numerous other areas. These would include systems science, computer science, information science, and cognitive sciences, as well as artificial intelligence, quantitative management, mathematical social sciences, and virtually all areas of engineering and natural sciences. The book may also be useful for researchers in these areas.

Although a solid background in mathematical analysis is required for understanding the material presented, the book is otherwise self-contained. This is achieved by the inclusion of needed prerequisites regarding classical sets, classical measures, and fuzzy sets, as given in Chapter 2. In general, the book is written in the textbook style, characterized by generous use of examples and exercises. Each chapter concludes with notes containing relevant historical, bibliographical, and other remarks relating to the covered material, which are useful for further study of generalized measure theory and its applications. Compared with our previous book, the bibliography of *Generalized Measure Theory* is substantially expanded. Two glossaries are included for convenience of the reader, Glossary of Key Concepts (Appendix A) and Glossary of Symbols (Appendix B).

Omaha, Nebraska, USA Zhenyuan Wang
Binghamton, New York, USA George J.Klir

Contents

List of Figures

List of Tables

Chapter 1
Introduction

Generalized measure theory, which is the subject of this book, emerged from the well-established *classical measure theory* by the process of generalization. As is well known, classical measures are nonnegative real-valued set functions, each defined on a specific class of subsets of a given universal set, that satisfy certain axiomatic requirements. One of these requirements, crucial to classical measures, is known as the requirement of *additivity*. This requirement is basically that the measure of the union (finite or countably infinite) of any recognized family of sets that are pairwise disjoint be equal to the sum of measures of the individual sets in the union. In generalized measure theory, the additivity requirement is replaced with a considerably weaker requirement. Any real-valued set function μ on a given class of sets that vanishes on the empty set (i.e., $\mu(\emptyset) = 0$) is accepted in generalized measure theory as a measure. Clearly, various additional requirements are applied as needed to introduce special types of measures. One of these special types consists of classical (i.e., additive) measures. The meaning of the term "measure" in generalized measure theory is thus very much broader than its counterpart in classical measure theory. Since generalized measure theory deals with various types of measures, contrary to classical measure theory, each type is characterized by an adjective added to the term "measure." When we refer to measures of classical measure theory, we use either the term "classical measures" or the term "additive measures."

Classical (additive) measures have their roots in metric geometry, which is characterized by assigning numbers to lengths, areas, or volumes. In antiquity this assignment process—or *measurement*—was first conceived simply as a comparison with a standard unit and the requirement that the assigned numbers be invariant under displacement of the respective geometric objects. Soon, however, *the problem of incommensurables* (exemplified by the problem of measuring the length of the diagonal of a square whose sides each measure one unit) revealed that measurement is more complicated than this simple, intuitively suggestive process. It became clear that measurement must inevitably involve infinite sets and infinite processes.

Prior to the emergence and sufficient development of the calculus, the problem of incommensurables had caused a lot of anxiety since there were no satisfactory tools to deal with it. *Integral calculus*, based upon the Riemann

Z. Wang, G.J. Klir, *Generalized Measure Theory*,
DOI: 10.1007/978-0-387-76852-6_1, © Springer Science+Business Media, LLC 2009

integral, which became well developed in the second half of the nineteenth century, was the first tool to deal with the problem. Certain measurements contingent upon the existence of associated limits could finally be determined by using appropriate techniques of integration. However, it was increasingly recognized in the 1870s and 1880s that the Riemann integral also had a number of deficiencies. One deficiency of the Riemann integral is its applicability only to functions that are continuous except at a finite number of points. This means that the class of Riemann integrable functions is overly restrictive. Another deficiency is that the fundamental operations of differentiation and integration are, in general, not reversible within the context of Riemann's theory of integration. One additional deficiency is connected with limiting operations and can be described as follows. If functions $f_1, f_2, \ldots$ are Riemann integrable on interval $X = [a, b]$ and $\lim f_n(x) = f(x)$ everywhere in X, then it is not, in general, true that the Riemann integral of $f(x)$ is equal to the limit of the Riemann integrals of $f_n(x)$.

In the late nineteenth century, there was a growing need for more precise mathematical analysis, a need induced primarily by the rapidly advancing science and technology. As a result, new questions regarding measurement emerged. Considering, for example, the set of all real numbers between 0 and 1, which may be viewed as points on a real line, mathematicians asked: When we remove the end points 0 and 1 from this set, what is the measure of the remaining set (or the length of the remaining open interval on the real line)? What is the measure of the set obtained from the given set by removing some rational numbers, say 1, 1/2, 1/3, 1/4, and so on? What is the measure of the set obtained by removing all rational numbers?

Questions like these and many more difficult questions were carefully examined by Émile Borel (1871–1956), a French mathematician. He developed a theory [Borel, 1898] to deal with these questions, which was an important step toward a more general theory that we now refer to as the *classical measure theory*.

Borel's theory deals with the σ-algebra (the class of sets closed under the set union of countably many sets and the set complement) that is generated by the family of all open (or semi-open) intervals of real numbers (or within some interval $[a, b]$ of real numbers). Borel defines a measure that associates a positive real number with each bounded subset in the σ-algebra, which, in the case of an interval, is exactly equal to the length of the interval. The measure is additive in the sense that its value for a bounded union of a sequence of pairwise disjoints sets is equal to the sum of the values associated with the individual sets.

In the second half of the nineteenth century there was a growing interest in studying arbitrary real-valued functions, particularly in the context of integration. This involved some strange classes of functions, such as functions that are nowhere continuous or continuous functions that are nowhere differentiable. The existence of such strange classes of functions was already well established at that time. The first example of a continuous function that is nowhere differentiable was apparently constructed already in 1830 by Bernard Bolzano

(1781–1848). These functions eventually became significant within a relatively new area: fractal geometry.

Borel did not connect his theory with the theory of integration. This was done a few years later (between 1899 and 1902) by Henri Lebesgue (1875–1941), another French mathematician. In a paper published in 1901, he defined an integral, more general than the Riemann integral, which is based on a measure that subsumes the Borel measure as a special case. These new concepts of a measure and an integral (further developed in Lebesgue's doctoral dissertation and published in the Italian journal *Annali di Matematica* in 1902), which are now referred to as the *Lebesgue measure* and the *Lebesgue integral*, are the cornerstones of classical measure theory [Halmos, 1950]. The significance of Lebesgue's work is that he connected, in a natural way, measures of sets with measures of functions.

Perhaps the best nontechnical exposition of the motivation behind the Lebesgue measure and the Lebesgue integral, and a discussion of their physical meaning, was prepared by Lebesgue himself; it is available in a book edited by K. O. May, which also contains a biographical sketch of Lebesgue and a list of his key publications [Lebesgue, 1966].

Classical measure theory is closely connected with *probability theory*. A probability measure, as any other classical measure, is a set function that assigns measure 0 to the empty set and a nonnegative number to any other set, and that is additive. However, a probability measure requires, in addition, that measure 1 be assigned to the universal set in question. Hence, probability theory may be viewed as a part of classical measure theory.

The concept of a *probability measure* (or simply a *probability*) was formulated axiomatically by Andrei N. Kolmogorov (1903–1987), a Russian mathematician, in a book written in German that was published in 1933. An English translation of the book was published almost 20 years later [Kolmogorov, 1950]. Kolmogorov's concept of probability is sometimes called a *quantitative* or *numerical probability* to distinguish it from other types of probability, such as *classificatory* or *comparative probabilities* [Fine, 1973; Walley and Fine, 1979; Walley, 1991]. Nevertheless, the term "probability theory" with no additional qualifications refers normally to the theory based upon Kolmogorov's axioms.

After more than 50 years of the existence and steady development of the classical measure theory, the additivity requirement of classical measures became a subject of controversy. Some mathematicians felt that additivity is too restrictive in some application contexts. For example, it is too restrictive to capture adequately the full scope of measurement. While additivity characterizes well many types of measurements under idealized, error-free conditions, it is not fully adequate to characterize most measurements under real, physical conditions, when measurement errors are unavoidable. Moreover, some measurements, for example, those involving subjective judgments or nonrepeatable experiments, are intrinsically nonadditive.

Numerous arguments have been or can be raised against the necessity and adequacy of the additivity axiom of probability theory. One such argument was

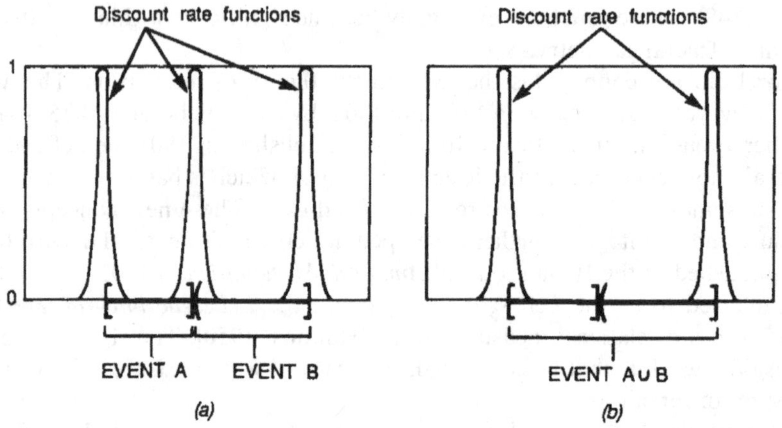

Fig. 1.1 An example illustrating the violation of the additivity axiom of probability theory

presented by Viertl [1987]. It is based on the fact that all measurements are inherently imprecise due to finite resolution of measuring instruments and unavoidable measurement errors. Consider, for example, two disjoint events A and B defined in terms of adjoining intervals of real numbers, as shown in Fig. 1.1a. Observations in close neighborhoods (within a measurement error) of the endpoint of each event are unreliable and should be properly discounted, for example, according to the discount rate functions shown in Fig. 1.1a. That is, observations in the neighborhood of the endpoints should carry less evidence than those outside these neighborhoods. When measurements are taken for the union of the two events, as shown in Fig. 1.1b, one of the discount rate functions is not applicable. Hence, the same observations produce more evidence for the single event $A \cup B$ than they do for the two disjoint events A and B. This implies that the degree of support for $A \cup B$ (probability of $A \cup B$) should be greater than the sum of the respective degrees of support for A and B (probabilities of A and B). The additivity axiom is thus violated.

The earliest challenge to classical measure theory came from a theory proposed by a French mathematician, Gustave Choquet, for which he coined the name *theory of capacities*. This theory is based on a potentially infinite family of distinct types of nonadditive measures that are linearly ordered by their generalities. They range from capacities of order 2 (the most general type) to capacities of order infinity (the least general type). Each Choquet capacity is a real-valued function defined on a class of subsets of a given universal set (with an appropriate algebraic structure) that is monotone increasing with respect to set inclusion and, depending on its type, satisfies one additional axiomatic requirement. For each given Choquet capacity there exists a unique *alternating capacity*. These two capacities are always dual in the sense defined in Chapter 3. Choquet also developed integrals applicable to his capacities, which are now routinely referred to as *Choquet integrals* (Chapter 11).

Choquet developed his theory of capacities at the University of Kansas in Lawrence, where he spent an entire academic year (1953–54) as Visiting Research Professor of Mathematics. The theory was an outcome of a research project entitled "Research on Modern Potential Theory and Dirichlet Problem," which was sponsored by the US Air Force. It was initially published in May 1954 as Technical Note No. 1 (on the project) by the Department of Mathematics of the university, and it was soon republished as [Choquet, 1953–54]. The theory is also covered in Volume I of Choquet's lecture notes for a course on mathematical analysis he gave at Princeton University [Choquet, 1969].

In his writings, Choquet emphasizes that the theory of capacities is closely connected with *potential theory* [Dellacherie and Meyer, 1978; Du Plessis, 1970; Helms, 1963] and that the former theory emerged from the latter. However, the applicability of his capacities extends far beyond potential theory. For example, these capacities play an important role in formalizing imprecise probabilities of various types [Klir, 2006].

Another approach to developing generalized measures was taken in the 1970s by Michio Sugeno, a distinguished Japanese scholar. He tried to compare membership functions of fuzzy sets with probabilities [Sugeno, 1974, 1977]. Since no direct comparison is possible, Sugeno conceived of the generalization of classical measures into nonclassical (nonadditive) measures as an analogy of the generalization of classical (crisp) sets into fuzzy sets. Using this analogy he coined for the nonclassical (nonadditive) measures the term "fuzzy measures."

Fuzzy measures, according to Sugeno, are obtained by replacing the additivity requirement of classical measures with the weaker requirements of *increasing monotonicity* (with respect to set inclusion) and *continuity*. The requirement of continuity was later found to be too restrictive and was replaced with a weaker requirement of *semicontinuity*. For example, lower and upper probability measures in the various theories of imprecise probabilities, which are introduced in Section 15.3, are only semicontinuous.

The term "fuzzy measure" in the sense Sugeno introduced it has been accepted by most researchers working in the area of generalized measures, including the authors of this book. We published a book entitled *Fuzzy Measure Theory* [Wang and Klir, 1992], and we used the term in our other publications. Unfortunately, the term is confusing since there is no fuzziness involved in so-called "fuzzy measures." These are just special nonnegative real-valued set functions that are defined on specified classes of classical sets, not on classes of fuzzy sets. However, these functions, as well as any other types of set functions involved in generalized measure theory, can be fuzzified (defined on fuzzy sets), as is shown in Chapter 14. This would result then in the term "fuzzified fuzzy measures," which is even more confusing. These are the main considerations that led us to abandon the confusing term "fuzzy measures" and replace it with a more descriptive term "monotone measures." We are glad to observe that some other authors are starting to abandon the other term as well.

We should mention at this point that monotone measures, as we define them in this book, do not require continuity. If the requirement of continuity is added, then we refer to the measures as *continuous monotone measures.* It is this class of measures that is equivalent to the class of fuzzy measures, as defined in our previous book. If the requirement of only lower or upper semicontinuity is added to monotonicity, we call the measures *lower-semicontinuous monotone measures* or *upper-semicontinuous monotone measures,* respectively.

In the second half of the twentieth century, some researchers recognized that the required precision of classical probabilities is not realistic in some applications. This stimulated interest in investigating imprecise probabilities. It seems that the notion of imprecise probabilities was first introduced and investigated by Dempster [1967a]. He was concerned with convex sets of probability measures rather than single probability measures. For each given convex set of probability measures, Dempster introduced two types of nonadditive measures, which he called *lower and upper probabilities*. These measures are superadditive and subadditive, respectively, in the sense introduced in Chapter 3, and allow us to represent probabilities imprecisely by intervals of real numbers rather than by precise real numbers.

Special types of lower and upper probabilities, referred to as *belief measures* and *plausibility measures*, were later introduced and thoroughly investigated by Shafer [1976]. The theory based on these two nonadditive measures is usually called *Dempster–Shafer theory* or *evidence theory*. Since belief measures are always smaller than or equal to the corresponding plausibility measures, the intervals between belief and plausibility values may be viewed as ranges of admissible probabilities. The Dempster–Shafer theory may thus be viewed as a theory that is capable of dealing with interval-valued probabilities. It turns out that belief measures are Choquet capacities of order infinity, and plausibility measures are alternating capacities of order infinity.

Another theory based upon nonadditive measures, referred to as the *theory of graded possibilities*, emerged from the concept of a *fuzzy set*, which was proposed by Zadeh [1965]. A fuzzy set is a set whose boundary is not required to be sharp. That is, the change from nonmembership to membership is allowed to be gradual rather than abrupt. This gradual change is expressed by a *membership function* of the fuzzy set, which assigns to each individual of a given universal set its degree of membership in the fuzzy set. If these degrees are expressed by values in the unit interval [0, 1], the fuzzy set is called *standard.* At this time standard fuzzy sets are the most common, and these are the only fuzzy sets that are considered in this book. A fuzzy set is called *normalized* if the supremum of its membership function is 1. It is clear that classical sets are special fuzzy sets in which each degree of membership is either 0 or 1. These special fuzzy sets are usually referred to as *crisp sets*.

Given a standard fuzzy set that is normalized, Zadeh [1978] defines a *possibility function* associated with the fuzzy set as numerically equal to its membership function. Then, he defines a *possibility measure* by taking the supremum of the possibility function in each crisp set or, more generally, fuzzy set of concern.

If the fuzzy set is not normalized the possibility function must be defined in a more general way [Klir, 1999]. The fuzzy-set interpretation is only one of several other established interpretations of the theory of graded possibilities.

It turns out that possibility measures also emerge from Dempster–Shafer theory. They are plausibility measures with a special mathematical structure. In this context, possibility measures are usually called *consonant plausibility measures* [Dubois and Prade, 1988; Klir, 2006]. Their dual measures, which are special (consonant) belief measures, are called *necessity measures*.

Similar to Choquet capacities, monotone measures are too loose to allow us to develop a theory that would capture their full generality and, yet, are of pragmatic utility. On the other hand, some very special types of general measures appear to be unnecessarily restrictive in some application contexts. These considerations led to a more systematic investigation of useful structural characteristics of set functions, primarily by Wang [1984, 1985a], as presented in Chapter 5. These characteristics are essential for capturing mathematical properties of measurable functions on generalized measure spaces (Chapter 7), and that, in turn, is requisite for developing a *theory of generalized integrals* (Chapters 8–12).

There have been many additional developments pertaining to various aspects of generalized measure theory that we do not deem necessary to cover in this introduction. Since most of these developments are rather technical and involve special terminology, we leave their historical and bibliographical coverage to Notes accompanying the individual chapters.

Notes

1.1 An overview of relevant concepts and results of classical measure theory is given in Section 2.2. For further study we recommend the classic text by Halmos [1950]. An excellent text on classical measure theory by Billingsley [1986] is recommended to readers that are interested particularly in probability measures.

1.2 Among many other books on classical measure theory, let us mention a few that are significant in various respects. The book by Caratheodory [1963], whose original German version was published in 1956, is one of the earliest and most highly influential books on classical measure theory. Books by Chae [1995], Temple [1971], and Weir [1973] provide pedagogically excellent introductions to classical measure theory; they require only some basic knowledge of calculus and algebra as prerequisites. The book by Constantinescu and Weber [1985], suitable for a mathematically mature reader, attempts to unify abstract and topological approaches. Other valuable books are by Berberian [1965], Kingman and Taylor [1966], and Wheeden and Zygmund [1977]. The book by Faden [1977] is an extensive treatise on the use of measure theory, particularly in the area of economics, which also contains a good introduction to measure theory itself.

1.3 The various shortcomings of classical probabilities and the reasons why nonadditive measures are needed to overcome these shortcomings are thoroughly discussed by Walley [1991]. These shortcomings have also been discussed within the area of generalized information theory [Klir, 2006]. It has been demonstrated that classical probability measures can capture only one of several recognized types of uncertainty.
1.4 The history of classical measure theory and Lebesgue's integral is carefully traced in a fascinating book by Hawkins [1975]. He describes how modern mathematical concepts regarding these theories (involving concepts such as a function, continuity, convergence, measure, integration, and the like) developed (primarily in the nineteenth century and the early twentieth century) through the work of many mathematicians, including Cauchy (1789–1857), Fourier (1768–1830), Dirichlet (1805–1859), Weierstrass (1815–1897), Riemann (1826–1866), Borel (1871–1956), Cantor (1845–1918), Hankel (1839–1873), Jordan (1838–1922), Volterra (1860–1897), Peano (1858–1932), Lebesgue (1875–1941), Radon (1887–1956), Riecz (1880–1956), Vitali (1875–1932), Egoroff (1869–1931), Fubini (1879–1943), Young (1863–1942), Dini (1845–1918), and many others. The book by Saks [1937] is an excellent overview of the development of the classical theory of integration.
1.5 For the history of probability theory, we recommend a book by Hacking [1975] and a paper by Shafer [1978]. From the standpoint of generalized measure theory it is most interesting that Bernoulli (1654–1705) and, later, Lambert (1728–1777) were already concerned with a calculus of probabilities that are not additive and, consequently, are imprecise. Their work, unfortunately, was forgotten for more than two centuries.
1.6 The significance of the contribution by Kolmogorov [1950] to the transition from a classical foundation of probability, which emerged early in the eighteenth century, to a measure-theoretic foundation is thoroughly discussed by Shafer and Vovk in an extensive report entitled "The Origins and Legacy of Kolmororov's *Grundbegriffe*" on the following website: http://www.probabilityandfinance.com. This report, which was prepared in 2005 and consists of more than 100 pages, contains a large list of references pertaining to the discussed issue. A shorter version of this report is [Shafer and Vovk, 2006]. The evolution from Lebesgue's measure and integral to Kolmogorov's axiomatic formulation of probability theory is also well described in papers by Doob [1994] and Bingham [2000].
1.7 The literature dealing with classical probability theory is abundant. Perhaps the most comprehensive examination of foundations of classical probability theory was made by Fine [1973].
1.8 A very valuable resource regarding classical as well as generalized measures and the associated integration is a Handbook edited by Pap [2002a].
1.9 In our previous book [Wang and Klir, 1992] fuzzy measures are defined as monotone and continuous, nonnegative, real-valued set functions that vanish at the empty set. In the literature the term "fuzzy measure" is usually used in this sense, but in some publications the continuity is not required.

Chapter 2
Preliminaries

2.1 Classical Sets

2.1.1 Set Inclusion and Characteristic Function

Let X be a nonempty set. Unless otherwise stated, all sets that we consider are subsets of X. Set X is called a *universe of discourse* or a *universal set*. The elements of X are called *points*. Universal set X may contain finite, countably infinite, or uncountably infinite number of points. A set that consists of a finite number of points $x_1, x_2, \ldots, x_n$ (or, a countably infinite number of points $x_1, x_2, \ldots$) may be denoted by $\{x_1, x_2, \ldots, x_n\}$($\{x_1, x_2, \ldots\}$, respectively). A set containing no point is called the *empty set* and is denoted by Ø.

If x is a point of X and E is a subset of X, the notation

$$x \in E$$

means that x belongs to E, i.e., x is an element of E; and the statement that x does not belong to E is denoted by

$$x \notin E.$$

Thus, for every point x of X we have

$$x \in X$$

and

$$x \notin \emptyset.$$

A set of sets is called a *class*. If E is a set and $\mathbf{C}$ is a class, then

$$E \in \mathbf{C}$$

means that set E belongs to class $\mathbf{C}$.

Z. Wang, G.J. Klir, *Generalized Measure Theory*,
DOI: 10.1007/978-0-387-76852-6_2, © Springer Science+Business Media, LLC 2009

If, for each $x, \pi(x)$ is a proposition concerning x, then the symbol

$$\{x|\pi(x)\}$$

denotes the set of all those points x for which $\pi(x)$ is true; that is,

$$x_0 \in \{x|\pi(x)\} \Leftrightarrow \pi(x_0) \text{is true.}$$

If the point x is replaced with set E, such a symbol may be used to indicate a class. For example,

$$\{E|x \in E\}$$

denotes the class of those sets that contain the point x.

Example 2.1. Let $X = \{1, 2, \ldots\}$. Then, $A = \{x|x \text{ is odd and less than} 10\} = \{1, 3, 5, 7, 9\}$.

Example 2.2. Let X be the set of all real numbers, which is often referred to as the real line or one-dimensional Euclidean space. The class $\{(a,b)| -\infty < a < b < \infty\}$ is the class consisting of all open intervals on the real line.

If E and F are sets, the notation

$$E \subset F \text{ or } F \supset E$$

means that E is a subset of F, i.e., every point of E belongs to F. In this case, we say that F *includes* E, or that E *is included* by F. For every set E we have

$$\emptyset \subset E \subset X.$$

Two sets E and F are called *equal* iff

$$E \subset F \text{ and } F \subset E;$$

that is, they contain exactly the same points. This is denoted by

$$E = F.$$

The symbols $\subset$ or $\supset$ also may be used for classes. If $\mathbf{E}$ and $\mathbf{F}$ are classes, then

$$\mathbf{E} \subset \mathbf{F}$$

means that every set of $\mathbf{E}$ belongs to $\mathbf{F}$, that is, $\mathbf{E}$ is a subclass of $\mathbf{F}$.

If $E_1, E_2, \ldots, E_n$ are nonempty sets, then

$$E = \{(x_1, x_2, \ldots, x_n)|x_i \in E_i,\ i = 1, 2, \ldots, n\}$$

is called an *n-dimensional product set* and is denoted by

$$E = E_1 \times E_2 \times \ldots \times E_n.$$

Similarly, if $\{E_t | t \in T\}$ is a family of nonempty sets, where T is an infinite index set, then

$$E = \{x_t, t \in T \,|\, x_t \in E_t \text{ for each } t \in T\}$$

is called an *infinite-dimensional product set.*

Example 2.3. Let X_1 and X_2 be one-dimensional Euclidean spaces. Then $X = X_1 \times X_2 = \{(x_1, x_2) | x_1 \in (-\infty, \infty), x_2 \in (-\infty, \infty)\}$ is the two-dimensional Euclidean space. The set $\{(x_1, x_2) | x_1 > x_2\}$ is a half (open) plane under the line $x_2 = x_1$, while the set $\{(x_1, x_2) | x_1^2 + x_2^2 < r^2\}$ is the open circle centering at the origin with a radius r, where $r > 0$.

Example 2.4. Let $X_t = \{0, 1\}, t \in \{1, 2, \ldots\}$. The space

$$\begin{aligned} X &= X_1 \times X_2 \times \ldots \times X_n \times \ldots \\ &= \{(x_1, x_2, \ldots, x_n, \ldots) | x_t \in \{0, 1\} \text{ for each } t \in \{1, 2, \ldots\}\} \end{aligned}$$

is an infinite-dimensional product space. Each point $(x_1, x_2, \ldots, x_n, \ldots)$ in this space corresponds to the binary number 0. $x_1 x_2 \ldots x_n \ldots$ in $[0, 1]$. Such a correspondence is not one to one, but it is onto.

If E is a set, the function χ_E, defined for all $x \in X$ by

$$\chi_E(x) = \begin{cases} 1 & \text{if } x \in E \\ 0 & \text{if } x \notin E, \end{cases}$$

is called the *characteristic function* of set E. The correspondence between sets and their characteristic functions is one to one, that is,

$$E = F \Leftrightarrow \chi_E(x) = \chi_F(x), \ \forall x \in X.$$

It is easy to see that

$$E \subset F \Leftrightarrow \chi_E(x) \leq \chi_F(x), \ \forall x \in X,$$

and that

$$\chi_X \equiv 1, \chi_\varnothing \equiv 0.$$

2.1.2 Operations on Sets

Let **C** be any class of subsets of X. The set of all those points of X that belong to at least one set of the class **C** is called the *union* of the sets of **C**. This is denoted by

$$\bigcup \mathbf{C}.$$

If to every t of a certain index set T there corresponds a set E_t, then the union of the sets of class

$$\{E_t | t \in T\}$$

may be also denoted by

$$\bigcup_{t \in T} E_t \quad \text{or} \quad \bigcup_t E_t.$$

Especially, when

$$\mathbf{C} = \{E_1, E_2\},$$

then $\bigcup \mathbf{C}$ is denoted by

$$E_1 \cup E_2;$$

and if

$$\mathbf{C} = \{E_1, E_2, \ldots, E_n\} \qquad (\mathbf{C} = \{E_1, E_2, \ldots\})$$

then $\bigcup \mathbf{C}$ is denoted by

$$E_1 \cup E_2 \cup \ldots \cup E_n \quad \text{or} \quad \bigcup_{i=1}^{n} E_i \quad \left(\bigcup_{i=1}^{\infty} E_i, \text{ respectively} \right).$$

The set of all those points of X which belong to every set of the class **C** is called the *intersection* of the sets of **C**. This is denoted by $\bigcap \mathbf{C}$. Symbols similar to those used for unions are available, such as $\bigcap_{t \in T} E_t$ (or $\bigcap_t E_t$), $E_1 \cap E_2$, $E_1 \cap E_2 \cap \ldots \cap E_n$ (or $\bigcap_{i=1}^{n} E_i$), and $\bigcap_{i=1}^{\infty} E_i$. If F is a set, the class $\{E \cap F | E \in \mathbf{C}\}$ is denoted by $\mathbf{C} \cap F$.

Example 2.5. Let $X = \{a, b, c, d\}, \mathbf{C} = \{\{a\}, \{b, c\}, \{b, d\}, \{c, d\}\}, F = \{a, b\}$. Then $\mathbf{C} \cap F = \{\{a\}, \{b\}, \emptyset\}$.

Example 2.6. Let $X = (-\infty, \infty)$, $\mathbf{C} = \{[a, b] | -\infty < a \le b < \infty\}$, $F = [0, 1]$. Then, $\mathbf{C} \cap F = \{[a, b] | 0 \le a \le b \le 1\}$, that is, the class of all closed subintervals of the unit closed interval.

It is convenient to adopt the conventions that

$$\bigcup_{t \in T} E_t = \emptyset$$

and

$$\bigcap_{t \in T} E_t = X$$

when T is empty.

Proposition 2.1. *The following statements are equivalent:*

(1) $E \subset F$;
(2) $E \cup F = F$;
(3) $E \cap F = E$.

Two sets E and F are called *disjoint* iff

$$E \cap F = \emptyset.$$

A class $\mathbf{C}$ is called *disjoint* iff every two distinct sets of $\mathbf{C}$ are disjoint; in this case we refer to the union of the sets of $\mathbf{C}$ as a disjoint union.

If E is a set, the set of all those points of X that do not belong to E is called the *complement* of E. This is denoted by $\overline{E}$.

Proposition 2.2. *The set operations union, intersection, and complement have the following properties:*

Involution:

$$\overline{\overline{E}} = E$$

Commutativity:

$$E \cup F = F \cup E$$
$$E \cap F = F \cap E$$

Associativity:

$$\bigcup_{t \in T} \left(\bigcup_{s \in S_t} E_s \right) = \bigcup_{s \in \cup_{t \in T} S_t} E_s$$

$$\bigcap_{t \in T} \left(\bigcap_{s \in S_t} E_s \right) = \bigcap_{s \in \cup_{t \in T} S_t} E_s$$

Distributivity:

$$F \cap \left(\bigcup_{t \in T} E_t \right) = \bigcup_{t \in T} (F \cap E_t)$$

$$F \cup \left(\bigcap_{t \in T} E_t \right) = \bigcap_{t \in T} (F \cup E_t)$$

Idempotence:

$$E \cup E = E$$

$$E \cap E = E$$

Absorption: $E \cup (E \cap F) = E$

$E \cap (E \cup F) = E$

Absorption of complement: $E \cup (\overline{E} \cap F) = E \cup F$

$E \cap (\overline{E} \cup F) = E \cap F$

Absorption by X and Ø: $E \cup X = X$

$E \cap \emptyset = \emptyset$

Identity: $E \cup \emptyset = E$

$E \cap X = E$

Law of contradiction: $E \cap \overline{E} = \emptyset$

Law of excluded middle: $E \cup \overline{E} = X$

DeMorgan's laws: $\overline{\bigcup_{t \in T} E_t} = \bigcap_{t \in T} \overline{E_t}$

$\overline{\bigcap_{t \in T} E_t} = \bigcup_{t \in T} \overline{E_t}$

where S_t, T are index sets.

From the above a duality is suggestive. In general, we have the following principle of duality: Any valid identity among sets obtained by unions, intersections, and complements, remains valid if the symbols

$$\cap, \subset, \text{ and } \emptyset$$

are interchanged with

$$\cup, \supset, \text{ and } X,$$

respectively (and if the equality and complementation are left unchanged).

If E and F are sets, the set of all those points of E that do not belong to F is called the difference of E and F. This is denoted by

$$E - F.$$

If $E \supset F$, the difference $E - F$ is called *proper*. Clearly,

$$E - F = E \cap \overline{F}.$$

The *symmetric difference* of E and F, in symbols

$$E \,\Delta\, F,$$

is defined by

$$E \Delta F = (E - F) \cup (F - E).$$

Let $\{E_1, E_2, \ldots\}$ (or $\{E_n\}$, briefly) be a sequence of sets. The set of all those points of X that belong to E_n for infinitely many values of n is called the *superior limit* of $\{E_n\}$, and is denoted by

$$\limsup_n E_n \text{ or } \overline{\lim_n} E_n;$$

the set of all points of X that belong to E_n for all but a finite number of values of n is called the *inferior limit* of $\{E_n\}$, and denoted by

$$\liminf_n E_n \text{ or } \underline{\lim_n} E_n.$$

Proposition 2.3.

$$\limsup_n E_n = \bigcap_{n=1}^{\infty} \bigcup_{i=n}^{\infty} E_i;$$

$$\liminf_n E_n = \bigcup_{n=1}^{\infty} \bigcap_{i=n}^{\infty} E_i.$$

Example 2.7. Let $X = \{a, b\}$ and let a set sequence $\{E_n\}$ be defined as follows: $E_n = \begin{cases} \{a\} & \text{if } n \text{ is even} \\ \{b\} & \text{if } n \text{ is odd.} \end{cases}$ Then, $\limsup_n E_n = X$ and $\liminf_n E_n = \emptyset$.

Example 2.8. Let $X = (-\infty, \infty)$ and let a set sequence $\{E_n\}$ be defined as follows: $E_1 = [0, 1)$, $E_2 = [0, 1/2)$, $E_3 = [1/2, 1)$, $E_4 = [0, 1/4)$, $E_5 = [1/4, 1/2)$, $E_6 = [1/2, 3/4)$, $E_7 = [3/4, 1)$, $E_8 = [0, 1/8)$, ... Then, $\limsup_n E_n = [0, 1)$ and $\liminf_n E_n = \emptyset$.

Proposition 2.4. $\liminf_n E_n \subset \limsup_n E_n$.

If

$$\limsup_n E_n = \liminf_n E_n$$

we use the notation

$$\lim_n E_n$$

for this set and say that the limit of $\{E_n\}$ exists and that this set is the *limit* of $\{E_n\}$. Sometimes we write $E_n \to E$ when $\lim_n E_n = E$.

Example 2.9. Let $X = \{1, 2, \dots\}$ and let $\{E_n\}$ be a set sequence in which $E_n = \{n\}, n = 1, 2, \dots$. Then, we have

$$\limsup_n E_n = \liminf_n E_n = \emptyset.$$

Hence, the limit of $\{E_n\}$ exists, and $\lim_n E_n = \emptyset$.

We say that $\{E_n\}$ is *increasing* if

$$E_n \subset E_{n+1}, \quad \forall n = 1, 2, \dots,$$

and $\{E_n\}$ is *decreasing* if

$$E_n \supset E_{n+1}, \quad \forall n = 1, 2, \dots.$$

Both increasing and decreasing sequences are called *monotone*.

Proposition 2.5. *For any monotone sequence $\{E_n\}$, $\lim_n E_n$ exists and equals*

$$\bigcup_n E_n \quad \text{or} \quad \bigcap_n E_n$$

according as $\{E_n\}$ is increasing or decreasing, respectively.

Usually, we write $E_n \nearrow E$ when $\{E_n\}$ is increasing and $\lim_n E_n = E$, whereas we write $E_n \searrow E$ when $\{E_n\}$ is decreasing and $\lim_n E_n = E$.

Example 2.10. Let $X = (-\infty, \infty)$. If $\{E_n\}$ is a set sequence in which $E_n = [1/n, 1], n = 1, 2, \dots$, then $\{E_n\}$ is increasing, and $E_n \nearrow \bigcup_n E_n = (0, 1]$. If $\{F_n\}$ is a set sequence in which $F_n = (-(1 + 1/n), 1 + 1/n), n = 1, 2, \dots$, then $\{F_n\}$ is decreasing, and $F_n \searrow \bigcap_n F_n = [-1, 1]$.

The discussion of monotone sequences $\{E_n\}$ can be generalized to families of sets $\{E_t | t \in T\}$, where T is an interval (finite or infinite) of real numbers. If for any $t, t' \in T, E_t \subset E_{t'}$ whenever $t \le t'$, then $\{E_t\}$ is increasing, and

$$\lim_{t \to t_0-} E_t = \bigcup_{t < t_0, t \in T} E_t,$$

$$\lim_{t \to t_0+} E_t = \bigcap_{t > t_0, t \in T} E_t;$$

if for any $t, t' \in T, E_t \supset E_{t'}$ whenever $t \le t'$, then $\{E_t\}$ is decreasing, and

$$\lim_{t \to t_0-} E_t = \bigcap_{t < t_0, t \in T} E_t,$$

$$\lim_{t \to t_0+} E_t = \bigcup_{t > t_0, t \in T} E_t,$$

where symbols $\lim_{t \to t_0 -}$ and $\lim_{t \to t_0 +}$ denote the left limit at t_0 and the right limit at t_0, respectively.

The following proposition gives the correspondence between the operations of sets and the operations of characteristic functions of sets.

Proposition 2.6.

(1)
$$\chi_E = \sup_{t \in T} \chi_{E_t}, \quad \textit{where } E = \bigcup_{t \in T} E_t;$$

in particular,

$$\chi_{E \cup F} = \max(\chi_E, \chi_F);$$

(2)
$$\chi_E = \inf_{t \in T} \chi_{E_t}, \quad \textit{where } E = \bigcap_{t \in T} E_t;$$

in particular,

$$\chi_{E \cap F} = \min(\chi_E, \chi_F);$$

(3)
$$\chi_{\overline{E}} = 1 - \chi_E;$$

(4)
$$\chi_{E-F} = \chi_E - \min(\chi_E, \chi_F) = \min(\chi_E, 1 - \chi_F) = \max(0, \chi_E - \chi_F);$$

(5)
$$\chi_{E \Delta F} = |\chi_E - \chi_F|;$$

(6)
$$\chi_{\limsup_n E_n} = \limsup_n \chi_{E_n},$$

$$\chi_{\liminf_n E_n} = \liminf_n \chi_{E_n},$$

and if $\lim_n E_n$ *exists, then*

$$\chi_{\lim_n E_n} = \lim_n \chi_{E_n}.$$

2.1.3 *Classes of Sets*

Definition 2.1. The class of all subsets of X is called the *power set* of X, and is denoted by

$$\mathbf{P}(X).$$

Definition 2.2. A nonempty class $\mathbf{R}$ is called a *ring*, iff $\forall E, F \in \mathbf{R}$,

$$E \cup F \in \mathbf{R} \text{ and } E - F \in \mathbf{R}.$$

In other words, a ring is a nonempty class that is closed under the formation of unions and differences. Because of the associativity of the set union a ring is also closed under the formation of finite unions.

Proposition 2.7. *The empty set* $\emptyset$ *belongs to every ring.*

Theorem 2.1. *Any ring is closed under the formation of symmetric differences and intersections; and, conversely, a nonempty class that is closed under the formation of symmetric differences and intersections is a ring.*

Proof. From

$$E \Delta F = (E - F) \cup (F - E)$$

and

$$E \cap F = (E \cup F) - (E \Delta F),$$

we obtain the first conclusion. The converse conclusion issues from

$$E \cup F = (E \Delta F) \ \Delta \ (F \cap F)$$

and

$$E - F = (E \Delta F) \cap E.$$

□

Theorem 2.2. *A nonempty class that is closed under the formation of intersections, proper differences, and disjoint unions is a ring.*

Proof. The conclusion follows from

$$E\Delta F = [E - (E \cap F)] \cup [F - (E \cap F)]$$

and Theorem 2.1. □

Example 2.11. The class of all finite subsets of X is a ring.

Example 2.12. Let X be the real line, that is

$$X = (-\infty, \infty) = \{x | -\infty < x < \infty\}.$$

The class of all finite unions of bounded, left closed, and right open intervals, that is, the class of all sets which have the form

$$\bigcup_{i=1}^{n} \{x | -\infty < a_i \leq x < b_i < \infty\},$$

is a ring.

Definition 2.3. A nonempty class **R** is called an *algebra* iff

(1) $\forall E, F \in \mathbf{R}$,

$$E \cup F \in \mathbf{R};$$

(2) $\forall E \in \mathbf{R}$,

$$\overline{E} \in \mathbf{R}.$$

In other words, an algebra is a nonempty class that is closed under the formation of unions and complements. Obviously, in this definition, "$\cup$" can be replaced by "$\cap$".

Theorem 2.3. *An algebra is a ring containing X and, conversely, a ring that contains X is an algebra.*

Proof. Let $\mathbf{R}$ be an algebra. Since

$$E - F = E \cap \overline{F} = (\overline{\overline{E} \cup F}),$$

and, if $E \in \mathbf{R}$, then

$$X = E \cup \overline{E} \in \mathbf{R},$$

we have the first part of the theorem. Conversely, if $\mathbf{R}$ is a ring containing X, then $\forall E \in \mathbf{R}$,

$$\overline{E} = X - E \in \mathbf{R},$$

and the second part follows. □

Example 2.13. The class of all finite sets and their complements is an algebra.

The property described by this example can be generalized into the following proposition.

Proposition 2.8. *If* $\mathbf{R}$ *is a ring, then* $\mathbf{R} \cup \{E | \overline{E} \in \mathbf{R}\}$ *is an algebra.*

Definition 2.4. A nonempty class $\mathbf{S}$ is called a *semiring* iff

(1) $\forall E,\ F \in \mathbf{S}$,

$$E \cap F \in \mathbf{S};$$

(2) $\forall E,\ F \in \mathbf{S}$ satisfying $E \subset F$, there exists a finite class $\{C_0, C_1, \ldots, C_n\}$ of sets in $\mathbf{S}$, such that

$$E = C_0 \subset C_1 \subset \ldots \subset C_n = F$$

and

$$D_i = C_i - C_{i-1} \in \mathbf{S} \qquad \text{for } i = 1, 2, \ldots, n.$$

Every ring is a semiring, and the empty set belongs to any semiring.

Example 2.14. The class consisting of all singletons of X and the empty set is a semiring.

Example 2.15. Let X be the real line. The class of all bounded, left closed, and right open intervals is a semiring.

Definition 2.5. A nonempty class $\mathbf{F}$ is called a *σ-ring* iff

(1) $\forall E,\ F \in \mathbf{F}$,

$$E - F \in \mathbf{F};$$

(2) $\forall E_i \in \mathbf{F}, i = 1, 2, \ldots,$

$$\bigcup_{i=1}^{\infty} E_i \in \mathbf{F}.$$

Any σ-ring is a ring which is closed under the formation of countable unions.

Proposition 2.9. *Any σ-ring is closed under the formation of countable intersections; and, therefore, if* $\mathbf{F}$ *is a σ-ring and a set sequence* $\{E_n\} \subset \mathbf{F}$, *then*

$$\limsup_n E_n \in \mathbf{F} \text{ and } \liminf_n E_n \in \mathbf{F}.$$

Example 2.16. The class of all countable sets is a σ-ring.

Definition 2.6. A *σ-algebra* (or say, *σ-field*) is a *σ-ring* that contains X.

Example 2.17. The class of all countable sets and their complements is a σ-algebra.

Proposition 2.10. *If* $\mathbf{F}$ *is a σ-ring, then* $\mathbf{F} \cup \{E | \overline{E} \in \mathbf{F}\}$ *is a σ-algebra.*

Definition 2.7. A nonempty class $\mathbf{M}$ is called a *monotone class* iff, for every monotone sequence $\{E_n\} \subset \mathbf{M}$, we have

$$\lim_n E_n \in \mathbf{M}.$$

Proposition 2.11. *Any σ-ring is a monotone class.*

Proposition 2.12. *If a ring is also a monotone class, then it is a σ-ring.*

Example 2.18. Let X be the real line. The class of all intervals (the empty set and singletons may be regarded as intervals: $\emptyset = (a, a], \{a\} = [a, a]$) is a monotone class.

Definition 2.8. A nonempty class $\mathbf{F}_p$ is called a *plump class* iff $\forall \{E_t | t \in T\} \subset \mathbf{F}_p$,

$$\bigcup_t t \in \mathbf{F}_p \quad \text{and} \quad \bigcap_t E_t \in \mathbf{F}_p,$$

where T is an arbitrary index set.

Proposition 2.13. *Any plump class is a monotone class.*

Example 2.19. Let X be the unit closed interval [0,1]. The class of all sets that have the form $[0, a)$, or the form $[0, a]$, where $a \in [0, 1]$, is a plump class.

The relations among the above-mentioned concepts of classes are illustrated in Fig. 2.1.

Proposition 2.14. *Let E be a fixed set. If $\mathbf{C}$ is a σ-ring (respectively, ring, semiring, monotone class, plump class), then so is $\mathbf{C} \cap E$.*

Theorem 2.4. *Let $\mathbf{C}$ be a class. There exists a unique ring $\mathbf{R}_0$ such that it is the smallest ring including $\mathbf{C}$; that is,*

$$\mathbf{R}_0 \supset \mathbf{C}$$

and for any ring $\mathbf{R}$,

$$\mathbf{R} \supset \mathbf{C} \Rightarrow \mathbf{R} \supset \mathbf{R}_0.$$

$\mathbf{R}_0$ is called the ring generated by $\mathbf{C}$ and is denoted by $\mathbf{R}(\mathbf{C})$.

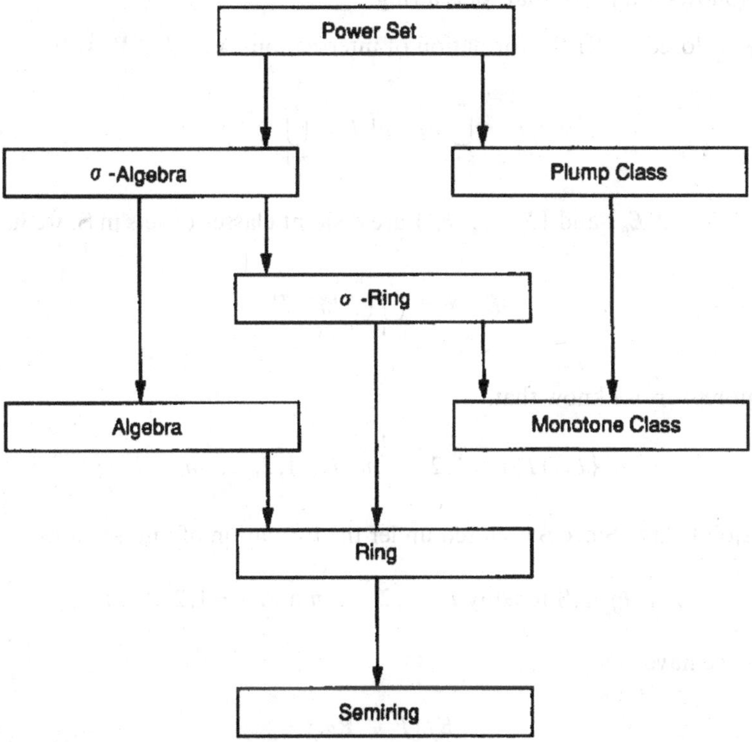

Fig. 2.1 The ordering of classes of sets

Proof. $\mathbf{P}(X)$ is a ring including $\mathbf{C}$. The intersection of all rings including $\mathbf{C}$ is also a ring including $\mathbf{C}$, and it is the desired ring $\mathbf{R}_0$. The uniqueness is evident. □

In the same way, we can also give the concepts of σ-ring, monotone class, and plump class generated by $\mathbf{C}$, and use $\mathbf{F}(\mathbf{C})$, $\mathbf{M}(\mathbf{C})$, and $\mathbf{F}_p(\mathbf{C})$ to denote them, respectively.

Example 2.20. Let X be an infinite set. If $\mathbf{C}$ is the class of all singletons, then $\mathbf{R}(\mathbf{C})$ is the class of all finite sets, and $\mathbf{F}(\mathbf{C})$ is the class of all countable sets.

Example 2.21. Let X be the real line. If $\mathbf{C}$ is the class of all finite open intervals, then $\mathbf{M}(\mathbf{C})$ is the class of all intervals, and $\mathbf{F}_p(\mathbf{C}) = \mathbf{P}(X)$.

Proposition 2.15. *If* $\mathbf{C}_1 \subset \mathbf{C}_2$, *then* $\mathbf{K}(\mathbf{C}_1) \subset \mathbf{K}(\mathbf{C}_2)$, *where* $\mathbf{K}$ *may be taken as* $\mathbf{R}$, $\mathbf{F}$, $\mathbf{M}$, *or* $\mathbf{F}_p$.

Theorem 2.5. *Let* $\mathbf{S}$ *be a semiring. Then,* $\mathbf{R}(\mathbf{S})$ *is the class of all finite, disjoint unions of sets in* $\mathbf{S}$.

Proof. Denote the class of all finite, disjoint unions of sets in $\mathbf{S}$ by $\mathbf{R}_0$. Clearly,

$$\mathbf{R}_0 \supset \mathbf{S}.$$

What follows is a proof that $\mathbf{R}_0$ is a ring.

(1) $\mathbf{R}_0$ is closed under the formation of intersections: $\forall E,\ F \in \mathbf{R}_0$ with

$$E = \bigcup_{i=1}^{n} E_i \text{ and } F = \bigcup_{j=1}^{m} F_j,$$

where $\{E_1, \ldots, E_n\}$ and $\{F_1, \ldots, F_m\}$ are disjoint classes of sets in $\mathbf{S}$, we have

$$E \cap F = \bigcup_{i=1}^{n} \bigcup_{j=1}^{m} E_i \cap F_j$$

and, moreover, we know that

$$\{E_j \cap F_j | i = 1, 2, \ldots, n;\ j = 1, 2, \ldots, m\}$$

is a disjoint class. Since $\mathbf{S}$ is closed under the formation of intersections,

$$E_i \cap F_j \in \mathbf{S} \text{ for any } i = 1, 2, \ldots, n \text{ and } j = 1, 2, \ldots, m.$$

Hence, we have

$$E \cap F \in \mathbf{R}_0.$$

(2) $\mathbf{R}_0$ is closed under the formation of proper differences: For any E and F given in (1), if $F \subset E$, the difference $E - F$ may be expressed by a finite, disjoint union of sets having the form

$$E_i - \bigcup_{j=1}^{m} F_j.$$

Each $E_i - \bigcup_{j=1}^{m} F_j$ may also be expressed by a finite, disjoint union of the sets in $\mathbf{S}$. Thus, we have

$$E - F \in \mathbf{R}_0.$$

(3) It is evident that $\mathbf{R}_0$ is closed under the formation of disjoint unions. By Theorem 2.2, we know that $\mathbf{R}_0$ is a ring.

Finally, since $\mathbf{R}$ is closed under the formation of finite unions, if $\mathbf{R}$ is a ring containing $\mathbf{S}$, it should contain every finite union of sets in $\mathbf{S}$. Hence, $\mathbf{R} \supset \mathbf{R}_0$. This completes the proof. □

Theorem 2.6. $\mathbf{F}(\mathbf{S}) = \mathbf{F}(\mathbf{R}(\mathbf{S}))$.

Proof. On the one hand, since $\mathbf{S} \subset \mathbf{R}(\mathbf{S})$, by Proposition 2.15, we have

$$\mathbf{F}(\mathbf{S}) \subset \mathbf{F}(\mathbf{R}(\mathbf{S})).$$

On the other hand, since $\mathbf{F}(\mathbf{S}) \supset \mathbf{S}$ and $\mathbf{F}(\mathbf{S})$ is a ring, we have $\mathbf{F}(\mathbf{S}) \supset \mathbf{R}(\mathbf{S})$. Furthermore, since $\mathbf{F}(\mathbf{S})$ is a σ-ring, we have

$$\mathbf{F}(\mathbf{S}) \supset \mathbf{F}(\mathbf{R}(\mathbf{S})).$$

Consequently, we have

$$\mathbf{F}(\mathbf{S}) = \mathbf{F}(\mathbf{R}(\mathbf{S})).$$ □

Example 2.22. Let X be the real line and let $\mathbf{S}$ be the semiring given in Example 2.15. Then $\mathbf{F}(\mathbf{S})$ is called the *Borel field* on the real line, and it is usually denoted by $\mathbf{B}$. The sets in $\mathbf{B}$ are called *Borel sets*. We have seen the process of constructing $\mathbf{R}(\mathbf{S})$ from $\mathbf{S}$ by Theorem 2.5, and $\mathbf{R}(\mathbf{S})$ is just the ring given in Example 2.12. But the process for constructing $\mathbf{B}$ from $\mathbf{R}(\mathbf{S})$ is quite complex. $\mathbf{B}$ is also the σ-ring generated by the class of all open intervals, by the class of all closed intervals, by the class of all left open and right closed intervals, by the class of all left closed and right open intervals, or by the class of all intervals, respectively.

Theorem 2.7. *If* **C** *is a class, then*

$$\mathbf{F}_p(\mathbf{C}) = \left\{ \bigcup_{t \in T} \bigcap_{s \in S_t} E_S | E_S \in \mathbf{C}, S_t \text{ and } T \text{ are arbitrary index sets} \right\}.$$

Proof. Denote the right part of this equality by **E**.

(1) $\mathbf{E} \supset \mathbf{C}$ because S_t and T may be taken as singletons.
(2) By an application of the associativity of the set union, we know that **E** is closed under the formation of arbitrary unions.
(3) Because an arbitrary intersection of arbitrary unions of sets in a class **C** may be expressed by an arbitrary union of arbitrary intersections of sets in that class **C**, and because arbitrary intersections are associative, **E** is closed under the formation of arbitrary intersections.

Thus, **E** is a plump class including **C** and, therefore, $\mathbf{E} \supset \mathbf{F}_p(\mathbf{C})$. Conversely, any plump class including **C** includes **E**; hence, $\mathbf{F}_p(\mathbf{C}) \supset \mathbf{E}$. Consequently, $\mathbf{F}_p(\mathbf{C}) = \mathbf{E}$. □

Theorem 2.8. *For any class* **C** *and any set* A,

$$\mathbf{F}(\mathbf{C}) \cap A = \mathbf{F}(\mathbf{C} \cap A).$$

Similar conclusions about rings, monotone classes, and plump classes are true, as well.

Proof.

(1) $\mathbf{F}(\mathbf{C}) \cap A$ is a σ-ring and includes $\mathbf{C} \cap A$; so

$$\mathbf{F}(\mathbf{C}) \cap A \supset \mathbf{F}(\mathbf{C} \cap A) .$$

(2) Let

$$\mathbf{E} = \{E | E \cap A \in \mathbf{F}(\mathbf{C} \cap A), E \in \mathbf{F}(\mathbf{C})\}.$$

E is a σ-ring, and $\mathbf{E} \supset \mathbf{C}$. So $\mathbf{E} \supset \mathbf{F}(\mathbf{C})$, that is, $\forall E \in \mathbf{F}(\mathbf{C})$,

$$E \cap A \in \mathbf{F}(\mathbf{C} \cap A).$$

This shows that

$$\mathbf{F}(\mathbf{C}) \cap A \subset \mathbf{F}(\mathbf{C} \cap A).$$

Consequently,

$$\mathbf{F}(\mathbf{C}) \cap A = \mathbf{F}(\mathbf{C} \cap A).$$

The rest may be proved in the same way. □

Example 2.23. Let **B** be the Borel field on the real line. $\mathbf{B} \cap [0, 1]$ is called the Borel field on the unit interval. It is the σ-ring generated by the class of all intervals in [0, 1].

Theorem 2.9. *If* **R** *is a ring, then*

$$\mathbf{M}(\mathbf{R}) = \mathbf{F}(\mathbf{R}).$$

Proof. From Proposition 2.11, we know that **F(R)** is a monotone class. Since $\mathbf{F}(\mathbf{R}) \supset \mathbf{R}$, we have

$$\mathbf{F}(\mathbf{R}) \supset \mathbf{M}(\mathbf{R}).$$

If **M(R)** is a σ-ring, then we have

$$\mathbf{M}(\mathbf{R}) \supset \mathbf{F}(\mathbf{R}),$$

and, therefore, the proof would be complete.

To complete the proof, we need to prove that **M(R)** is a σ-ring. For any set F, let **K**(F) be the class of all those sets E for which $E - F$, $F - E$, and $E \cup F$ are all in **M(R)**. It is easy to see, by the symmetry of the positions of E and F in the definition of $\mathbf{K}(F)$, that

$$E \in \mathbf{K}(F) \Leftrightarrow F \in \mathbf{K}(E).$$

If $\{E_n\}$ is a monotone sequence of sets in $\mathbf{K}(F)$, then we have

$$\lim_n E_n - F = \lim_n (E_n - F) \in \mathbf{M}(\mathbf{R}),$$

$$F - \lim_n E_n = \lim_n (F - E_n) \in \mathbf{M}(\mathbf{R}),$$

$$F \cup \lim_n E_n = \lim_n (F \cup E_n) \in \mathbf{M}(\mathbf{R}),$$

that is, $\lim_n E_n \in \mathbf{K}(F)$. So, if $\mathbf{K}(F)$ is not empty, then it is a monotone class. $\forall F \in \mathbf{R}$, if $E \in \mathbf{R}$, then $E \in \mathbf{K}(F)$; that is, $\mathbf{R} \subset \mathbf{K}(F)$. It follows that

$$\mathbf{M}(\mathbf{R}) \subset \mathbf{K}(F), \quad \forall F \in \mathbf{R}.$$

Hence, $\forall E \in \mathbf{M}(\mathbf{R}), \forall F \in \mathbf{R}$, we have $E \in \mathbf{K}(F)$; therefore, by symmetry, $F \in \mathbf{K}(E)$; that is,

$$\mathbf{R} \subset \mathbf{K}(E),$$

for any$E \in \mathbf{M}(\mathbf{R})$. Noting again that $\mathbf{K}(E)$ is a monotone class, we have

$$\mathbf{M}(\mathbf{R}) \subset \mathbf{K}(E), \qquad \forall E \in \mathbf{M}(\mathbf{R}).$$

This shows that **M(R)** is a ring. From Proposition 2.12, we know that **M(R)** is a σ-ring. □

Corollary 2.1. *A monotone class including a ring includes the σ-ring generated by this ring.*

2.1.4 Atoms and Holes

Let **C** be an arbitrary nonempty class of subsets of X.

Definition 2.9. For any point $x \in X$, the set $\bigcap\{E|x \in E \in \mathbf{C}\}$ is called the *atom* of **C** at x, and denoted by $A(x/\mathbf{C})$. If there is no confusion, it will be called the atom at x, or atom for short, and denoted by $A(x)$. The class of all atoms of **C** is denoted by **A** [**C**], that is,

$$\mathbf{A}[\mathbf{C}] = \{A(x/\mathbf{C})|x \in X\}.$$

Clearly, for every $x \in X, x \in A(x)$. So, every atom is nonempty.
When $\bigcup \mathbf{C} \neq X$, then $A(x/\mathbf{C}) = X$ for any $x \notin \bigcup \mathbf{C}$. Thus, if we write

$$\mathbf{A}^{-}[\mathbf{C}] = \{A(x/\mathbf{C})|x \in \bigcup \mathbf{C}\},$$

then we have

$$\mathbf{A}[\mathbf{C}] - \mathbf{A}^{-}[\mathbf{C}] \subset \{X\}.$$

Proposition 2.16. *If $x \in E \in \mathbf{C}$, then $A(x) \subset E$.*

Example 2.24. Let $X = \{a, b, c\}, \mathbf{C} = \{A, B, C\}$, where $A = \{a\}$, $B = \{a, b\}$, $C = \{b, c\}$. Then, A, $\{b\}$, and C are atoms. That is, $A = A(a)$, $\{b\} = A(b)$, $C = A(c)$. From this example, we can see that it is not necessary that all sets in **C** be atoms of **C**, and that all atoms of **C** belong to **C**. But, if **C** is closed under the formation of arbitrary intersections, then we have

$$\mathbf{A}[\mathbf{C}] \subset \mathbf{C};$$

that is, in this case, $\forall A(x)$,

$$A(x) = \bigcap\{E|x \in E \in \mathbf{C}\} \in \mathbf{C}.$$

Example 2.25. If $\mathbf{C} = \mathbf{P}(X)$, then $\mathbf{A}[\mathbf{C}] = \{\{x\}|x \in X\}$.

Proposition 2.17. $\bigcup \mathbf{A}^{-}[\mathbf{C}] = \bigcup \mathbf{C}$.

Theorem 2.10. *Any set in **C** may be expressed by a union of atoms of **C**; moreover, any intersection of sets in **C** may be expressed by a union of atoms of **C**.*

Proof. It is sufficient to prove the second conclusion.

Let $\{E_t | t \in T\}$ be a family of sets in **C**. We have

$$\bigcap_{t\in T} E_t = \bigcup \left\{ A(x) | x \in \bigcap_{t\in T} E_t \right\}$$

In fact, on the one hand, by Proposition 2.16, for any $x \in \bigcap_{t\in T} E_t$, and any $t \in T$,

$$A(x) \subset E_t.$$

So, for any $x \in \bigcap_{t\in T} E_t$,

$$A(x) \subset \bigcap_{t\in T} E_t,$$

and it follows that

$$\bigcup \left\{ A(x) | x \in \bigcap_{t\in T} E_t \right\} \subset \bigcap_{t\in T} E_t.$$

On the other hand, since $x \in A(x)$, we have

$$\bigcap_{t\in T} E_t = \left\{ x | x \in \bigcap_{t\in T} E_t \right\} \subset \bigcup \left\{ A(x) | x \in \bigcap_{t\in T} E_t \right\}.$$

The proof is thus complete. □

Theorem 2.11. *Any intersection of atoms may be expressed by a union of atoms.*

Proof. Since any atom of **C** is an intersection of sets in **C**, by Theorem 2.10 and the associativity of intersections, we obtain the conclusion. □

Example 2.26. Let $X = \{a, b, c, d\}$, $\mathbf{C} = \{A, B, C, D\}$, where $A = \{a, c, d\}$, $B = \{b, c, d\}$, $C = \{c\}$, $D = \{d\}$. Then, $A(a) = A$, $A(b) = B$, $A(c) = C$, $A(d) = D$. We have

$$A(a) \cap A(b) = A(c) \cup A(d).$$

Theorem 2.12. *If* $A' \in \mathbf{A}[\mathbf{C}], x \in A'$, *then* $A(x) \subset A'$

Proof. Let

$$A' = A(x') = \bigcap \{E | x' \in E \in \mathbf{C}\} = \bigcap_{t\in T} E_t,$$

where $E_t \in \mathbf{C}$, T is an index set. Since $x \in A'$, we have $x \in E_t$ for all $t \in T$. Therefore, by Proposition 2.16, $A(x) \subset E_t$ for all $t \in T$. Consequently, we have

$$A(x) \subset A'. \qquad \square$$

Theorem 2.13. $A(x/\mathbf{C}) = A(x/\mathbf{A}[\mathbf{C}])$ *for any* $x \in X$, *and* $\mathbf{A}[\mathbf{C}] = \mathbf{A}[\mathbf{A}[\mathbf{C}]]$.

Proof. $\forall x \in X$, if $x \in B$ for some $B \in \mathbf{A}[\mathbf{C}]$, we have, by Theorem 2.12,

$$A(x/\mathbf{C}) \subset B,$$

and, therefore,

$$A(x/\mathbf{C}) \subset \bigcap\{B|x \in B \in \mathbf{A}[\mathbf{C}]\}.$$

Reviewing $x \in A(x/\mathbf{C}) \in \mathbf{A}[\mathbf{C}]$, we have

$$A(x/\mathbf{C}) \supset \bigcap\{B|x \in B \in \mathbf{A}[\mathbf{C}]\}.$$

Thus,

$$A(x/\mathbf{C}) = \bigcap\{B|x \in B \in \mathbf{A}[\mathbf{C}]\} = A(x/\mathbf{A}[\mathbf{C}]).$$

Consequently, we have

$$\mathbf{A}[\mathbf{C}] = \mathbf{A}[\mathbf{A}[\mathbf{C}]]. \qquad \square$$

Theorem 2.14. $\mathbf{A}[\mathbf{C} \cup \mathbf{A}[\mathbf{C}]] = \mathbf{A}[\mathbf{C}]$.

Proof. $\forall x \in X$,

$$\begin{aligned} A(x/\mathbf{C} \cup \mathbf{A}[\mathbf{C}]) &= \bigcap\{E|x \in E \in \mathbf{C} \cup \mathbf{A}[\mathbf{C}]\} \\ &= \left(\bigcap\{E|x \in E \in \mathbf{C}\}\right) \cap \left(\bigcap\{E|x \in E \in \mathbf{A}[\mathbf{C}]\}\right). \\ &= A(x/\mathbf{C}) \cap A(x/\mathbf{A}[\mathbf{C}]) = A(x/\mathbf{C}) \end{aligned}$$

Thus,

$$\mathbf{A}[\mathbf{C} \cup \mathbf{A}[\mathbf{C}]] = \mathbf{A}[\mathbf{C}]. \qquad \square$$

Theorem 2.15. *If* $\mathbf{C}' = \{\bigcup_{t \in T} E_t | E_t \in \mathbf{C}, t \in T,\ T \text{ is an arbitrary index set}\}$, *then* $\mathbf{A}[\mathbf{C}'] = \mathbf{A}[\mathbf{C}]$.

Proof. $\bigcup \mathbf{C}' = \bigcup \mathbf{C}$. $\forall x \in \bigcup \mathbf{C}$, by absorption, we have

$$\begin{aligned} A(x/\mathbf{C}') &= \bigcap \{E \mid x \in E \in \mathbf{C}'\} \\ &= \bigcup \left\{ \bigcup_{t \in T} E_t \,\middle|\, x \in \bigcup_{t \in T} E_t, E_t \in \mathbf{C}, t \in T, T \text{ is an arbitrary set} \right\}. \\ &= \bigcap \{E \mid x \in E \in \mathbf{C}\} = A(x/\mathbf{C}) \end{aligned}$$

Thus, we have

$$\mathbf{A}[\mathbf{C}'] = \mathbf{A}[\mathbf{C}]. \qquad \square$$

Theorem 2.16. *If* $\mathbf{C}$ *is closed under the formation of difference, then* $\mathbf{A}^-[\mathbf{C}]$ *is a partition of* $\bigcup \mathbf{C}$ *(Definition 2.18).*

Proof. Since $\bigcup \mathbf{A}^-[\mathbf{C}] = \bigcup \mathbf{C}$, we only need to prove that the different atoms in $\mathbf{A}^-[\mathbf{C}]$ must be disjoint, that is, $\forall A(x), A(y) \in \mathbf{A}^-[\mathbf{C}]$,

$$A(x) \neq A(y) \Rightarrow A(x) \cap A(y) = \emptyset.$$

If both $x \in A(y)$ and $y \in A(x)$, then, by Theorem 2.12, we have $A(x) = A(y)$. So, when $A(x) \neq A(y)$, we can suppose $x \notin A(y)$ without any loss of generality. In this case, if there exists $z \in A(x) \cap A(y)$, we have the result that, from $x \notin A(y)$ and $z \in A(y)$, there exists $E \in \mathbf{C}$ such that $x \notin E$, but $z \in E$. Thus, if we take $F \in \mathbf{C}$, satisfying $x \in F$ and set $G = F - E$, then $x \in G \in \mathbf{C}$, but $z \notin G$. This contradicts the fact that $z \in A(x)$. Therefore, we have $A(x) \cap A(y) = \emptyset$. □

Corollary 2.2. *If* $\mathbf{F}$ *is an algebra, then* $\mathbf{A}[\mathbf{F}]$ *is a partition of* X.

The following theorem provides an expression of $\mathbf{F}_p(\mathbf{C})$ by the atoms of $\mathbf{C}$.

Theorem 2.17. $\mathbf{F}_p(\mathbf{C}) = \{\bigcup_{t \in T} A_t \mid A_t \in \mathbf{A}[\mathbf{C}], T$ *is an arbitrary index set*$\}$.

Proof. By Theorem 2.7, Theorem 2.10, and the associativity of set unions, the conclusion immediately follows. □

Theorem 2.18. $\mathbf{A}[\mathbf{F}_p(\mathbf{C})] = \mathbf{A}[\mathbf{C}]$.

Proof. It follows directly from Theorems 2.13, 2.15, and 2.17. □

Theorem 2.19. $\mathbf{F}_p(\mathbf{C}) = \mathbf{F}_p(\mathbf{A}[\mathbf{C}])$.

Proof. From the definition of the atom and Theorem 2.10, the equality is easily obtained. □

A concept of AU-class is interrelated closely with the concept of the atom.

Definition 2.10. The *AU-class* is a nonempty class $\mathbf{C}$ with anticlosedness under the formation of unions, that is, $\forall \mathbf{C}' \subset \mathbf{C}$,

$$\bigcup \mathbf{C}' \in \mathbf{C} \Rightarrow \bigcup \mathbf{C}' \in \mathbf{C}'.$$

By the convention for operations of union and intersection (introduced in Section 2.1.2), if $\mathbf{C}'$ is an empty class, then $\bigcup \mathbf{C}' = \emptyset$. Hence, if $\emptyset \in \mathbf{C}$, and $\mathbf{C}$ is an AU-class, it should follow that $\emptyset \in \mathbf{C}'$. This is a contradiction. So, no AU-class contains the empty set $\emptyset$.

Proposition 2.18. *If* $\mathbf{C}$ *is an AU-class, then all nonempty subclasses of* $\mathbf{C}$ *are AU-classes as well.*

Theorem 2.20. $\mathbf{A}[\mathbf{C}]$ *is an AU-class.*

Proof. Let $\{A(x)|x \in D\}$ be a family of atoms of $\mathbf{C}$. Denote

$$B = \bigcup \{A(x)|x \in D\} = \bigcup_{x \in D} A(x).$$

If $B \in \mathbf{A}[\mathbf{C}]$, then $\exists x_0 \in B$ such that $B = A(x_0)$. From $x_0 \in \bigcup_{x \in D} A(x)$, we have $x_0 \in A(x_0')$ for some $x_0' \in D$. By applying Theorem 2.12, it follows that

$$A(x_0') \supset A(x_0) = B.$$

The inverse inclusion relation is evident. Consequently, we have

$$B = A(x_0') \in \{A(x)|x \in D \subset \bigcup \mathbf{C}\}.$$

This shows that $\mathbf{A}[\mathbf{C}]$ is an AU-class. □

In general, if $\mathbf{C}$ is an AU-class, a set in $\mathbf{C}$ may not be an atom of $\mathbf{C}$.

Example 2.27. X and $\mathbf{C}$ are given as in Example 2.24. It is easy to verify that $\mathbf{C}$ is an AU-class, but B is not an atom of $\mathbf{C}$.

However, we have the following property.

Theorem 2.21. *Let* $\mathbf{C}$ *be an AU-class. If* $\mathbf{C} \supset \mathbf{A}[\mathbf{C}]$, *then we have*

$$\mathbf{C} = \mathbf{A}[\mathbf{C}].$$

Proof. If $\mathbf{C} \neq \mathbf{A}[\mathbf{C}]$, then there exists a nonempty set $E \in \mathbf{C}$, but $E \notin \mathbf{A}[\mathbf{C}]$. By Theorem 2.10 there exists a family of atoms $\{A_t| t \in T\}$ such that $E = \bigcup_{t \in T} A_t$. Since $\mathbf{C}$ is an AU-class, $\exists t_0 \in T$ such that $E = A_{t_0} \in \mathbf{A}[\mathbf{C}]$. This contradicts $E \notin \mathbf{A}[\mathbf{C}]$. □

A dual concept to the "atom" is the "hole."

Definition 2.11. Let $\hat{\mathbf{C}} = \{\overline{E} | E \in \mathbf{C}\}$. For any point $x \in X$, the set

$$\bigcup\{E | x \in \overline{E} \in \hat{\mathbf{C}}\}$$

is called the *hole* of **C** at x, denoted by $H(x/\mathbf{C})$, or $H(x)$ for short. The class of all holes of **C** is denoted by **H**[**C**].

We can also write

$$H(x/\mathbf{C}) = \bigcup\{E | x \notin E \in \mathbf{C}\}.$$

It is evident that, for any $x \in X, x \notin H(x/\mathbf{C})$. So, X is not a hole.

The relation between hole and atom is given in the following proposition.

Proposition 2.19. $H(x/\mathbf{C}) = \overline{A(x/\hat{\mathbf{C}})}$.

Example 2.28. We use X and **C** given in Example 2.24. In this case, $\overline{A} = \{b, c\} = C, \overline{B} = \{c\},\ \overline{C} = \{a\} = A$. Consequently, $H(a) = C,\ H(b) = A,\ H(c) = B$.

Example 2.29. If $\mathbf{C} = \mathbf{P}(X)$, then $\mathbf{H}[\mathbf{C}] = \{\overline{\{x\}} | x \in X\}$.

Definition 2.12. The *AI-class* is a nonempty class **C** with anticlosedness under the formation of intersections, that is, $\forall \mathbf{C}' \subset \mathbf{C}$,

$$\bigcap \mathbf{C}' \in \mathbf{C} \Rightarrow \bigcap \mathbf{C}' \in \mathbf{C}'.$$

All properties of the AU-class can be easily converted into analogous properties of the AI-class by replacing atoms with holes [Liu and Wang 1985, 1987].

2.1.5 S-Compact Space

Let **C** be a nonempty class of subsets of X. Usually, we also use the term "space" to mean (X, **C**). Especially, when **C** is a σ-algebra (or σ-ring), denoted by **F**, we call (X, **F**) a *measurable space*, and the sets in **F** are called *measurable sets*. We say (X, **C**) or (X, **F**) is to be finite, countable, or uncountable if X is finite, countable, or uncountable, respectively.

Definition 2.13. (X, **C**) is said to be *S-precompact* iff for any sequence of sets in **C** there exists some convergent subsequence, that is, $\forall \{E_n\} \subset \mathbf{C}, \exists \{E_{n_i}\} \subset \{E_n\}$ such that,

$$\limsup_i E_{n_i} = \liminf_i E_{n_i};$$

(X, **C**) is said to be *S-compact* iff it is S-precompact and the limit of the above-mentioned subsequence belongs to **C**, that is, $\forall \{E_n\} \subset \mathbf{C}, \exists \{E_{n_i}\} \subset \{E_n\}$ such that $\lim_i E_{n_i}$ exists and

$$\lim_i E_{n_i} \in \mathbf{C}.$$

Obviously, any S-precompact measurable space is S-compact.

Example 2.30. Any finite space is S-compact. In fact, if $(X, \mathbf{C})$ is a finite space, then $\mathbf{C}$ is finite too. So, from any sequence of sets in $\mathbf{C}$, we can always pick out a subsequence in which all sets are identical; therefore, this subsequence converges to the same set as that in the subsequence.

From the above example we can also see that, although X is not finite, $(X, \mathbf{C})$ is S-compact so long as $\mathbf{C}$ is finite.

Example 2.31. If $\mathbf{C}$ is a nest (or, say, a chain; in this case it is fully ordered by the inclusion relation between sets), then $(X, \mathbf{C})$ is S-precompact. To show this, it is sufficient to prove the following lemma.

Lemma 2.1. *If* $\mathbf{C}$ *is an infinite nest, then there exists a monotone subsequence of sets in* $\mathbf{C}$.

Proof. According to the order given by the inclusion relation, if there exists $\mathbf{D} \subset \mathbf{C}$ that does not have the greatest element, then we can pick out an increasing sequence of elements (that is, sets) in $\mathbf{D}$ (and therefore, in $\mathbf{C}$). Otherwise, any subset of $\mathbf{C}$ has its greatest element. Thus, we take the greatest element of $\mathbf{C}$ as E_1, the greatest element of $\mathbf{C} - \{E_1\}$ as E_2, the greatest element of $\mathbf{C} - \{E_1, E_2\}$ as $E_3, \ldots$. Finally, we obtain a decreasing subsequence $\{E_n\}$ of $\mathbf{C}$. □

In the following, we give an example of the non-S-precompact space, in which the universe of discourse X is an uncountable set.

Example 2.32. Let X_0 be a set that contains at least two points, $X = X_1 \times X_2 \times \ldots \times X_n \times \ldots$ be an infinite-dimensional product space, where $X_i = X_0,\ i = 1, 2, \ldots$, and $\mathbf{C} = \mathbf{P}(X)$. Take $a \in X_0$ arbitrarily and denote

$$A_n = X_1 \times X_2 \times \ldots \times X_{n-1} \times \{a\} \times X_{n+1} \times \ldots.$$

A_n is an nth dimensional cylinder set based on $\{a\}$. Then, for such a set sequence $\{A_n\}$ there exists no subsequence that is convergent. In fact, for any given subsequence $\{A_{n_i}\} \subset \{A_n\}$, we take $b \in X_0 - \{a\}$ arbitrarily, and set

$$x_k = \begin{cases} a & \text{if } k = n_{2i}, i = 1, 2, \ldots \\ b & \text{else.} \end{cases}$$

Denote $x = (x_1, x_2, \ldots)$; then $x \in A_{n_{2i}}$, but $x \notin A_{n_{2i-1}}, i = 1, 2, \ldots$.
So,

$$x \in \limsup_i A_{n_i},$$

but

$$x \quad \liminf_i A_{n_i}.$$

That is, the subsequence $\{A_{n_i}\}$ does not converge. Therefore, $(X, \mathbf{C})$ is not S-precompact.

For a countable space we have an affirmative conclusion.

Theorem 2.22. *If X is countable, then $(X, \mathbf{C})$ is S-precompact.*

Proof. Denote $X = \{x_1, x_2, \ldots\}$. Any subset E of X corresponds uniquely to a binary number

$$\begin{aligned} b(E) &= {}_1b \times (1/2) + {}_2b \times (1/2)^2 + \ldots + {}_nb \times (1/2)^n + \ldots \\ &= 0.{}_1b_2b \ldots {}_nb \ldots \end{aligned}$$

in [0,1], where

$${}_ib = \begin{cases} 1 & \text{if } x_i \in E \\ 0 & \text{if } x_i \notin E. \end{cases}$$

We should note that such a correspondence is not one to one; for example, $\{x_1\}$ corresponds to 0.1, $\overline{\{x_1\}}$ corresponds to 0.0111..., but 0.1 = 0.0111....

For an arbitrarily given set sequence $\{E_n\} \subset \mathbf{C}$, $\{E_n\}$ corresponds to a number sequence $\{b_n\} \subset [0,1]$ with $E_n \mapsto b_n$. Since $\{b_n\}$ is bounded, there exists a convergent subsequence $\{b_{n_i}\}$. If all $b_{n_i}, i = 1, 2, \ldots$, are constant, then the conclusion of this theorem is obviously true. Otherwise, we can suppose, with no loss of generality, that $\{b_{n_i}\}$ is strictly decreasing, and $b_{n_i} \to b \in [0,1]$. If we adopt the restriction that b is represented by a binary number with infinitely many zeros after its decimal point, then b corresponds uniquely to a set E by the converse of the above-mentioned correspondence. It is not difficult to see that $\overline{E}$ must be an infinite set. Arbitrarily fixing a bit ${}_jb$ of b, we have ${}_jb_{n_i} = {}_jb$ when i is large enough. That is to say, there exist at most finitely many sets in $\{E_{n_i}\}$ that do not contain x_j when $x_j \in E$; and there exist at most finitely many sets in $\{E_{n_i}\}$ that contain x_j when $x_j \in \overline{E}$. This shows that $x_j \in \liminf_i E_{n_i}$ when $x_j \in E$ and $x_j \in \liminf_i \overline{E_{n_i}} = \overline{\limsup_i E_{n_i}}$ when $x_j \in \overline{E}$, namely,

$$\liminf_i E_{n_i} \supset E \text{ and } \overline{\limsup_i E_{n_i}} \supset \overline{E}.$$

The latter implies that

$$\limsup_i E_{n_i} \subset E.$$

So,

$$\limsup_i E_{n_i} \subset E \subset \liminf_i E_{n_i}.$$

This means that $\lim_i E_{n_i}$ exists.

Thus, we have proved that $(X, \mathbf{C})$ is S-precompact. □

If we consider a measurable space $(X, \mathbf{F})$ with $X \in \mathbf{F}$, then, by Theorem 2.16, $\mathbf{A}[\mathbf{F}]$ is a partition of X. The quotient space $(X_{\mathbf{A}}, \mathbf{F}_{\mathbf{A}})$ induced by $\mathbf{A}[\mathbf{F}]$ from $(X, \mathbf{F})$

(Definition 2.19) is called the *reduced space* of (*X*, **F**). $\mathbf{F_A}$ and **F** are isomorphic. So, we can get a further theorem as follows.

Theorem 2.23. *If the reduced space of* (*X*, **F**) *is countable, then* (*X*, **F**) *is S-compact.*

Proof. The conclusion of this theorem follows from Theorem 2.22 and the fact that the *S*-precompact measurable space is *S*-compact. □

Theorem 2.24. *If* **F** *is a σ-algebra containing only countably many sets* (*that is,* **F** *is a countable class*), *then* (*X*, **F**) *is S-compact.*

Proof. Since F is a countable class of sets and **F** is closed under the formation of countable intersections, every atom $A(x/\mathbf{F}) \in \mathbf{F}$. So, A[F] is a countable class, too. This shows that the reduced space $(X_A, \mathbf{F}_A)$ of (*X*, **F**) is countable. Therefore, by Theorem 2.23, (*X*, **F**) is *S*-compact. □

2.1.6 Relations, Posets, and Lattices

Definition 2.14. Let *E* and *F* be nonempty sets. A *relation R* from *E* to *F* is a subset of $E \times F$. If $(a, b) \in R$, we say "*a* is related to *b*" and write *aRb*; if $(a, b) \notin R$, we say "*a* is not related to *b*" and write $a \not R b$. In the special case when $R \subset E \times E$, we use "on *E*" instead of "from *E* to *E*."

Example 2.33. Let $X = \{a, b, c\}, E = \{a, b\}$, and $B = \{0, 1\}$. The characteristic function χ_E of *E* is a relation (denoted by R_E) from *X* to *B*. We have $aR_E1, bR_E1, cR_E0, a\not R_E0, b\not R_E0, c\not R_E1$.

Example 2.34. Let $X = (-\infty, \infty)$. The symbol $<$ with the common meaning "less than" is a relation on *X*, and it is a subset of $X \times X : R = \{(x, y)|x<y\}$. We have, for example $(1, 2) \in R, (-5, 5) \in R, (2, 1) \notin R$, and $(1, 1) \notin R$.

Example 2.35. Let *X* be a nonempty set. The inclusion of sets $\subset$ is a relation on **P**(*X*); that is, $\{(E, F)|E \subset F\}$ is a subset of $\mathbf{P}(X) \times \mathbf{P}(X)$.

Example 2.36. Let *E* be any nonempty set. The *identity relation* on *E*, denoted by Δ_E, is the set of all pairs in $E \times E$ with equal elements:

$$\Delta_E = \{(a, a)|a \in E\}.$$

Example 2.37. Let $X = \{0, 1, 2, \ldots\}$. We can define a relation R_3 on *X* as follows: aR_3b iff $a = b$ (mod 3); that is, *a* and *b* have the same remainder when they are divided by 3.

Definition 2.15. Let *R* be a relation from *E* to *F*. The *inverse* of *R*, denoted by R^{-1}, is the relation from *F* to *E* which consists of those ordered pairs (*b*, *a*) for which *aRb*; that is $R^{-1} = \{(b, a)|(a, b) \in R\}$.

It is easy to see that

$$aRb \Leftrightarrow bR^{-1}a$$

and, therefore, we have the following proposition.

Proposition 2.20. $(R^{-1})^{-1} = R$.

Example 2.38. Let R be the relation given in Example 2.34. Its inverse, $R^{-1} = \{(x,y)|y<x\} = \{(x,y)|x>y\}$, has the meaning "greater than" and is denoted by the symbol $>$.

Definition 2.16. A relation R on a set E is called:

(a) *reflexive* iff aRa for each $a \in E$;
(b) *symmetric* iff aRb implies bRa for any $a, b \in E$;
(c) *transitive* iff aRb and bRc implies aRc for any $a, b, c \in E$.

Definition 2.17. A relation R on a set E is called an *equivalence relation* iff R is reflexive, symmetric, and transitive.

Example 2.39. The identity relation Δ, as defined in Example 2.36, is reflexive, symmetric, and transitive; hence, it is an equivalence relation.

Example 2.40. The relation defined in Example 2.34 ("less than," $<$) is neither reflexive nor symmetric, but it is transitive.

Example 2.41. Let $X = (-\infty, \infty)$. The relation described by the phrase "less than or equal to," which is usually denoted by the symbol $\leq$, is reflexive and transitive but it is not symmetric.

Example 2.42. The relation R_3 defined in Example 2.37 is reflexive, symmetric, and transitive; consequently, it is an equivalence relation.

Definition 2.18. A disjoint class $\{E_1, E_2, \ldots, E_n\}$ of nonempty subsets of E is called a *partition* of E iff $\bigcup_{i=1}^{n} E_i = E$.

Example 2.43. Let $X = \{a, b, c, d, e, f, g\}$, and let

(1) $A_1 = \{a, c, e\}, A_2 = \{b\}, A_3 = \{d, g\}$
(2) $B_1 = \{a, e, g\}, B_2 = \{c, d\}, B_3 = \{b, e, f\}$
(3) $C_1 = \{a, b, e, g\}, C_2 = \{c\}, C_3 = \{d, f\}$
(4) $D_1 = X$
(5) $E_1 = \{a\}, E_2 = \{b\}, E_3 = \{c\}, E_4 = \{d\}, E_5 = \{e\}, E_6 = \{f\}, E_7 = \{g\}$.

Then, classes $\{C_1, C_2, C_3\}, \{D_1\}$, and $\{E_1, E_2, E_3, E_4, E_5, E_6, E_7\}$ are partitions of X, but $\{A_1, A_2, A_3\}$ and $\{B_1, B_2, B_3\}$ are not.

Example 2.44. Let $X = [0, \infty)$. The class $\{[n-1, n)|n = 1, 2, \ldots\}$ is a partition of X.

Definition 2.19. Let R be an equivalence relation on E. For each $x \in E$, the set $[x] = \{y|xRy\}$ is called an *equivalence class* of E (in fact, it is a subset of E). The class of all equivalence classes of E induced by R, denoted by E/R, is called the *quotient* of E by R, that is, $E/R = \{[x]|x \in E\}$.

Proposition 2.21. *Let R be an equivalence relation on a set E. Then*

$$[x] = [y] \Leftrightarrow xRy$$

for any $x, y \in E$, and E/R is a partition of E.

Example 2.45. For the relation R_3 defined in Example 2.37, the quotient X/R_3 is formed by the following three distinct equivalence classes:

$$E_0 = \{0, 3, 6, 9, \ldots\}$$
$$E_1 = \{1, 4, 7, 10, \ldots\}$$
$$E_2 = \{2, 5, 8, 11, \ldots\}$$

$\{E_0, E_1, E_2\}$ is a partition of $X = \{0, 1, 2, \ldots\}$.

Definition 2.20. A relation R on set E is called *antisymmetric* iff aRb and bRa imply $a = b$ for any $a, b \in E$.

Example 2.46. The relations given in Example 2.34, 2.35, and 2.41 are antisymmetric.

Definition 2.21. Let R be a relation on a set E. If R is reflexive, antisymmetric, and transitive, then R is called a *partial ordering on E*, and (E, R) is called a *partially ordered set* (or, *poset*).

Example 2.47. Referring to Example 2.35, the pair $(\mathbf{P}(X), \subset)$ is a partially ordered set.

Example 2.48. Referring to Example 2.41, the pair $(X, \leq)$ is a partially ordered set.

Example 2.49. Let $\overline{F}$ be the set of all generalized real-valued functions on $(-\infty, \infty)$. We define a relation $\leq$ on $\overline{F}$ as follows: $f \leq g$ iff $f(x) \leq g(x)$ for all $x \in (-\infty, \infty)$. The relation $\leq$ is a partial ordering on $\overline{F}$ and, therefore, $(\overline{F}, \leq)$ is a partially ordered set.

From now on we use $(P, \leq)$ to denote a partially ordered set.

Definition 2.22. Let $(P, \leq)$ be a partially ordered set and let $E \subset P$. An element a in P is called an *upper bound* of E iff $x \leq a$ for all $x \in E$. An upper bound a of E is called the *least upper bound* of E (or *supremum* of E) iff $a \leq b$ for any upper bound b of E. The least upper bound of E is denoted by $\sup E$ or $\vee E$. An element a in P is called a *lower bound* of E iff $a \leq x$ for all $x \in E$. A lower bound a of E is called the *greatest lower bound* of E (or *infimum* of E) iff $b \leq a$ for any lower bound b of E. The greatest lower bound of E is denoted by $\inf E$ or $\wedge E$.

When E consists of only two elements, say x and y, we may write $x \vee y$ instead of $\vee \{x, y\}$ and $x \wedge y$ instead of $\wedge \{x, y\}$.

Proposition 2.22. *If the least upper bound (or the greatest lower bound) of a set $E \subset P$ exists, then it is unique.*

Definition 2.23. A partially ordered set $(P, \leq)$ is called an *upper semilattice* (or *lower semilattice*) iff $x \vee y$ (or $x \wedge y$, respectively) exists for any $x, y \in P$. $(P, \leq)$. is called a *lattice* iff it is both upper semilattice and lower semilattice.

Example 2.50. The partially ordered set $(\mathbf{P}(X), \subset)$ is a lattice. For any sets $E, F \subset X, E \cup F = \sup\{E, F\}$ and $E \cap F = \inf\{E, F\}$.

Definition 2.24. A partially ordered set $(P, \leq)$ is called a *fully ordered set* or a *chain* iff either $x \leq y$ or $y \leq x$ for any $x, y \in P$.

Example 2.51. The partially ordered set $(X, \leq)$ of Example 2.41 is a fully ordered set.

Example 2.52. The partially ordered set $(\overline{F}, \leq)$, of Example 2.49 is not a fully ordered set.

Example 2.53. The partially ordered set $(\mathbf{P}(X), \subset)$ is not a fully ordered set if X consists of more than one point.

The fully ordered set $((-\infty, \infty), \leq)$ has many convenient properties. One of them, which is often used in this text, is expressed by the following proposition.

Proposition 2.23. *Let E be a set of real numbers. If E has an upper bound (or a lower bound), then $\sup E$ (or $\inf E$) exists; furthermore, for any given $\varepsilon > 0$, there exists $x = x(\varepsilon) \in E$ such that $\sup E \leq x + \varepsilon$ (or $x - \varepsilon \leq \inf E$, respectively).*

2.2 Classical Measures

Let X be a nonempty set, $\mathbf{C}$ be a nonempty class of subsets of X, and $\mu : \mathbf{C} \to [0, \infty]$ be a nonnegative, extended real valued set function defined on $\mathbf{C}$.

Definition 2.25. A set E in $\mathbf{C}$ is called the *null set* (with respect to μ) iff $\mu(E) = 0$.

Definition 2.26. μ is *additive* iff

$$\mu(E \cup F) = \mu(E) + \mu(F)$$

whenever

$$E \in \mathbf{C}, F \in \mathbf{C}, E \cup F \in \mathbf{C}, \text{ and } , E \cap F = \emptyset.$$

Definition 2.27. μ is *finitely additive* iff

$$\mu\left(\bigcup_{i=1}^{n} E_i\right) = \sum_{i=1}^{n} \mu(E_i)$$

for any finite, disjoint class $\{E_1, E_2, \ldots, E_n\}$ of sets in **C** whose union is also in **C**.

Definition 2.28. μ is *countably additive* iff

$$\mu\left(\bigcup_{i=1}^{\infty} E_i\right) = \sum_{i=1}^{\infty} \mu(E_i)$$

for any disjoint sequence $\{E_n\}$ of sets in **C** whose union is also in **C**.

Definition 2.29. μ is *subtractive* iff

$$E \in \mathbf{C}, F \in \mathbf{C}, E \subset F, F - E \in \mathbf{C}, \text{ and } \mu(E) < \infty$$

imply

$$\mu(F - E) = \mu(F) - \mu(E).$$

Theorem 2.25. *If μ is additive, then it is subtractive.*

Definition 2.30. μ is called a *measure* on **C** iff it is countably additive and there exists $E \in \mathbf{C}$ such that $\mu(E) < \infty$.

Example 2.54. If $\mu(E) = 0, \forall E \in \mathbf{C}$, then μ is a measure on **C**.

Example 2.55. Let **C** contain at least one finite set. If $\mu(E) = |E|, \forall E \in \mathbf{C}$, where $|E|$ is the number of those points that belong to E, then μ is a measure on **C**.

Theorem 2.26. *If μ is a measure on* **C** *and* $\emptyset \in \mathbf{C}$, *then* $\mu(\emptyset) = 0$. *Moreover, μ is finitely additive.*

Definition 2.31. Let μ be a measure on **C**. A set E in **C** is said to have a *finite measure* iff $\mu(E) < \infty$; E is said to have a *σ-finite measure* iff there exists a sequence $\{E_n\}$ of sets in **C** such that

$$E \subset \bigcup_{n=1}^{\infty} E_n \text{ and } \mu(E_n) < \infty, n = 1, 2, \ldots.$$

μ is *finite* (or *σ-finite*) on **C** iff every $\mu(E)$ is finite (or σ-finite, respectively) for every $E \in \mathbf{C}$.

Definition 2.32. Let μ be a measure on **C**. μ is *complete* iff

$$E \in \mathbf{C}, F \subset E, \text{and } \mu(E) = 0$$

imply

$$F \in \mathbf{C}.$$

In other words, a measure on **C** is complete if and only if any subset of a null set belongs to **C**.

Definition 2.33. μ is *monotone* iff

$$E \in \mathbf{C}, F \in \mathbf{C}, \textit{and}, E \subset F$$

imply

$$\mu(E) \leq \mu(F).$$

In the following, we take a semiring **S**, a ring **R**, and a σ-ring **F**, respectively, as the class **C**, and μ is always a nonnegative, extended real-valued set function on this class.

Theorem 2.27. *Let* **S** *be a semiring. If* μ *is additive on* **S**, *then it is finitely additive and monotone.*

Definition 2.34. μ is *subadditive* iff

$$\mu(E) \leq \mu(E_1) + \mu(E_2)$$

whenever

$$E \in \mathbf{C}, \ E_1 \in \mathbf{C}, \quad E_2 \in \mathbf{C}, \ \text{and } E = E_1 \cup E_2.$$

Definition 2.35. μ is *finitely subadditive* iff

$$\mu(E) \leq \sum_{i=1}^{n} \mu(E_i)$$

for any finite class $\{E_1, E_2, \ldots, E_n\}$ of sets in **C** such that $E = \bigcup_{i=1}^{n} E_i \in \mathbf{C}$.

Definition 3.36. μ is *countably subadditive* iff

$$\mu(E) \leq \sum_{i=1}^{\infty} \mu(E_i)$$

for any sequence $\{E_i\}$ of sets in **C** such that $E = \bigcup_{i=1}^{\infty} E_i \in \mathbf{C}$.

Theorem 2.28. *If* μ *is countably subadditive and* $\mu(\emptyset) = 0$, *then* μ *is finitely subadditive.*

Definition 2.37. Let $E \in \mathbf{C}$. μ is *continuous from below at* E iff

$$\{E_n\} \subset \mathbf{C},\ E_1 \subset E_2 \subset \ldots,\ \text{and } \lim_n E_n = E$$

imply

$$\lim_n \mu(E_n) = \mu(E);$$

μ is *continuous from above at* E iff

$$\{E_n\} \subset \mathbf{C},\ E_1 \supset E_2 \supset \ldots, \mu(E_1) < \infty,$$

and

$$\lim_n E_n = E$$

imply

$$\lim_n \mu(E_n) = \mu(E).$$

μ is *continuous from below* (*on* $\mathbf{C}$) iff it is continuous from below at every set in $\mathbf{C}$; μ is *continuous from above* (*on* $\mathbf{C}$) iff it is continuous from above at every set in $\mathbf{C}$; μ is *continuous* iff it is both continuous from below and continuous from above.

Theorem 2.29. *If* μ *is a measure on a semiring* $\mathbf{S}$, *then* μ *is countably subadditive and continuous.*

Definition 2.38. Let $\mathbf{C}_1$ and $\mathbf{C}_2$ be classes of subsets of X, $\mathbf{C}_1 \subset \mathbf{C}_2$, and μ_1 and μ_2 be set functions on $\mathbf{C}_1$ and $\mathbf{C}_2$, respectively. μ_2 is called an *extension* of μ_1 iff $\mu_1(E) = \mu_2(E)$ whenever $E \in \mathbf{C}_1$.

Let **S** be a semiring, **R(S)** be the ring generated by **S**. Since any set in **R(S)** can be expressed by a disjoint finite union of sets in **S**, we have the following extension theorem for a measure on **S**.

Theorem 2.30. *If* μ *is a measure on* **S**, *then there is a unique measure* $\overline{\mu}$ *on* **R(S)** *such that* $\overline{\mu}$ *is an extension of* μ. *If* μ *is finite or* σ*-finite, then so is* $\overline{\mu}$.

The extension of μ (on **S**) may also be denoted by $\overline{\mu}$ [on **R(S)**] without any confusion.

Example 2.56. Let $X = (-\infty, \infty)$. $\mathbf{S} = \{[a, b) | -\infty < a \le b < \infty\}$ is a semiring. Define a set function μ on **S** by

$$\mu([a, b)) = b - a.$$

μ is countably additive, and, therefore, μ is a finite measure on **S**. μ can be extended to a finite measure on **R(S)**, the class of all finite, disjoint unions of bounded, left closed, and right open intervals. More generally, if g is a finite, increasing, and left continuous real-valued function of a real variable, then

$$\mu_g([a,b)) = g(b) - g(a), \forall [a,b) \in \mathbf{S}$$

determines a finite measure μ_g on **S**, and it can be extended onto **R(S)**.

Example 2.57. Let the ring **R** consist of all finite subsets of X and f be an extended real-valued, nonnegative function on X. If we define μ by

$$\mu(\{x_1, x_2, \ldots, x_n\}) = \sum_{i=1}^{n} f(x_i) \text{ for any } \{x_1, x_2, \ldots, x_m\} \in \mathbf{R} \text{ and } \mu(\emptyset) = 0,$$

then μ is a measure on **R**. In fact, the class **S** consisting of all singletons of X and the empty set $\emptyset$ is a semiring. If we define μ on **S** by

$$\mu(\{x\}) = f(x) \text{ for any } x \in X \text{ and } \mu(\emptyset) = 0,$$

then μ is a measure on **S**, and the above-mentioned measure μ on **R** is just the extension of this measure on **S**.

Theorem 2.31. *If μ is a measure on a ring* **R**, *then it is continuous.*

Theorem 2.32. *Let μ be additive on a ring* **R** *and $\mu(\emptyset) = 0$. If μ is either continuous from below, or continuous from above at the empty set $\emptyset$ and finite, then it is σ-additive on* **R**.

It should be noted that, on a semiring, an analogous conclusion of Theorem 2.32 is not true.

Example 2.58. Let $X = \{x | 0 \le x \le 1, x \text{ is a rational number}\}$, $\mathbf{S} = \{\{x | a \le x \le b,\ x \text{ is a rational number}\} \,|\, 0 \le a \le b \le 1,\ a \text{ and } b \text{ are rational numbers}\}$. If we define μ on **S** by

$$\mu(\{x | a \le x \le b, x \text{ is a rational number}\}) = b - a,$$

then μ is finitely additive and continuous, but it is not countably additive.

Definition 2.39. A nonempty class **C** is *hereditary* iff

$$F \in \mathbf{C}$$

whenever

$$E \in \mathbf{C} \text{ and } F \subset E.$$

A hereditary class is a σ-ring if and only if it is closed under the formation of countable unions.

Example 2.59. The classes given in Examples 2.11, 2.14, and 2.16 are hereditary, and the last one is a hereditary σ-ring.

The hereditary σ-ring generated by a class $\mathbf{C}$, i.e., the smallest hereditary σ-ring containing $\mathbf{C}$, is denoted by $\mathbf{H}_\sigma(\mathbf{C})$

Theorem 2.33. $\mathbf{H}_\sigma(\mathbf{C})$ *is the class of all sets that can be covered by countably many sets in* $\mathbf{C}$.

Example 2.60. Let $X = (-\infty, \infty)$ and $\mathbf{C}$ be the class of all bounded intervals in X. Then $\mathbf{H}_\sigma(\mathbf{C}) = \mathbf{P}(X)$.

If $\mathbf{C}$ is a nonempty class closed under the formation of countable unions, then $\mathbf{H}_\sigma(\mathbf{C})$ is just the class of all sets that are subsets of some set in $\mathbf{C}$.

Definition 2.40. Let $\mathbf{H}_\sigma$ be a hereditary σ-ring, μ^* be an extended, real-valued, nonnegative set function on $\mathbf{H}_\sigma$. μ^* is called an *outer measure* iff it is monotone, countably subadditive, and such that $\mu^*(\emptyset) = 0$.

The same terminology concerning finiteness, σ-finiteness, and extension is used for outer measures as for measures.

Example 2.61. Let X be a finite set and $X \times X$ be a product space. $\mathbf{P}(X \times X)$ is a hereditary σ-ring. Define μ^* on $\mathbf{P}(X \times X)$ by

$$\mu^*(E) = |\mathrm{Proj}(E)|, \forall E \in \mathbf{P}(X \times X),$$

where $\mathrm{Proj}(E) = \{x|(x, y) \in E\}$. Then μ^* is a finite outer measure on $\mathbf{P}(X \times X)$.

Theorem 2.34. *If* μ *is a measure on a ring* $\mathbf{R}$, *then the set function* μ^* *on* $\mathbf{H}_\sigma(\mathbf{R})$ *defined by*

$$\mu^*(E) = \inf\left\{\sum_{n=1}^{\infty} \mu(E_n)|E_n \in \mathbf{R},\, n = 1, 2, \ldots; E \subset \bigcup_{n=1}^{\infty} E_n\right\}$$

is an extension of μ *to an outer measure on* $\mathbf{H}_\sigma(\mathbf{R})$; *if* μ *is* σ-*finite, then so is* μ^*.

This outer measure μ^* is called the outer measure induced by the measure μ.

Definition 2.41. Let μ^* be an outer measure on a hereditary σ-ring $\mathbf{H}_\sigma$. A set $E \in \mathbf{H}_\sigma$ is μ^*-*measurable* iff

$$\mu^*(A) = \mu^*(A \cap E) + \mu^*(A \cap \overline{E}), \forall A \in \mathbf{H}_\sigma.$$

Theorem 2.35. *If* μ^* *is an outer measure on a hereditary* σ-*ring* $\mathbf{H}_\sigma$ *and if* $\overline{\mathbf{F}}$ *is the class of all* μ^* *-measurable sets, then* $\overline{\mathbf{F}}$ *is a* σ-*ring, and the set function* $\overline{\mu}$ *defined for every* $E \in \overline{\mathbf{F}}$ *by* $\overline{\mu}(E) = \mu^*(E)$ *is a complete measure on* $\overline{\mathbf{F}}$.

This measure $\overline{\mu}$ is called the measure induced by the outer measure μ^*.

Theorem 2.36. *Let μ be a measure on a ring* **R**, *μ^* be the outer measure induced by μ. Then every set in* **R** *is μ^*-measurable, and therefore* $\mathbf{F}(\mathbf{R}) \subset \overline{\mathbf{F}}$

Theorem 2.37. *If μ is a σ-finite measure on a ring* **R**, *then so is the measure $\overline{\mu}$ on* **F(R)**, *and $\overline{\mu}$ is the unique extension of μ on* **R** *to* **F(R)**.

Theorem 2.38. *If μ is a measure on a σ-ring* **F**,

$$\mathbf{F}' = \{E\Delta N | E \in \mathbf{F},\ N \subset F \text{ for some } F \in \mathbf{F} \text{ with } \mu(F) = 0\},$$

then $\mathbf{F}'$ *is a σ-ring, and set function μ' defined for every $E \in F$* by $\mu'(E\Delta N) = \mu(E)$ *is a complete measure on* $\mathbf{F}'$.

This measure μ' is called the completion of μ.

Theorem 2.39. *If μ is a σ-finite measure on a ring* **R**, *then* $\mathbf{F}' = \overline{\mathbf{F}}$, *and μ' is just identical with $\overline{\mu}$*

Theorem 2.40. *If μ is a σ-finite measure on a ring* **R**, *then for every $\varepsilon > 0$ and every set $E \in$* **F(R)** *that has finite measure there exists a set $E_0 \in$* **R** *such that*

$$\overline{\mu}(E \,\Delta\, E_0) \leq \varepsilon.$$

Example 2.62. In Example 2.56, a finite measure μ on **R(S)** satisfying $\mu([a, b)) = b - a$ for any $[a - b) \in \mathbf{S} = \{[a, b)| - \infty < a \leq b < \infty\}$ is obtained. This measure μ can be extended uniquely to a σ-finite measure on a σ-ring $\mathbf{B} = \mathbf{F}(\mathbf{R}(\mathbf{S})) = \mathbf{F}(\mathbf{S})$, the class of all Borel sets (this class is also a σ-field, so-called Borel field on the real line). The complete measure $\overline{\mu}$ on $\overline{\mathbf{B}}$ is called a *Lebesgue measure* (the incomplete measure μ on **B** is usually called a Lebesgue measure as well), and the sets in $\overline{\mathbf{B}}$ are called Lebesgue measurable sets of the real line. More generally, if g is a finite, increasing, and left continuous real-valued function of a real variable, the measure μ_g on **R(S)** obtained in Example 2.56 can be extended uniquely to a complete measure $\overline{\mu}_g$ on a σ-field $\overline{\mathbf{F}}_g$ containing the Borel field, and the measure $\overline{\mu}_g$ is called a *Lebesgue-Stieltjes measure* induced by g. In particular, if g is a probability distribution function, then g can uniquely determine a probability measure on the Borel field **B** on the real line. At last, it should be noted that not all subsets of $X = (-\infty,\ \infty)$ are Lebesgue measurable.

2.3 Fuzzy Sets

Let X be a nonempty set considered as the universe of discourse. A *standard fuzzy set* in X (that is, in fact, a standard fuzzy subset of X) is characterized by a membership function $m : X \rightarrow [0, 1]$. A standard fuzzy set is called *normalized* if

$$\sup_{x \in X} m(x) = 1.$$

We use the same symbols, capital letters $A, B, C, \ldots$, to denote both standard fuzzy sets and ordinary (crisp) sets in classical set theory. Membership functions of standard fuzzy sets $A, B, C, \ldots$, are denoted by $m_A, m_B, m_C, \ldots$. If A denotes a standard fuzzy set, then $m_A(x)$ is called the *grade of membership* of x in A. The class of all standard fuzzy sets is denoted by $\tilde{\mathbf{P}}(X)$. Since any ordinary set E in $\mathbf{P}(X)$ can be defined by its characteristic function $\chi_E : X \to \{0, 1\}$, it is a special standard fuzzy set and, therefore, $\mathbf{P}(X) \subset \tilde{\mathbf{P}}(X)$. In this book, only standard fuzzy sets are considered, and we refer to them from now on as, simply, fuzzy sets.

Definition 2.42. If $m_A(x) \leq m_B(x)$ for any $x \in X$, we say that fuzzy set A is *included* in fuzzy set B, and we write $A \subset B$ or, equivalently, $B \supset A$. If $A \subset B$ and $B \subset A$, we say that A and B are *equal* (or, A equals B), which we write as $A = B$.

Definition 2.43. Let A and B be fuzzy sets. The *standard union* of A and B, $A \cup B$, is defined by

$$m_{A\cup B}(x) = m_A(x) \vee m_B(x), \quad \forall x \in X,$$

where $\vee$ denotes the maximum operator.

Definition 2.44. Let A and B be fuzzy sets. The *standard intersection* of A and B, $A \cap B$, is defined by

$$m_{A\cap B}(x) = m_A(x) \wedge m_B(x), \quad \forall x \in X,$$

where $\wedge$ denotes the minimum operator.

Similar to the way operations on ordinary sets are treated, we can generalize the standard union and the standard intersection for an arbitrary class of fuzzy sets: If $\{A_t | t \in T\}$ is a class of fuzzy sets, where T is an arbitrary index set, then $\cup_{t\in T} A_t$ is the fuzzy set having membership function $\sup_{t\in T} m_{A_t}(x), x \in X$, and $\cap_{t\in T} A_t$ is the fuzzy set having membership function $\inf_{t\in T} m_{A_t}(x), x \in X$.

Definition 2.45. Let A be a fuzzy set. The *standard complement* of A, $\overline{A}$, is defined by

$$m_{\overline{A}}(x) = 1 - m_A(x), \ \forall x \in X.$$

Two or more of the three basic operations can also be combined. For example, the difference $A - B$ of fuzzy sets A and B can be expressed as $A \cap \overline{B}$, so that

$$m_{A-B}(x) = \min[m_A(x), 1 - m_B(x)]$$

for all $x \in X$. Since only standard operations on fuzzy sets are used in this book, we omit from now on the adjective "standard" if there is no confusion.

Example 2.63. Let $X = \{a, b, c\}$ and let fuzzy sets A, B, and C be defined by the following membership functions

$$m_A(x) = \begin{cases} 0.4 & \text{if } x = a \\ 0.7 & \text{if } x = b \\ 0 & \text{if } x = c, \end{cases}$$

$$m_B(x) = \begin{cases} 0.6 & \text{if } x = a \\ 1 & \text{if } x = b \\ 0.2 & \text{if } x = c, \end{cases}$$

$$m_C(x) = \begin{cases} 0.1 & \text{if } x = a \\ 0 & \text{if } x = b \\ 1 & \text{if } x = c. \end{cases}$$

Then, $A \subset B$,

$$m_{A \cup C}(x) = \begin{cases} 0.4 & \text{if } x = a \\ 0.7 & \text{if } x = b \\ 1 & \text{if } x = c, \end{cases}$$

$$m_{A \cap C}(x) = \begin{cases} 0.1 & \text{if } x = a \\ 0 & \text{if } x = b \\ 0 & \text{if } x = c, \end{cases}$$

and

$$m_{\overline{B}}(x) = \begin{cases} 0.4 & \text{if } x = a \\ 0 & \text{if } x = b \\ 0.8 & \text{if } x = c. \end{cases}$$

Example 2.64. Fuzzy sets can be used to represent fuzzy concepts. Let X be a reasonable age interval of human beings: [0, 100]. Assume that the concept of "young" is represented by a fuzzy set Y whose membership function is

$$m_Y(x) = \begin{cases} 1 & \text{if } x \leq 25 \\ (40 - x)/15 & \text{if } 25 < x < 40 \\ 0 & \text{if } x \geq 40 \end{cases}$$

and the concept of "old" is represented by a fuzzy set O whose membership function is

$$m_O(x) = \begin{cases} 0 & \text{if } x \leq 50 \\ (x-50)/15 & \text{if } 50 < x < 65 \\ 1 & \text{if } x \geq 65. \end{cases}$$

Then, the grade of membership of 28 years of age in Y is 0.8 while that of 45 years of age in O is 0. Consider now the complement of Y and O whose membership functions are

$$m_{\overline{Y}}(x) = \begin{cases} 0 & \text{if } x \leq 25 \\ (x-25)/15 & \text{if } 25 < x < 40 \\ 1 & \text{if } x \geq 40, \end{cases}$$

and

$$m_{\overline{O}}(x) = \begin{cases} 1 & \text{if } x \leq 50 \\ (65-x)/15 & \text{if } 50 < x < 65 \\ 0 & \text{if } x \geq 65. \end{cases}$$

These fuzzy sets represent the concepts of "not young" and "not old," respectively. Fuzzy set $\overline{Y} \cap \overline{O}$ whose membership function is

$$m_{\overline{Y} \cap \overline{O}}(x) = \begin{cases} 0 & \text{if } x \leq 25 \\ (x-25)/15 & \text{if } 25 < x < 40 \\ 1 & \text{if } 40 \leq x \leq 50 \\ (65-x)/15 & \text{if } 50 < x < 65 \\ 0 & \text{if } x \geq 65, \end{cases}$$

represents the concept of "neither young nor old," that is, "middle-aged" (Fig. 2.2a–2.2e). Furthermore, we have $O \subset \overline{Y}$ that is, "old" implies "not young."

Theorem 2.41. *The standard operations of union, intersection, and complement of fuzzy sets have the following properties, where S_t and T are index sets:*

Involution: $\overline{\overline{A}} = A$

Commutativity: $A \cup B = B \cup A$

$A \cap B = B \cap A$

Associativity: $$\bigcup_{t \in T} \left(\bigcup_{s \in S_t} A_s \right) = \bigcup_{s \in \bigcup_{t \in T} S_t} A_s$$

$$\bigcap_{t \in T} \left(\bigcap_{s \in S_t} A_s \right) = \bigcap_{s \in \bigcup_{t \in T} S_t} A_s$$

Distributivity:

$$B \cap \left(\bigcup_{t \in T} A_t \right) = \bigcup_{t \in T} (B \cap A_t)$$

$$B \cup \left(\bigcap_{t \in T} A_t \right) = \bigcap_{t \in T} (B \cup A_t)$$

Idempotence:

$$A \cup A = A$$

$$A \cap A = A$$

Absorption:

$$A \cup (A \cap B) = A$$

$$A \cap (A \cup B) = A$$

Absorption by X *and* Ø:

$$A \cup X = X$$

$$A \cap \emptyset = \emptyset$$

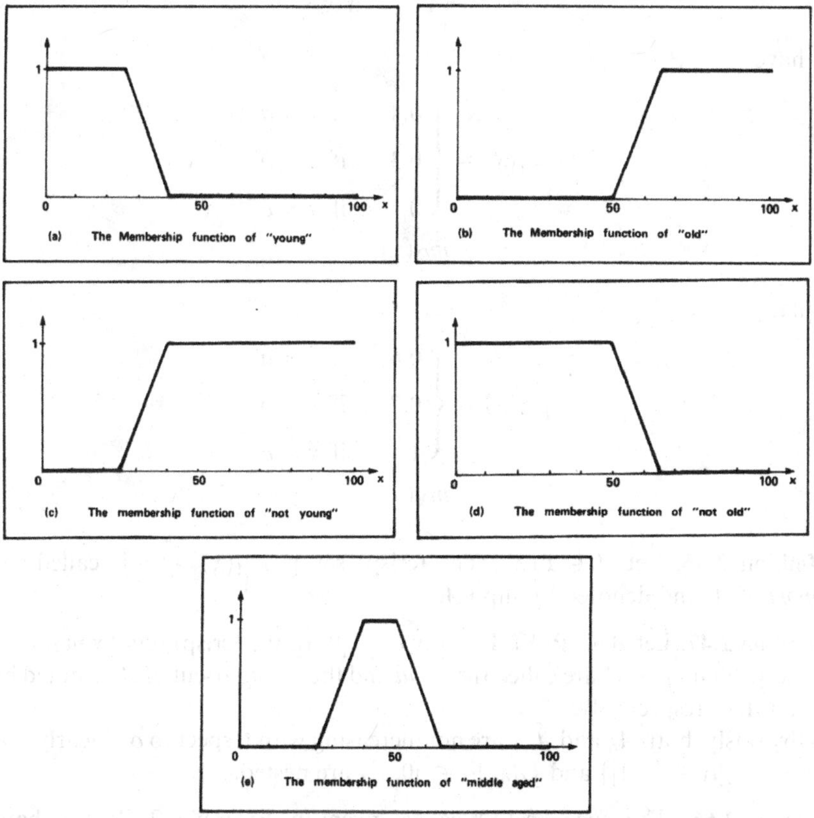

Fig. 2.2 Membership functions of fuzzy sets defined on the interval [0,100] and representing linguistic terms pertaining to age of human beings

Identity:

$$A \cup \emptyset = A$$

$$A \cap X = A$$

De Morgan's laws:

$$\overline{\bigcup_{t \in T} A_t} = \bigcap_{t \in T} \overline{A}_t$$

$$\overline{\bigcap_{t \in T} A_t} = \bigcup_{t \in T} \overline{A}_t$$

In comparison with crisp set operations (see Proposition 2.2), the law of contradiction and the law of excluded middle are not true for fuzzy sets. This is illustrated by the following example.

Example 2.65. X and A are given in Example 2.63. The complement $\overline{A}$ of A has a membership function

$$m_{\overline{A}}(x) = \begin{cases} 0.6 & \text{if } x = a \\ 0.3 & \text{if } x = b \\ 1 & \text{if } x = c. \end{cases}$$

We have

$$m_{A \cap \overline{A}}(x) = \begin{cases} 0.4 & \text{if } x = a \\ 0.3 & \text{if } x = b \\ 0 & \text{if } x = c \end{cases}$$
$$\neq m_{\emptyset}(x).$$

Similarly,

$$m_{A \cup \overline{A}}(x) = \begin{cases} 0.6 & \text{if } x = a \\ 0.7 & \text{if } x = b \\ 1 & \text{if } x = c \end{cases}$$
$$\neq m_X(x).$$

Definition 2.46. Let $A \in \tilde{\mathbf{P}}(X)$. The (crisp) set $\{x | m_A(x) > 0\}$ is called the *support* of A, and denoted by supp A.

Definition 2.47. Let $A \in \tilde{\mathbf{P}}(X)$. For any $\alpha \in [0, 1]$, the (crisp) sets $\{x | m_A(x) \geq \alpha\}$ and $\{x | m_A(x) > \alpha\}$ are called the *α-cut* and the *strong* α-cut of A, denoted by A_α and $A_{\alpha+}$, respectively.

Obviously, both A_α and $A_{\alpha+}$ are nonincreasing with respect to α. Clearly, the classes $\{A_\alpha | \alpha \in [0, 1]\}$ and $\{A_{\alpha+} | \alpha \in [0, 1]\}$ are nested.

Example 2.66. The fuzzy set Y is as given in Example 2.64. We have $Y_{0.2} = [0, 37]$ and $Y_{0.6} = [0, 31]$.

Theorem 2.42. *Let $\{A_t | t \in T\} \subset \tilde{\mathbf{P}}(X)$. Then, for any $\alpha \in [0, 1]$,*

$$\left(\bigcup_{t \in T} A_t\right)_{\alpha+} = \bigcup_{t \in T} (A_t)_{\alpha+}$$

and

$$\left(\bigcap_{t \in T} A_t\right)_{\alpha} = \bigcap_{t \in T} (A_t)_{\alpha}.$$

Theorem 2.43. *Let $A \in \tilde{\mathbf{P}}(X)$. Then*

$$\overline{A}_{\alpha} = \overline{(A_{(1-\alpha)+})}.$$

If we use a symbol $\alpha \cdot E$ to denote the fuzzy set whose membership function is

$$m(x) = \begin{cases} \alpha & \text{if } x \in E \\ 0 & \text{if } x \notin E \end{cases}$$

for any $\alpha \in [0, 1]$ and any crisp set $E \in \mathbf{P}(X)$, then each fuzzy set can be fully characterized by its α-cuts, as expressed by the following theorem.

Theorem 2.44. (Decomposition Theorem). *For any $A \in \tilde{\mathbf{P}}(X)$,*

$$A = \bigcup_{\alpha \in [0,1]} \alpha \cdot A_{\alpha}.$$

Definition 2.48. Let $X = (-\infty, \infty)$. A normalized fuzzy set $A \in \tilde{\mathbf{P}}(X)$ is called a *fuzzy number* if A_{α} is a finite closed interval for each $\alpha \in (0, 1]$.

Definition 2.49. A *rectangular fuzzy number* is a fuzzy number with membership function having a form as

$$m(x) = \begin{cases} 1 & \text{if } x \in [a_l, a_r] \\ 0 & \text{otherwise,} \end{cases}$$

where $a_l, a_r \in R$ with $a_l \leq a_r$.

A rectangular fuzzy number is identified with the corresponding vector $\langle a_l, a_r \rangle$ and is an interval number, essentially. Any crisp real number a can be regarded as a special rectangular fuzzy number with $a_l = a_r = a$.

Definition 2.50. A *triangular fuzzy number* is a fuzzy number with membership function

$$m(x) = \begin{cases} 1 & \text{if } x = a_0 \\ \dfrac{x - a_l}{a_0 - a_l} & \text{if } x \in [a_l, a_0) \\ \dfrac{x - a_r}{a_0 - a_r} & \text{if } x \in (a_0, a_r] \\ 0 & \text{otherwise,} \end{cases}$$

where $a_l, a_0, a_r \in R$, with $a_l \leq a_0 \leq a_r$.

A triangular fuzzy number is identified with the corresponding vector $\langle a_l, a_0, a_r \rangle$. A triangular fuzzy number is called *symmetric* if $a_0 - a_l = a_r - a_0$. Any crisp real number a can be regarded as a special triangular fuzzy number with $a_l = a_0 = a_r = a$.

Example 2.67. Let $X = (-\infty, \infty)$. Fuzzy sets A and B with

$$m_A(x) = \begin{cases} 0 & \text{if } x < 6 \text{ or } x > 12 \\ (x-6)/3 & \text{if } 6 \leq x \leq 9 \\ (12-x)/3 & \text{if } 9 < x \leq 12, \end{cases}$$

$$m_B(x) = \begin{cases} 0 & \text{if } x < 2 \text{ or } x > 4 \\ x-2 & \text{if } 2 \leq x \leq 3 \\ 4-x & \text{if } 3 < x \leq 4 \end{cases}$$

are triangular fuzzy numbers (Fig. 2.3).

Definition 2.51. A *trapezoidal fuzzy number* is a fuzzy number with membership function

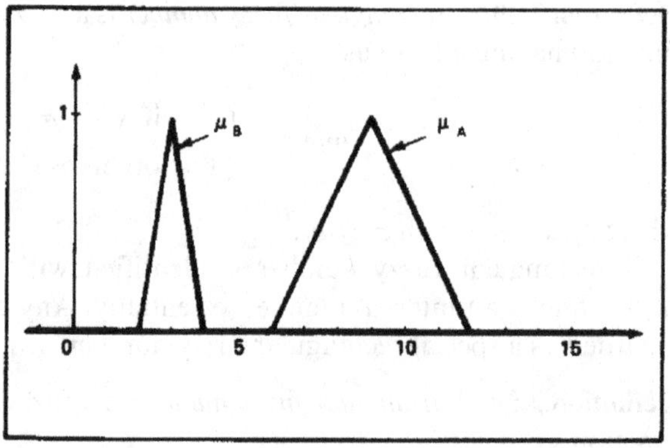

Fig. 2.3 Membership functions of triangular fuzzy numbers A and B

$$m(x) = \begin{cases} 1 & \text{if } x = [a_b, a_c] \\ \dfrac{x - a_l}{a_b - a_l} & \text{if } x \in [a_l, a_b) \\ \dfrac{x - a_r}{a_c - a_r} & \text{if } x \in (a_c, a_r] \\ 0 & \text{otherwise,} \end{cases}$$

where $a_l, a_b, a_c, a_r \in R$, with $a_l \leq a_b \leq a_c \leq a_r$.

A trapezoidal fuzzy number is identified with the corresponding vector $\langle a_l, a_b, a_c, a_r \rangle$. A trapezoidal fuzzy number is called *symmetric* if $a_b - a_l = a_r - a_c$. Any rectangular fuzzy number $\langle a_l, a_r \rangle$can be regarded as a special trapezoidal fuzzy number with $a_l = a_b$ and $a_c = a_r$. Similarly, any triangular fuzzy number $\langle a_l, a_0, a_r \rangle$ can be regarded as a special trapezoidal fuzzy number with $a_b = a_c = a_0$. Of course, any crisp real number a can be regarded as a special trapezoidal fuzzy number with $a_l = a_b = a_c = a_r = a$.

Example 2.68. Fuzzy sets Y, M, and O discussed in Examples 2.64 are trapezoidal fuzzy numbers.

Definition 2.52. Let $X = (-\infty, \infty)$. A fuzzy set $A \in \tilde{\mathbf{P}}(X)$ is called convex, if for any $x_1, x_2, x_3 \in X$,

$$m_A(x_2) \geq m_A(x_1) \wedge m_A(x_3)$$

where $x_1 \leq x_2 \leq x_3$.

Theorem 2.45. *Any fuzzy number is a convex fuzzy subset of* $(-\infty, \infty)$, *and its membership function is upper semicontinuous.*

The following extension principle introduced by Zadeh [1975] is a useful tool for extending nonfuzzy mathematical concepts to fuzzy sets (to fuzzify classical mathematical concepts).

Extension Principle. Let $X_1, X_2, \ldots, X_n$, and Y be nonempty (crisp) sets, $X = X_1 \times X_2 \times \cdots \times X_n$ be the product set of $X_1, X_2 \ldots, X_n$, and f be a mapping from X to Y. Then, for any given n fuzzy sets $A_i \in \tilde{\mathbf{P}}(X_i), i = 1, 2, \ldots, n$, we can induce a fuzzy set $B \in \tilde{\mathbf{P}}(Y)$ through f such that

$$m_B(y) = \sup_{y=f(x_1,x_2,\ldots,x_n)} \min[m_{A_1}(x_1), m_{A_2}(x_2), \ldots, m_{A_n}(x_n)],$$

where we use the convention that

$$\sup_{x \in \emptyset}\{x | x \in [0, \infty]\} = 0$$

when $f^{-1}(y) = \emptyset$.

As a special case, if $*$ is a binary operator on points in the universe of discourse X, then, by using the extension principle, we can obtain a binary operator $*$ (we use the same symbol) on fuzzy sets in $\tilde{\mathbf{P}}(X)$:

$$m_{A*B}(z) = \sup_{x*y=z} [m_A(x) \wedge m_B(y)], \forall z \in X,$$

where $A, B \in \tilde{\mathbf{P}}(X)$.

Now, we can use the extension principle to define addition, subtraction, multiplication, and division operations on fuzzy numbers, which are generalizations of the corresponding operations on real numbers.

Definition 2.53. Let A and B be fuzzy numbers. Then $A + B$, $A - B$, $A \cdot B$ and A/B are defined by

$$m_{A+B}(z) = \sup_{x+y=z} [m_A(x) \wedge m_B(y)],$$

$$m_{A-B}(z) = \sup_{x-y=z} [m_A(x) \wedge m_B(y)],$$

$$m_{A \cdot B}(z) = \sup_{x \cdot y=z} [m_A(x) \wedge m_B(y)],$$

and

$$m_{A/B}(z) = \sup_{x/y=z, y \neq 0} [m_A(x) \wedge m_B(y)] \text{ (when } 0 \notin \text{supp } B)$$

for any $z \in X$, respectively.

Example 2.69. Fuzzy numbers A and B are given in Example 2.67. Then we have

$$m_{A+B}(x) = \begin{cases} 0 & \text{if } x < 8 \text{ or } x > 16 \\ (x-8)/4 & \text{if } 8 \leq x \leq 12 \\ (16-x)/4 & \text{if } 12 < x \leq 16 \end{cases}$$

(Fig. 2.4), and

$$m_{A-B}(x) = \begin{cases} 0 & \text{if } x < 2 \text{ or } x > 10 \\ (x-2)/4 & \text{if } 2 \leq x \leq 16 \\ (10-x)/4 & \text{if } 6 < x \leq 10 \end{cases}$$

(Fig. 2.5). Viewing the real number 3 as a fuzzy number, we have $A = 3 \cdot B$ and $B = A/3$.

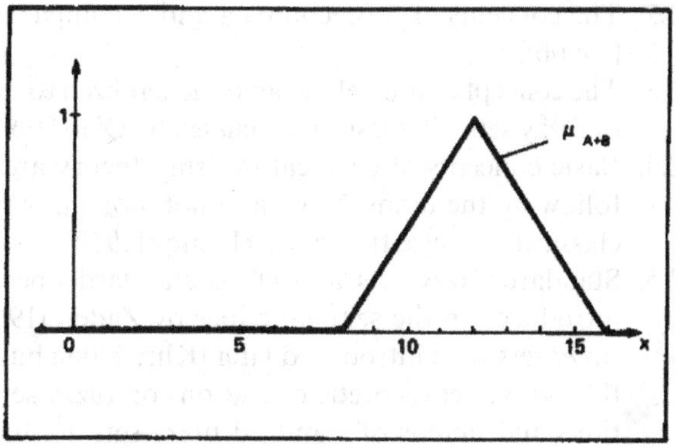

Fig. 2.4 Membership function of $A + B$

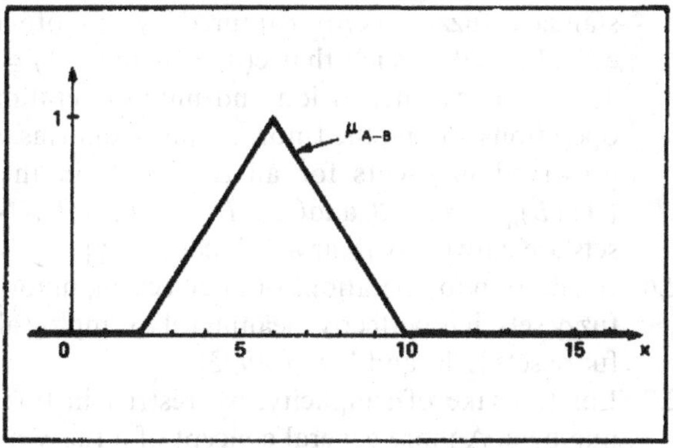

Fig. 2.5 Membership function of A–B

Definition 2.54. A *fuzzy partition* of X is a class of nonempty fuzzy sets defined on X, $\{A_i | i \in I\}$, such that

$$\sum_{i \in I} m_{A_i}(x) = 1$$

for all $x \in X$.

Clearly, any fuzzy set on X and its standard complement is a fuzzy partition of X. The three fuzzy sets that represent the concepts of *young, old,* and *middle-aged* in Example 2.64 form a fuzzy partition of the interval [0, 100].

Notes

2.1. The basic knowledge on sets and classes can be found in numerous books, including the classic book by Halmos [1950]. For a complete and up-to-date coverage of classical set theory, we recommend the book by Jech [2003].

2.2. The concepts of S-precompact and S-compact were introduced by Wang [1990b].
2.3. The concept of σ-algebra can be generalized to fuzzy σ-field, which is a class of fuzzy sets. This issue is discussed by Qiao [1990], as well as in Chapter 14.
2.4. Basic concepts of classical measure theory are introduced in Section 2.2 following the terminology and notation employed in the classic book on classical measure theory by Halmos [1950].
2.5. Standard fuzzy sets as well as standard operations on fuzzy sets were introduced in the seminal paper by Zadeh [1965]. Several other types of fuzzy sets were introduced later [Klir, 2006], but they are not considered in this book. Set theoretic operations on fuzzy sets are not unique. Intersections and unions of standard fuzzy sets are mathematically captured by operations known as triangular norms and conorms (or t-norms and t-conorms) [Klement et al., 2000, Klir and Yuan, 1995]. Complements of standard fuzzy sets are captured by monotone nonincreasing functions $c\colon [0,1] \to [0,1]$ such that $c(0) = 1$ and $c(1) = 0$ [Klir and Yuan, 1995]. The standard intersection and union operations are the only *cutworthy* operations among the t-norms and t-conorms, which means that they are preserved in α-cuts for all $\alpha \in (0,1]$ in the classical sense. That is, $(A \cap B)_\alpha = A_\alpha \cap B_\alpha \text{and} (A \cup B)_\alpha = A_\alpha \cup B_\alpha$. No complements of fuzzy sets are cutworthy [Klir and Yuan, 1995].
2.6. In addition to operations of intersection, union, and complementation on fuzzy sets, it is perfectly meaningful to employ also *averaging operations* on fuzzy sets [Klir and Yuan, 1995].
2.7. For the sake of simplicity, we restrict in this book to triangular fuzzy numbers. A more general concept of a fuzzy number (sometimes called a fuzzy interval) involves nonlinear functions and its α-cut for $\alpha = 1$might be an interval of real numbers [Klir and Yuan, 1995].
2.8. Arithmetic operations on fuzzy numbers introduced in Definition 2.53 form a basis for the so-called *standard fuzzy arithmetic*, which is based on the assumption that there are no constraints among the fuzzy numbers involved. If this assumption is not warranted, the constraints must be taken into account. Principles of *constrained fuzzy arithmetic* are discussed in [Klir, 1997, 2006; Klir and Pan, 1998].
2.9. The literature on fuzzy set theory has been rapidly growing, especially during the last twenty years or so. Two important handbooks, edited by Ruspini et al. [1998] and Dubois and Prade [2000], are recommended as convenient sources of information on virtually any aspect of fuzzy set theory. From among the growing number of textbooks on fuzzy set theory, any of the following general textbooks is recommended for further study: [Klir and Yuan, 1995a], [Lin and Lee, 1996], [Nguyen and Walker, 1997], [Pedrycz and Gomide, 1998], and Zimmermann [1996]. Another valuable resource is the following pair of books that contain classical papers on fuzzy set theory by Lotfi A. Zadeh, the founder of fuzzy set theory: [Yager et al., 1987] and [Klir and Yuan, 1996].

Exercises

2.1. Let $X = (-\infty, \infty)$. Explain the following sets and classes in natural language:

(a) $\{X|0 \leq x \leq 1\}$;
(b) $\{X|x < 0\}$;
(c) $\{\{x\}|x \in X\}$;
(d) $\{E|E \subset X\}$.

2.2. Let $X_1 = X_2 = (-\infty, \infty), X = X_1 \times X_2$. Use shading to indicate the following sets on the Euclidean plane:

(a) $\{(x_1, x_2)|x_1 + x_2 > 1\}$;
(b) $\{(x_1, x_2)|x_1^2 \leq x_2\}$;
(c) $\{(x_1, x_2)|x_2 > 5\}$.

2.3. Prove the following equalities:

(a) $(E - G) \cap (F - G) = (E \cap F) - G$;
(b) $(E - F) - G = E - (F \cup G)$;
(c) $E - (F - G) = (E - F) \cup (E \cap G)$;
(d) $(E - F) \cap (G - H) = (E \cap G) - (F \cup H)$.

2.4. Prove the following equalities:

(a) $E \Delta F = F \Delta E$;
(b) $E \Delta (F \Delta G) = (E \Delta F) \Delta G$;
(c) $E \cap (F \Delta G) = (E \cap F) \Delta (E \cap G)$.

2.5. Prove that $\overline{\limsup_n E_n} = \liminf_n \overline{E_n}$.

2.6. Indicate the superior limit and the inferior limit of the set sequence $\{E_n\}$ where E_n is given as follows:

(a) $E_n = (n, n + 3/2)$;
(b) $E_n = [a_n, b_n]$ with $a_n = \min(0, (-2)^n); b_n = \max\ (0, (-2)^n)$;
(c) $E_n = \{n, n + 1, \ldots\}$;
(d) $E_n = \{x|nx$ is a natural number$\}$;
(e) $E_n = [1/n, n]$.

2.7. Which set sequence in Exercise 2.6 is monotone and for which does the limit exist?

2.8. Prove:

$$\overline{\lim_n}(E \cup F_n) = E \cup \overline{\lim_n}\, F_n,$$

$$\underline{\lim_n}(E - F_n) = E - \overline{\lim_n}\, F_n.$$

2.9. Prove Proposition 2.6 (4), (5), and (6).

2.10. Let $X = (-\infty, \infty) \times (-\infty, \infty) = \{(x, y) | -\infty < x < \infty, -\infty < y < \infty\}$. Prove that the class of all sets that have the form

$$\{(x, y) | -\infty < a_1 \leq x < b_1 < \infty, -\infty < a_2 \leq y < b_2 < \infty\}$$

is a semiring.

2.11. Prove Proposition 2.11.

2.12. Is a monotone class closed under the formation of limit operations of set sequence? Why or why not?

2.13. Prove that

$$\mathbf{F}_p(\mathbf{C}) = \left\{ \bigcap_{t \in T} \bigcup_{s \in S_t} E_s | E_s \in \mathbf{C}, S_t \text{ and } T \text{ are arbitrary index sets} \right\}$$

2.14. Categorize the class **C** given in the following descriptions as either a ring, semiring, algebra, σ-ring, σ-algebra, monotone class, or a plump class:

(a) $X = (-\infty, \infty)$, **C** is the class of all bounded, left open, and right closed intervals
(b) $X = \{1, 2, \ldots\}$, $\mathbf{C} = \{\{n, n+1, \ldots\} | n = 1, 2, \ldots\} \cup \{\emptyset\}$
(c) X is a nonempty set, E is a nonempty subset of X, $\mathbf{C} = \{F | E \subset F \subset X\}$
(d) X is a nonempty set, E is a nonempty subset of $X, E \neq X$, $C = \{F | F \subset E\}$
(e) X is a nonempty set, E is a nonempty subset of X, $\mathbf{C} = \{E\}$.

2.15 What are the rings (algebras, σ-rings, σ-algebras, monotone classes, plump classes, respectively) generated by the classes **C** given in Exercise 2.14?

2.16. Indicate what A [C] is for each of the following classes **C**:

(a) $X = \{1, 2, 3, 4, 5\}$, $\mathbf{C} = \{A, B, C, D, E\}$, where $A = \{1, 2, 3\}$, $B = \{1, 2, 4\}$, $C = \{1\}, D = \{2, 4\}$, $E = \emptyset$
(b) $X = \{1, 2, 3, 4, 5\}, \mathbf{C} = \{A, B, C, D, E\}$, where $A = \{1, 2, 3\}$, $B = \{1, 2, 4\}$, $C = \{1\}, D = \{1, 5\}, E = \{4, 5\}$
(c) $X = \{1, 2, 3, 4, 5\}$, $\mathbf{C} = \{A, B, C, D, E\}$, where $A = \{1, 2, 3\}$, $B = \{1, 2, 4\}, C = \{1\}, D = \{1, 5\}, E = \{1, 2\}$
(d) $X = \{1,2,3,4, 5\}$, $\mathbf{C} = \{A, B, C, D, E\}$, where $A = \{4, 5\}$, $B = \{3, 5\}$, $C = \{2, 3, 4, 5\}, D = \{2, 3, 4\}, E = \{3, 4, 5\}$
(e) $X = (-\infty, \infty), \mathbf{C} = \{\emptyset, B, \overline{B}, X\}$, where $B = [0, \infty)$
(f) $X = (-\infty, \infty)$, **C** is the class of all open intervals in X
(g) $X = (-\infty, \infty)$, **C** is the class of all closed intervals in X
(h) $X = (-\infty, \infty)$, $\mathbf{C} = \{A_n | n = 1, 2, \ldots\}$, where $A_n = [1 - 1/n, n]$, $n = 1, 2, \ldots$.
(i) $X = \{1, 2, \ldots\}$, $\mathbf{C} = \{A_n | n = 1, 2, \ldots\}$, where $A_n = \{n, n+1, \ldots\}$, $n = 1, 2, \ldots$.

2.17. In Exercise 2.16, which classes are closed under the formation of arbitrary intersections? Verify that $\mathbf{A}[\mathbf{C}] \subset \mathbf{C}$ for these classes $\mathbf{C}$.

2.18. In Exercise 2.16, which classes are AU-classes? Referring to Exercise 2.17, observe that Theorem 2.21 is applicable in some of these cases.

2.19. Prove that

$$A(x/\mathbf{C}_1 \cup \mathbf{C}_2) = A(x/\mathbf{C}_1) \cap A(x/\mathbf{C}_2) \text{ for any } x \in X.$$

May we regard Theorem 2.14 as a special case of this statement?

2.20 Prove that if

$$\mathbf{C}' = \left\{ \bigcap_{t \in T} E_t | E_t \in \mathbf{C},\ t \in T, T \text{ is an arbitrary index sets} \right\},$$

then $\mathbf{A}[\mathbf{C}'] = [\mathbf{C}]$. Can you find a class larger than $\mathbf{C}'$ for which this result still holds?

2.21. Determine the class $\mathbf{H}[\mathbf{C}]$ based upon each of the classes given in Exercise 2.16.

2.22. Prove that any set in $\mathbf{C}$ may be expressed by an intersection of the holes of $\mathbf{C}$, moreover, prove that any union of sets in $\mathbf{C}$ may be expressed by an intersection of the holes of $\mathbf{C}$.

2.23. Use one of the classes given in Exercise 2.16 to verify the conclusion given in Exercise 2.21.

2.24. Prove that

$$\mathbf{F}_p(\mathbf{C}) = \left\{ \bigcap_{t \in T} H_t | H_t \in \mathbf{H}[\mathbf{C}], T \text{ is an arbitrary index set} \right\}.$$

2.25. Prove that if $\mathbf{C}$ is closed under the formation of unions then $\mathbf{H}[\mathbf{C}] \subset \mathbf{C}$.

2.26. In Exercise 2.16, which classes are closed under the formation of unions? Verify that $\mathbf{H}[\mathbf{C}] \subset \mathbf{C}$ for these classes.

2.27. Prove that if $\mathbf{C}$ is an AI-class then $X \notin \mathbf{C}$.

2.28. Let $\mathbf{C}$ be an AI-class. Prove that if $\mathbf{C} \supset \mathbf{H}[\mathbf{C}]$ then $\mathbf{C} = H[\mathbf{C}]$.

2.29. In Exercise 2.16, which classes are AI-classes? Referring also to Exercise 2.26, verify, for some class(es) $\mathbf{C}$, the statement suggested in Exercise 2.28.

2.30. Let X be the set of all integers and $\mathbf{C} = \mathbf{P}(X)$. Is $(X, \mathbf{C})$ S-compact?

Take $E_n = \{x | 0 \leq (-1)^n x \leq n, x \in X\}, n = 1, 2, \ldots$. Can you find a convergent subsequence of $\{E_n\}$?

2.31. Prove that, if $(X, \mathbf{C})$ is S-precompact and $A \subset X$, then $(A, \mathbf{C} \cap A)$ is S-precompact.

2.32. Prove that if $(X, \mathbf{C})$ is S-precompact and $\mathbf{C}' \subset \mathbf{C}$, then $(X, \mathbf{C}')$ is S-precompact.

2.33. Let $X = \{1, 2, 3, 4\}$. Consider the following relation on X:
$R_1 = \{(1,1),(1,3)\}$
$R_2 = \{(2,2),(3,2),(4,1)\}$
$R_3 = \{(1,4),(2,3)\}$
$R_4 = \{(1,1),(4,4)\}$
$R_5 = \{(1,1),(2,2),(3,3),(4,4),(1,4)\}$
$R_6 = \{(1,2),(2,1),(2,3),(1,3),(3,1)\}$
$R_7 = X \times X$
$R_8 = \emptyset$.

Determine whether or not each relation is

(a) reflexive
(b) symmetric
(c) transitive.

2.34. Let $X = (-\infty, \infty)$ and $\cong$ be the relation on $X \times X$ defined by

$$(x_1, y_1) \cong (x_2, y_2) \text{ iff } x_1 - y_1 = x_2 - y_2.$$

(a) Prove that $\cong$ is an equivalence relation.
(b) Find the equivalence class of (2,1).
(c) Find the quotient $X/\cong$.

2.35. Let R be a relation on X. Prove that $R \subset \Delta$ iff R is both symmetric and antisymmetric.

2.36. Let $X = \{0, 1, 2, \ldots\}$. A relation $\leq$ on $X \times X$ is defined as follows:

$$(x_1, y_1) \leq (x_2, y_2) \text{ iff } x_1 \leq x_2 \text{ and } y_1 \leq y_2.$$

Prove that $(X \times X, \leq)$ is a lattice. Show that by replacing $X \times X$ with the two-dimensional Euclidean space $(-\infty, \infty) \times (-\infty, \infty)$ we still obtain a lattice.

2.37. Let $X = [0, 1]$ and let $\mathbf{C}$ consist of $\emptyset, X, A = [0, 0.25)$, $B = [0, 0.5)$, $C = [0, 0.75)$, and $D = [0.25, 0.75)$. Consider a set function μ defined on $\mathbf{C}$ as follows: $\mu(\emptyset) = 0, \mu(A) = 2, \mu(B) = 2, \mu(C) = 4, \mu(D) = 2, \mu(X) = 4$.

(a) Show that μ is additive on $\mathbf{C}$.
(b) Can μ be extended to an additive function on the ring generated by $\mathbf{C}$?

2.38. Assuming that a set function μ is finitely additive on a ring $\mathbf{R}$, show that

$$\mu(A \cup B) = \mu(A) + \mu(B) - \mu(A \cap B)$$

for all $A, B \in \mathbf{R}$.

2.39. Let $X = \{x_1, x_2, x_3\}$. μ is a set function defined for all singleton of X with $\mu(\{x_i\}) = 2^{-i}, i = 1, 2, 3$. Extend μ to be a measure on the power set of X.

2.40. Let $(X, \mathbf{F})$ be a measurable space, μ be a measure on $\mathbf{F}$. Show that

$$\mu\left(\bigcup_{i=1}^{n} E_i\right) = \sum_{I\subset\{1,\ldots,n\}, I\neq\emptyset} (-1)^{|I|+1}\mu\left(\bigcap_{i\in I} E_i\right)$$

and

$$\mu\left(\bigcap_{i=1}^{n} E_i\right) = \sum_{I\subset\{1,\ldots,n\}, I\neq\emptyset} (-1)^{|I|+1}\mu\left(\bigcup_{i\in I} E_i\right),$$

where $\{E_1, E_2, \ldots, E_n\}$ is a finite subclass of $\mathbf{F}$.

2.41 Prove Theorem 2.28.

2.42. Prove Theorem 2.31.

2.43. Let $(X, \mathbf{F})$ be a measurable space, and let μ be a measure on $\mathbf{F}$. For any $A \subset X$, define set function μ' by $\mu'(A) = \inf\{\mu(E) | A \subset E \subset X\}$.

Does μ' coincide with μ' on $\mathbf{F}$? Furthermore, is μ a measure on $\mathbf{P}(X)$? If yes, prove it; if not, construct a counterexample.

2.44. Consider the fuzzy sets A, B, and C defined on the set (interval) $X = [0, 10]$ by the following membership functions:

$$m_A(x) = \begin{cases} x^2 & \text{when } x \in [0,1] \\ (2-x)^2 & \text{when } x \in (1,2] \\ 0 & \text{otherwise} \end{cases}$$

$$m_B(x) = \begin{cases} x-2 & \text{when } x \in [2, 3] \\ 4-x & \text{when } x \in (3, 4] \\ 0 & \text{otherwise} \end{cases}$$

$$m_C(x) = \max\{0, 2(x-3) - (x-3)^2\}.$$

Determine:

(a) plots of the given membership functions and those representing standard complements of $A, B,$ and C, and C;

(b) the standard intersection and standard union of B and C;

(c) the α-cut representations of A, B, and C.

2.45. Viewing fuzzy sets A, B, C in Exercise 2.44 as fuzzy numbers on $\mathbf{R}$, determine:

(a) $A+B+C$
(b) $A-B-C$
(c) $AB+C$ and $AB-C$
(d) BC/A.

2.46. Show that under the standard operations fuzzy sets do not satisfy the law of excluded middle and the law of contradiction.
2.47. Show that under the standard operations fuzzy sets satisfy DeMorgan's laws.
2.48. Considering arithmetic operations on triangular fuzzy numbers, show that their:

(a) additions and subtractions are again triangular fuzzy numbers;
(b) multiplications and divisions may not be triangular fuzzy numbers.

2.49. Show that for any pair of fuzzy sets A and B on X, the concepts of set inclusion, standard intersection, and standard union are cutworthy (see Note 2.5).
2.50. Prove Theorem 2.42, which states that the operation of standard intersection and standard union on fuzzy sets are cutworthy and strong cutworthy, respectively.
2.51. Prove Theorem 2.43, which demonstrates that the standard complement of fuzzy sets is not cutworthy.
2.52. Explain why averaging operations are meaningful for fuzzy sets (even when they degenerate to crisp sets), while they are not meaningful for classical sets.

Chapter 3
Basic Ideas of Generalized Measure Theory

3.1 Generalizing Classical Measures

The principal feature of classical measures is the requirement of countable additivity. When this requirement is replaced with a set of requirements that, as a whole, are weaker than countable additivity, we obtain a class of measures that are more general than classical measures. The various classes of measures obtained in this way can be ordered, at least partially, by their generalities. Clearly, the weaker the set of requirements employed the more general is the class of measures characterized by these requirements.

In generalized measure theory the meaning of the term "measure" is thus considerably broader than its original meaning in classical measure theory. Since the subjects of generalized measure theory are various types of measures, including the classical ones, we need to distinguish each type by adding a suitable adjective to the term "measure." Thus, for example, when we want to refer to measures of classical measure theory, we use either the term "classical measures" or the term "additive measures."

Within generalized measure theory classical measures are obviously very special measures. In order to delimit the scope of measures in generalized measure theory, as understood in this book, we need to characterize the most general measures that we recognize. This is done in Definition 3.1. First, however, we need to introduce relevant notation and conventions.

Let X be a nonempty set, $\mathbf{C}$ be a nonempty class of subsets of X, and $\mu : \mathbf{C} \to [0, \infty]$ be a nonnegative and extended real-valued set function defined on $\mathbf{C}$. This notation is used throughout the book. However, symbol $\mathbf{C}$ usually signifies a class of subsets of X that is equipped with some mathematical structure, such as semiring, ring, algebra, σ-algebra (or, as a special case, Borel field), and the like. Similarly, symbol μ usually denotes a nonnegative and extended real-valued set function that possesses some additional properties, such as monotonicity with respect to set inclusion, continuity, semicontinuity, and the like. When appropriate, these general symbols are replaced with special symbols to signify the various special properties involved. Throughout the whole book, we use the following conventions:

$$\sup_{x \in \emptyset} \{x | x \in [0, \infty]\} = 0,$$

Z. Wang, G.J. Klir, *Generalized Measure Theory*,
DOI: 10.1007/978-0-387-76852-6_3, © Springer Science+Business Media, LLC 2009

$$\inf_{x \in \emptyset} \{x | x \in [0, 1]\} = 1,$$

$$0 \times \infty = \infty \times 0 = 0,$$

$$\frac{1}{\infty} = 0,$$

$$\infty - \infty = 0,$$

$$\sum_{i \in \emptyset} a_i = 0,$$

where $\{a_i\}$ is a real number sequence.

Definition 3.1. Set function μ: $\mathbf{C} \to [0, \infty]$ is called a *general measure* on $(X, \mathbf{C})$ iff $\mu(\emptyset) = 0$ when $\emptyset \in \mathbf{C}$.

The only requirement in Definition 3.1, one which characterizes general measures, is usually referred to as *vanishing at the empty set*. General measures on $(X, \mathbf{C})$ are thus nonnegative and extended real-valued set functions on $\mathbf{C}$ that vanish at the empty set.

We usually consider a monotone class, semiring, ring, algebra, σ-ring, σ-algebra, plump class, or power set of X as the class $\mathbf{C}$ on which μ is defined. We always call the pair $(X, \mathbf{F})$, where $\mathbf{F}$ denotes a σ-ring (or σ-algebra), a *measurable space*. The triple $(X, \mathbf{F}, \mu)$, where μ denotes a general measure, is then called a *general measure space*.

Example 3.1. Given a measurable space $(X, \mathbf{F})$, function μ defined by $\mu(E) = 1$ for a particular nonempty set $E \in \mathbf{F}$ and $\mu(A) = 0$ for all sets $A \in \mathbf{F}$ that are distinct from E is a general measure.

Example 3.2. Let $X = \{\text{a, b, c}\}$ and $\mathbf{C} = \mathbf{P}(X)$. Function μ defined by

$$\mu(A) = \begin{cases} 0 & \text{when } A = \emptyset \text{ or } |A| = 2 \\ 0.5 & \text{when } |A| = 1 \\ 1 & \text{when } A = X \end{cases}$$

is a general measure on $(X, \mathbf{P}(X))$.

The two properties required by general measures — nonnegativity and vanishing at the empty set — are obviously extremely weak requirements. In most applications various additional properties are needed. Although general measures are not completely devoid of applications (Note 3.1), their primary significance in this book is that they provide us with a broad framework under which all other types of measures, including the classical ones, are subsumed as special cases. The term *generalized measure theory* is used in this book for a theory that deals not only with additive measures, but also with their various generalizations, which are *nonadditive measures* of various types.

3.2 Monotone Measures

In this section, we introduce and examine a very broad class of measures that is obtained by restricting the class of general measures by the requirement of monotonicity with respect to set inclusion. Measures in this class are significant for two reasons: (i) they are quite general in the sense that they subsume all other types of measures, with the exception of general measures, as special cases; and (ii) in comparison with general measures the applicability of monotone measures is much greater.

Definition 3.2. Set function $\mu : \mathbf{C} \to [0, \infty]$ is called a *monotone measure* on $(X, \mathbf{C})$ iff it satisfies the following requirements:

(MM1) $\mu(\emptyset) = 0$ when $\emptyset \in \mathbf{C}$ (vanishing at $\emptyset$);
(MM2) $E \in \mathbf{C}, F \in \mathbf{C}$, and $E \subset F$ imply $\mu(E) \leq \mu(F)$ (monotonicity).

In the context of some applications, it is desirable that monotone measures also satisfy one or both of the following requirements:

(CB) $\{E_n\} \subset \mathbf{C}, E_1 \subset E_2 \subset \cdots$, and $\bigcup_{n=1}^{\infty} E_n \in \mathbf{C}$ imply

$$\lim_n \mu(E_n) = \mu\left(\bigcup_{n=1}^{\infty} E_n\right) \quad \text{(continuity from below);}$$

(CA) $\{E_n\} \subset \mathbf{C}, E_1 \supset E_2 \supset \cdots, \mu(E_1) < \infty$, and $\bigcap_{n=1}^{\infty} E_n \in \mathbf{C}$ imply

$$\lim_n \mu(E_n) = \mu\left(\bigcap_{n=1}^{\infty} E_n\right) \quad \text{(continuity from above).}$$

Monotone measures that satisfy requirement (CB) are called *semicontinuous from below* (or *lower-semicontinuous monotone measures*), and those that satisfy requirement (CA) are called *semicontinuous from above* (or *upper-semicontinuous monotone measures*). Monotone measures that satisfy both of the requirements are called *continuous monotone measures.*

In some application areas (for example, in the area of imprecise probabilities), it is useful to use monotone measures that are normalized in the following sense: A monotone measure μ on $(X, \mathbf{C})$ is said to be *normalized* iff $X \in \mathbf{C}$ and $\mu(X) = 1$.

The comments regarding the usual mathematical structure of class $\mathbf{C}$ that are made for general measures in Section 3.1 apply to monotone measures as well. Moreover, the terminology regarding finiteness, σ-finiteness, and extension introduced for classical measures in Section 2.2 is applicable to monotone measures and their semicontinuous or continuous subclasses as well.

We should mention at this point that measures defined by Definition 3.2, which are called monotone measures in this book, frequently have been discussed in the literature under the name "fuzzy measures." In fact, we used this term in our previous book [Wang and Klir, 1992]. However, we came to the conclusion that this term is confusing since it suggests that fuzzy sets are in some way involved in these measures, which, of course, they are not. The feature that distinguishes these measures from other types of measures is the property of monotonicity, expressed by the requirement (MM2) in Definition 3.2. The term "monotone measures" is thus far more expressive and transparent than the term "fuzzy measures."

Example 3.3. Let μ be the Dirac measure on $(X, \mathbf{F})$, i.e., for any $E \in \mathbf{F}$,

$$\mu(E) = \begin{cases} 1, & x_0 \in E \\ 0, & x_0 \notin E, \end{cases}$$

where x_0 is a fixed point in X. This set function μ is a probability measure and, of course, it is a normalized monotone measure.

Let us observe that any classical measure on a semiring is, in general, a monotone measure.

Example 3.4. $X = \{1, 2, \ldots, n\}, \mathbf{C} = \mathbf{P}(X)$. If

$$\mu(E) = \left(\frac{|E|}{n}\right)^2,$$

where $|E|$ is the number of those points that belong to E, then μ is a normalized monotone measure. In fact, since the space X is finite, the continuity (from above and below) is satisfied naturally.

Example 3.5. Let $X = \{1, 2, \ldots\}$ and $\mathbf{C} = \mathbf{P}(X)$. If

$$\mu(E) = |E| \cdot \sum_{i \in E} 2^{-i} \quad \forall E \in \mathbf{C},$$

then μ is a continuous monotone measure. In fact, it is clear that μ satisfies the conditions (MM1) and (MM2), and the continuity is guaranteed by the following lemma.

Lemma 3.1. *If μ_1 and μ_2 are continuous, nonnegative, extended, real-valued set functions on $(X, \mathbf{C})$, and $\mu_1 + \mu_2$ and $\mu_1 \times \mu_2$ are defined by*

$$(\mu_1 + \mu_2)(E) = \mu_1(E) + \mu_2(E)$$

and

$$(\mu_1 \times \mu_2)(E) = \mu_1(E) \times \mu_2(E)$$

for all $E \in \mathbf{C}$, *respectively, then both* $\mu_1 + \mu_2$ *and* $\mu_1 \times \mu_2$ *are continuous. If* μ_1 *and* μ_2 *are finite or* σ-*finite monotone measures (or semicontinuous monotone measures), then so are* $\mu_1 + \mu_2$ *and* $\mu_1 \times \mu_2$.

Example 3.6. Let $X_0 = \{1, 2, \ldots\}, X = X_0 \times X_0$, and $\mathbf{C} = \mathbf{P}(X)$. For any $E \in \mathbf{P}(X)$, let

$$\mu(E) = |\text{Proj } E|,$$

where

$$\text{Proj } E = \{x \mid (x, y) \in E\}.$$

Function μ satisfies the requirements (MMl), (MM2), and (CB), but it is not continuous from above. In fact, if $E_n = \{1\} \times \{n, n+1, \cdots\}$, then $E_1 \supset E_2 \supset \cdots$, and $\mu(E_n) = 1$ for any $n = 1, 2, \ldots$, but $\bigcap_{n=1}^{\infty} E_n = \emptyset$, and $\mu(\bigcap_{n=1}^{\infty} E_n) = 0$. So, the set function μ is a lower-semicontinuous monotone measure.

Example 3.7. Let $f(x)$ be a nonnegative, extended real-valued function defined on $X = (-\infty, \infty)$. If

$$\mu(E) = \sup_{x \in E} f(x) \; \forall E \in \mathbf{P}(X),$$

then μ satisfies the conditions (MMl), (MM2), and (CB). But, in general, it is not continuous from above. So μ is a lower-semicontinuous monotone measure on $(X, \mathbf{P}(X))$.

Example 3.8. Let the measurable space $(X, \mathbf{P}(X))$ be the same as that given in Example 3.7. if $f: X \to [0, 1]$ is such that $\inf_{x \in X} f(x) = 0$, then the set function μ which is determined by

$$\mu(E) = \inf_{x \notin E} f(x)$$

for every $E \in \mathbf{P}(X)$ is a normalized upper-semicontinuous monotone measure.

For each normalized monotone measure μ on $(X, \mathbf{C})$, where $\mathbf{C}$ is assumed to contain both $\emptyset$ and X and be closed under complementation, we define for all $A \in \mathbf{C}$ another monotone measure, ν, via the equation

$$\nu(A) = 1 - \mu(\bar{A}),$$

and we call ν the *dual monotone measure* of μ. Clearly, any monotone measure contains the same information as its dual measure, but the two measures represent the information differently. It is often preferable to work with pairs of dual measures rather than with single measures. The definition of duality applies not only to monotone measures, but to any special subclass of monotone measures as well. Observe that the dual of any additive measure is the measure itself. That is, additive measures are *autodual*. For any normalized monotone measure μ, its dual measure ν is also a normalized monotone measure. Moreover, the dual measure of any normalized, monotone, and lower-semicontinuous measure is a normalized, monotone, and upper-continuous measure. Similarly, the dual measure of any normalized, monotone, and upper-continuous measure is a normalized, monotone, and lower-continuous measure.

For any monotone measure μ, on $(X, \mathbf{C})$, if sets $A, B, A \cap B$, and $A \cup B$ are in $\mathbf{C}$, then

$$\mu(A \cap B) \leq \min\ (\mu(A),\ \mu(B)),$$

$$\mu(A \cup B) \geq \max\ (\mu(A),\ \mu(B)).$$

These inequalities follow from monotonicity of μ and from the simple facts that $A \cap B \subset A$ and $A \cap B \subset B$, and similarly, $A \cup B \supset A$ and $A \cup B \supset B$.

Observe that monotone measures, as introduced in Definition 3.2, are actually monotone *increasing* set functions. It is certainly possible to define monotone *decreasing* set functions as well, simply by replacing requirement (MM2) in Definition 3.2 with the following alternative requirement:

$$\text{(MM2}'\text{)}\ E \in \mathbf{C}, F \in \mathbf{C}, E \subset F \text{ imply } \mu(E) \geq \mu(F).$$

A fundamental difference between the two types of set functions is that the one satisfying requirement (MM2') is not a measure, as we understand the concept of a measure in this book, since it obviously cannot satisfy the requirement of general measures that they vanish at the empty set. Although monotone decreasing set functions emerge naturally on intuitive grounds in some applications (Note 3.2), it is possible, in general, to convert each of them to an associated monotone measure that contains the same information. Assume, for example, that σ is a decreasing set function on $(X, \mathbf{C})$ such that $\sigma(\emptyset) < \infty$. Then, we can define for all $A \in \mathbf{C}$ an associated monotone measure μ by the formula

$$\mu(A) = \sigma(\emptyset) - \sigma(A).$$

Clearly, set function μ contains the same information as set function σ. However, μ is a monotone measure while σ is not even a general measure and, consequently, it is outside the scope of generalized measure theory.

3.3 Superadditive and Subadditive Measures

It is easy to see that additivity and nonnegativity imply monotonicity, but not the other way around. However, monotone measures can be usefully classified from the standpoint of additivity into four classes: (i) additive measures; (ii) superadditive measures; (iii) subadditive measures; and (iv) monotone measures that do not belong to any of the other three classes. The superadditive and subadditive measures, which are quite important in generalized measure theory, are defined as follows.

Definition 3.3. A monotone measure μ on $(X, \mathbf{C})$ is *superadditive* iff

$$\mu(A \cup B) \geq \mu(A) + \mu(B)$$

whenever $A \cup B \in \mathbf{C}$, $A \in \mathbf{C}$, $B \in \mathbf{C}$, and $A \cap B = \emptyset$.

Definition 3.4. A monotone measure μ on $(X, \mathbf{C})$ is *subadditive* iff

$$\mu(A \cup B) \leq \mu(A) + \mu(B)$$

whenever $A \cup B \in \mathbf{C}$, $A \in \mathbf{C}$, and $B \in \mathbf{C}$.

Observe that superadditive measures are capable of expressing a cooperative action or synergy between sets in terms of the measured property, while subadditive measures are capable of expressing inhibitory effects or incompatibility between sets in terms of the measured property. Additive measures, on the other hand, are not able to express either of these interactive effects. They are applicable only to situations in which there is no interaction between sets as far as the measured property is concerned. In additive measures superadditivity and subadditivity are not distinguished; they both collapse into the property of additivity.

Inclusion relationship among additive measures and the four classes of measures introduced in Sections 3.1–3.3 is depicted in Fig. 3.1.

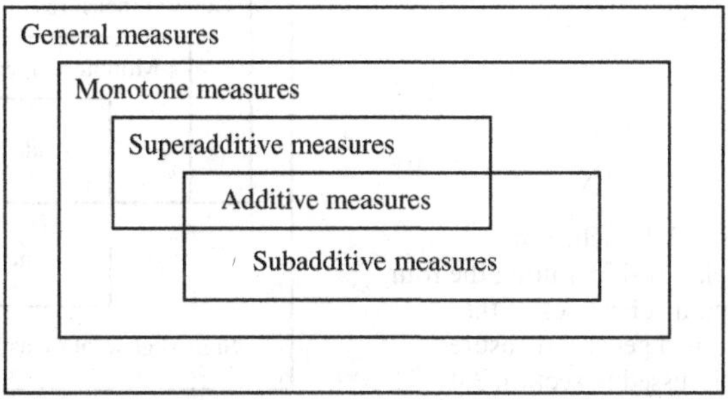

Fig. 3.1 Inclusion relationship among classes of measures considered in Sections 3.1–3.3

3.4 Signed General Measures

The notion of a *signed measure* refers in classical measure theory to an extended real-valued and countably additive set function μ on a measurable space (X, **C**) that assumes at most one of the values $+\infty$ and $-\infty$ and for which $\mu(\emptyset) = 0$ [Halmos, 1950]. Signed measures in classical measure theory are thus generalizations of classical measures. To emphasize that they are additive, it is more appropriate within the framework of generalized measure theory to call them *signed additive measures*.

In the same way as signed additive measures are obtained from additive measures, we can obtain *signed general measures* from general measures. Signed general measures vanish at the empty set, as general measures do, but they are not required to be nonnegative. Clearly, each additive measure is also a signed additive measure, but not the other way around. Each additive measure is also a monotone measure, but signed additive measures are generally not monotone. It is easy to show that a signed additive measure μ on (X, **C**) is nondecreasing iff it is an additive measure. This follows from the following two simple facts: (i) if μ is an additive measure, then we know that it is also a monotone measure; and (ii) if μ is a signed additive measure that is not an additive measure, then there exists a nonempty set $A \in \mathbf{C}$ for which $\mu(A) < 0$ and, considering the requirement that $\mu(\emptyset) = 0$, we have $\mu(A) < \mu(\emptyset)$ and $A \supset \emptyset$, which violates monotonicity. The same arguments and conclusions apply obviously to signed general measures and monotone measures as well. Inclusion relationship among four main subclasses of the class of signed general measures is depicted in Fig. 3.2.

Example 3.9. Let $X = \{a, b\}$. Set function $\mu : \mathbf{P}(X) \to (-\infty, \infty)$ is defined by $\mu(\emptyset) = 0, \mu(\{a\}) = 0.5, \mu(\{b\}) = -0.2$, and $\mu(X) = 1$. Then μ is a signed general measure on (X, **P**(X)).

We have the following decomposition theorem for signed general measures, which is similar to the *Jordan decomposition theorem* established in classical measure theory for signed measures. The proof of the theorem for signed

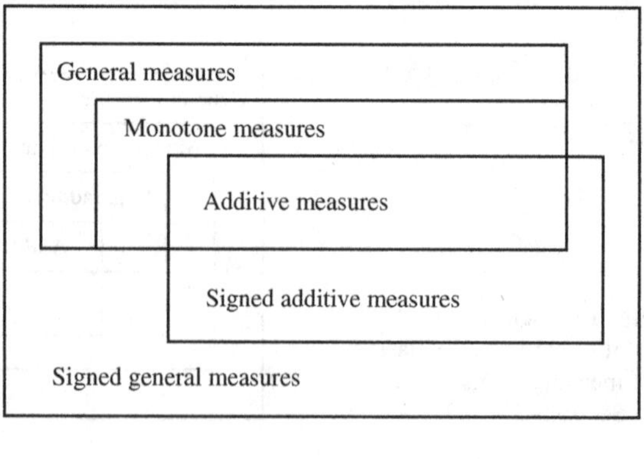

Fig. 3.2 Inclusion relationship among the four main subclasses of the signed general measures discussed in section 3.4

general measures is much simpler than that one for signed measures in classical measure theory.

Theorem 3.1. *Let μ be a signed general measure on $(X, \mathbf{C})$. Then μ can be expressed as the difference of two general measures, i.e., $\mu = \mu^+ - \mu^-$, where both μ^+ and μ^- are general measures on $(X, \mathbf{C})$.*

Proof. Define

$$\mu^+(E) = \begin{cases} \mu(E) & \text{if } \mu(E) \geq 0 \\ 0 & \text{otherwise} \end{cases}$$

and

$$\mu^-(E) = \begin{cases} -\mu(E) & \text{if } \mu(E) < 0 \\ 0 & \text{otherwise} \end{cases}$$

for every $E \in \mathbf{C}$. Then both μ^+ and μ^- are general measures on $(X, \mathbf{C})$ and $\mu(E) = \mu^+(E) - \mu^-(E)$ for every $E \in \mathbf{C}$. □

We may also call (μ^+, μ^-) a *Jordan decomposition* of signed general measure μ if $\mu = \mu^+ - \mu^-$, where both μ^+ and μ^- are general measures. Unlike the classical case, the Jordan decomposition of a signed general measure may not be unique.

Example 3.10. Consider the signed general measure μ given in Example 3.9. According to the proof of Theorem 3.1,

$$\mu^+(E) = \begin{cases} 1 & \text{if } E = X \\ 0.5 & \text{if } E = \{a\} \\ 0 & \text{otherwise} \end{cases}$$

and

$$\mu^-(E) = \begin{cases} 0.2 & \text{if } E = \{b\} \\ 0 & \text{otherwise} \end{cases}$$

for every $E \in \mathbf{P}(X)$. Then $\mu = \mu^+ - \mu^-$ and (μ^+, μ^-) is a Jordan decomposition of μ. However, if we take

$$\nu^+(E) = \begin{cases} 2 & \text{if } E = X \\ 1.5 & \text{if } E = \{a\} \\ 1 & \text{if } E = \{b\} \\ 0 & \text{if } E = \emptyset \end{cases}$$

and

$$\nu^-(E) = \begin{cases} 1 & \text{if } E = X \\ 1 & \text{if } E = \{a\} \\ 1.2 & \text{if } E = \{b\} \\ 0 & \text{if } E = \emptyset, \end{cases}$$

then (ν^+, ν^-) satisfies $\mu = \nu^+ - \nu^-$ and, therefore, it is a Jordan decomposition of μ as well.

Definition 3.5. A Jordan decomposition (μ^+, μ^-) of a signed general measure μ is called the *smallest Jordan decomposition* of μ if, for every $E \in \mathbf{C}$, $\mu^+(E) \leq \nu^+(E)$ and $\mu^-(E) \leq \nu^-(E)$ for any other Jordan decomposition (ν^+, ν^-) of μ.

The Jordan decomposition (μ^+, μ^-) of μ employed in the proof of Theorem 3.1 is the smallest one. This property can be seen easily in Example 3.10.

Notes

3.1. Aumann and Shapley [1974] use set functions that we call *signed general measures* for the study of non-atomic cooperative games. The term "non-atomic games" refers to games that involve very large number of players so that it is reasonable and convenient to represent them as points on a real line. These games are called "cooperative" since no individual player can affect the overall outcome and it is thus essential to cooperate and form coalitions. According to this interpretation, a general measure μ defined on a measurable space $(X, \mathbf{C})$, where X is the set of players and $\mathbf{C}$ is a class of coalitions of players, is a game. For each $A \in \mathbf{C}$, the number $\mu(A)$ is interpreted as the total payoff that the coalition A, if it forms, can obtain for its members. Other notable publications connecting generalized measure theory with the theory of cooperative games include [Shapley, 1953, 1971; Schmeidler, 1972; Owen, 1988; Branzei et al., 2005]. The utility of signed general measures or general measures in other areas is discussed in [Murofushi et al., 1994].

3.2. Special monotone decreasing set functions were introduced and discussed by the British economist George Shackle in several of his publications [Shackle, 1949, 1955, 1961]. He employed these functions to express degrees of potential surprise associated with judgments of individual human beings regarding future possibilities. The following is his own concise description of the idea of potential surprise:

> It is the degree of surprise to which we expose ourselves when we examine an imagined happening as to its possibility, in general or in the prevailing circumstances, and assess the obstacles, tensions and difficulties which arise in our minds when we try to imagine it occurring, that provide the indicator of degree of

possibility. This is the surprise we would feel, if the given thing did happen; it is potential surprise (Shackle, 1961, p. 68):.

As is shown in [Klir, 2002], each of these functions expressing degrees of potential surprise can be converted to an associated monotone measure—a possibility measure—that contains the same information. Shackle was aware of the two ways of expressing the same phenomenon. He preferred to work with set functions expressing potential surprise because they have a natural psychological interpretation. However, the associated possibility measures are more suitable for mathematical treatment.

3.3. The concept of monotone measures was first suggested and investigated by Sugeno [1974, 1977]. However, he introduced these measures under the name "fuzzy measures." Although this name is rather misleading, as there is no fuzziness in monotone measures, it has been quite a popular name in the literature. Sugeno's original definition of a fuzzy (i.e., monotone) measure is based on a measurable space $(X, \mathbf{B})$, where $\mathbf{B}$ is a Borel field. His definition differs from Definition 3.2 in the formulation of continuity. Sugeno does not distinguish continuity from below and continuity from above, and defines continuity as follows:
If $E_n \in \mathbf{B}$ for $1 \leq n < \infty$ and sequence $\{E_n\}$ is monotone, then $\lim_{n\to\infty} \mu(E_n) = \mu(\lim_{n\to\infty} E_n)$.

Exercises

3.1. Show that the measures in Examples 3.1 and 3.2 are not monotone.
3.2. Show that nonnegative monotone decreasing set functions are neither general measures nor signed general measures.
3.3. Consider a monotone decreasing set function σ on $(X, \mathbf{P}(X))$ where $X = \{a, b, c\}$, which is defined for all $A \in \mathbf{P}(X)$ by the formula $\sigma(A) = 1/|A|$. Determine an associated monotone measure μ that contains the same information as σ.
3.4. Determine whether each of the following set functions is a monotone measure, lower-semicontinuous monotone measure, upper-semicontinuous monotone measure, or continuous monotone measure.

(a) $X = (-\infty, \infty)$, $\mathbf{F}$ is the class of all Borel sets in $(-\infty, \infty)$, and $\mu(E) = c$ for any $E \in \mathbf{F}$, where c is a nonnegative constant.
(b) X is the set of all integers, $\mathbf{F} = \mathbf{P}(X)$, and $\mu(E) = \sum_{i\in E} i$ for any $E \in \mathbf{F}$.
(c) X is the set of all positive integers, $\mathbf{F} = \mathbf{P}(X)$, and $\mu(E) = \sum_{i\in E} i - (|E|^2/2)$ for any $E \in \mathbf{F}$.
(d) X is the set of all positive integers, $\mathbf{F} = \mathbf{P}(X)$, and

$$\mu(E) = \begin{cases} 1 & \text{if } E \neq \emptyset \\ 0 & \text{if } E = \emptyset \end{cases}$$

for any $E \in \mathbf{P}(X)$.

(e) $X = \{a, b, c, d\}, \mathbf{F} = \mathbf{P}(X)$, and

$$\mu(E) = \begin{cases} 1 & \text{if } E = X \\ 0 & \text{if } E = \emptyset \\ 1/3 & \text{otherwise.} \end{cases}$$

(f) $X = [0, 1) \cap R, \mathbf{C} = \mathbf{R}_{[0,1)} \cap R$, where R is the set of all rational numbers, $\mathbf{R}_{[0,1)}$ is the class of all finite unions of left closed right open intervals in [0, 1), and μ is defined on $\mathbf{C}$ by

$$\mu(A \cap R) = m(A)$$

for any $A \in \mathbf{R}_{[0,1)}$, where m is the Lebesgue measure.

3.5. Are there any normalized monotone measures in Exercise 3.4? If the answer is yes, find their dual monotone measures.

3.6. Determine whether each of the normalized monotone measures in Exercise 3.4 and its dual measure (Exercise 3.5) is superadditive, subadditive, or both.

3.7. Consider the measurable space $(X, \mathbf{P}(X))$, where $X = \{a, b, c\}$. Find an example of a set function μ on $\mathbf{P}(X)$ for each of the classes of measures or signed measures whose names are given in Figs. 3.1 and 3.2.

3.8. Prove Lemma 3.1.

Chapter 4
Special Areas of Generalized Measure Theory

4.1 An Overview

The term "generalized measure theory," as it is understood in this book, is delimited by two extremes—the classical theory of additive measures or signed additive measures at one extreme and the theory of general measures or signed general measures at the other extreme. There are of course many measure theories between these two extremes. They are based on measures that do not require additivity, but that are not fully general as well. Three major types of measures that are in this category are introduced in Chapter 3. They are monotone measures and their large subclasses: superadditive and subadditive measures. The purpose of this chapter is to further refine these large classes of measures by introducing their various subclasses. We focus on those subclasses that are well established in the literature.

In Section 4.2, we begin with an important family of measures that are referred to in the literature as *Choquet capacities of various orders*. Classes of measures captured by this family are significant as they are linearly ordered in terms of their interpretations and methodological capabilities. In some sense this family of measures is the core of generalized measure theory. Classes of measures in this family are benchmarks against which other classes of measures are compared in terms of their roles in generalized measure theory.

After introducing this important family of measures in Section 4.2, we return to classical measure theory and examine the various ways of how to generalize it. First, we introduce in Section 4.3 a simple generalization of classical measures via the so-called λ-measures. Next, we show in Section 4.4 that the class of λ-measures is a member of a broader class of measures that we call quasi-measures. Each member of this broader class of measures is connected to additive measures via a particular type of reversible transformation. After examining quasi-measures, we proceed in Section 4.5 to the strongest Choquet capacities (referred to as capacities of order ∞) and their dual measures (referred to *as alternating capacities of order* ∞). These pairs of dual measures, when normalized, form a basis for a well-developed and highly visible theory of uncertainty, which is usually referred to in the literature as the Dempster–Shafer theory. Another important and well-known theory of uncertainty, which is in

Z. Wang, G.J. Klir, *Generalized Measure Theory*,
DOI: 10.1007/978-0-387-76852-6_4, © Springer Science+Business Media, LLC 2009

some specific way connected with the Dempster–Shafer theory, is possibility theory. Nonadditive measures upon which possibility theory is based are introduced and examined in Section 4.6. Finally, some properties of finite monotone measures are presented in Section 4.7.

4.2 Choquet Capacities

Definition 4.1. Given a particular integer $k \geq 2$, a *Choquet capacity of order k* is a monotone measure μ on a measurable space $(X, \mathbf{F})$ that satisfies the inequalities

$$\mu\left(\bigcup_{j=1}^{k} A_j\right) \geq \sum_{\substack{K \subseteq N_k \\ K \neq \emptyset}} (-1)^{|K|+1} \mu\left(\bigcap_{j \in K} A_j\right) \tag{4.1}$$

for all families of k sets in $\mathbf{F}$, where $N_k = \{1, 2, \ldots, k\}$.

Since sets A_j in the inequalities (4.1) are not necessarily distinct, every Choquet capacity of order k is also of order $k' = k-1, k-2, \ldots, 2$. However, a capacity of order k may not be a capacity of any higher order ($k+1, k+2$, etc.). Hence, capacities of order 2, which satisfy the simple inequalities

$$\mu(A_1 \cup A_2) \geq \mu(A_1) + \mu(A_2) - \mu(A_1 \cap A_2) \tag{4.2}$$

for all pairs of sets in $\mathbf{F}$, are the most general capacities. The least general ones are those of order k for all $k \geq 2$. These are called *Choquet capacities of order* ∞ or *totally monotone measures*. They satisfy the inequalities

$$\begin{aligned} \mu(A_1 \cup A_2 \cup \cdots \cup A_k) \geq \sum_i \mu(A_i) - \sum_{i<j} \mu(A_i \cap A_j) + - \cdots \\ + (-1)^{k+1} \mu(A_1 \cap A_2 \cap \cdots \cap A_k) \end{aligned} \tag{4.3}$$

for every $k \geq 2$ and every family of k sets in $\mathbf{F}$.

It is trivial to see that the set of inequalities (4.2) contains all the inequalities required for superadditive measures in Definition 3.3 (when $A_1 \cap A_2 = \emptyset$), but contains additional inequalities (when $A_1 \cap A_2 \neq \emptyset$). Choquet capacities of order 2—the most general class of Choquet capacities—are thus a subclass of superadditive measures.

Definition 4.2. Given a particular integer $k \geq 2$, an *alternating Choquet capacity of order k* is a monotone measure μ on a measurable space $(X, \mathbf{F})$ that satisfies for all families of k sets in $\mathbf{F}$ the inequalities

$$\mu\left(\bigcap_{j=1}^{k} A_j\right) \leq \sum_{\substack{K \subseteq N_k \\ K \neq \emptyset}} (-1)^{|K|+1} \mu\left(\bigcup_{j \in K} A_j\right). \tag{4.4}$$

It is clear that the requirements for alternating capacities of some order $k \geq 2$ are weaker than those of orders $k+1, k+2, \ldots$. *Alternating capacities of order* 2, which are required to satisfy the inequalities

$$\mu(A_1 \cap A_2) \leq \mu(A_1) + \mu(A_2) - \mu(A_1 \cup A_2) \tag{4.5}$$

for all pairs of sets in **F**, are thus the most general alternating capacities. On the other hand, *alternating Choquet capacities of order* ∞, which are defined by the inequalities

$$\begin{aligned} \mu(A_1 \cap A_2 \cap \cdots \cap A_k) \leq \sum_i \mu(A_i) - \sum_{i<j} \mu(A_i \cup A_j) + - \cdots \\ + (-1)^{k+1} \mu(A_1 \cup A_2 \cup \cdots \cup A_k) \end{aligned} \tag{4.6}$$

for every $k \geq 2$ and every family of k sets in **F**, are the least general ones.

It is obvious that the set of inequalities (4.5) contains all the inequalities required in Definition 3.4 for subadditive measures (when $A_1 \cap A_2 = \emptyset$), but contains some additional inequalities (when $A_1 \cap A_2 \neq \emptyset$). Alternating Choquet capacities of order 2—the most general class of alternating Choquet capacities—are thus subadditive measures, but not the other way around.

Choquet capacities of order k are often referred to in the literature as k-*monotone measures* and, similarly, alternating Choquet capacities are often called k-*alternating measures*. These shorter names are adopted, by and large, in this book. For convenience, monotone measures that are not 2-monotone are often referred to as 1-*monotone measures*. Using this terminology the inclusion relationship among the introduced classes of k-monotone and k-alternating measures for $k \geq 1$is depicted in Fig. 4.1.

Theorem 4.1. *Let μ be a normalized 2-monotone measure on a measurable space $(X, \mathbf{F})$. Then the dual measure of μ, denoted by μ^*, is a normalized 2-alternating measure on $(X, \mathbf{F})$.*

Proof.

$$\begin{aligned} \mu^*(A_1 \cap A_2) &= 1 - \mu(\overline{A_1 \cap A_2}) \\ &= 1 - \mu(\overline{A_1} \cup \overline{A_2}) \\ &\leq 1 - \mu(\overline{A_1}) - \mu(\overline{A_2}) + \mu(\overline{A_1} \cap \overline{A_2}) \\ &= 1 - \mu(\overline{A_1}) + 1 - \mu(\overline{A_2}) - 1 + \mu(\overline{A_1} \cap \overline{A_2}) \\ &= 1 - \mu(\overline{A_1}) + 1 - \mu(\overline{A_2}) - 1 + \mu(\overline{A_1 \cup A_2}) \\ &= \mu^*(A_1) + \mu^*(A_2) - \mu^*(A_1 \cup A_2). \end{aligned}$$

□

This theorem can be easily generalized to normalized k-monotone measures for any $k \geq 2$. Observe, however, that the dual measure of a 1-monotone

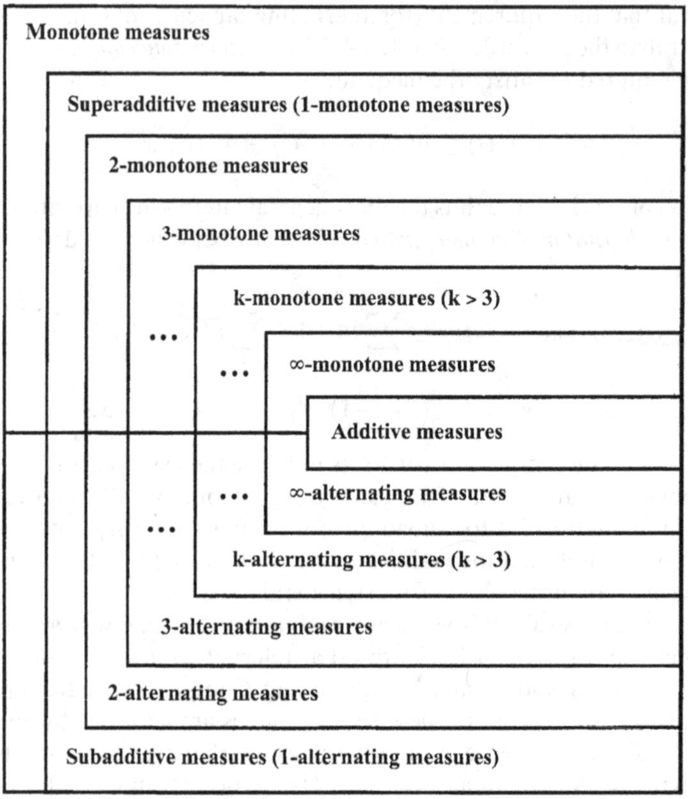

Fig. 4.1 Inclusion relationship among k-monotone and k-alternating measures for $k \geq 1$

measure that is superadditive is not necessarily subadditive, as is shown by the following counterexample.

Example 4.1. Let $X = \{a, b, c\}, \mathbf{F} = \mathbf{P}(X)$, and let μ be the 1-monotone measure on $(X,\ \mathbf{P}(X))$ defined in Table 4.1. This measure is clearly normalized and superadditive, but it is not 2-monotone due to the following two violations of the required inequalities (4.2):

$$\mu(X) = 1 < \mu(\{a, b\}) + \mu(\{b, c\} - \mu(\{b\}) = 1.4,$$
$$\mu(X) = 1 < \mu(\{a, c\}) + \mu(\{b, c\}) - \mu(\{c\}) = 1.1.$$

The dual measure of μ, denoted in Table 4.1 by μ^*, is not subadditive due to the following violation of the inequalities required for subadditive measures in Definition 3.4:

$$\mu(\{a, b\}) = 0.8 > \mu(\{a\}) + \mu(\{b\}) = 0.7,$$
$$\mu(\{a, c\}) = 1 > \mu(\{a\}) + \mu(\{c\}) = 0.6.$$

Table 4.1 Superadditive measure μ and its dual measure μ^* (Example 4.1)

$a\ b\ c$	$\mu(A)$	$\mu^*(A)$
A: 0 0 0	0.0	0.0
1 0 0	0.1	0.2
0 1 0	0.0	0.5
0 0 1	0.2	0.4
1 1 0	0.6	0.8
1 0 1	0.5	1.0
0 1 1	0.8	0.9
1 1 1	1.0	1.0

The whole family of k-monotone and k-alternating classes of measures plays an important role in generalized measure theory and, particularly, in its applications dealing with various types of uncertainty. Especially important are the classes of 2-monotone and 2-alternating measures, which are the most general classes in this family, and the classes of ∞-monotone and ∞-alternating measures. They represent important benchmarks from mathematical and computational points of view. These issues are discussed later in the book in various contexts.

Thus far, we have followed a top-down approach: we started by defining general measures and we proceeded to defining monotone measures, superadditive and subadditive measures, and, finally, k-monotone and k-alternating measures. In the rest of this chapter we switch to the complementary, bottom-up approach: we start with examining in detail some of the simplest generalizations of classical measures and we proceed then by enlarging the framework to discuss the various higher-level generalizations.

4.3 λ-Measures

Definition 4.3. A monotone measure μ satisfies the *λ-rule* (on **C**) iff there exists

$$\lambda \in \left(-\frac{1}{\sup \mu}, \infty\right) \cup \{0\},$$

where $\sup \mu = \sup_{E \in \mathbf{C}} \mu(E)$, such that

$$\mu(E \cup F) = \mu(E) + \mu(F) + \lambda \cdot \mu(E) \cdot \mu(F),$$

whenever

$$E \in \mathbf{C}, F \in \mathbf{C},\ \ E \cup F \in \mathbf{C}, \text{and } E \cap F = \emptyset.$$

μ satisfies the *finite λ-rule* (on **C**) iff there exists the above-mentioned λ such that

$$\mu\left(\bigcup_{i=1}^{n} E_i\right) = \begin{cases} \frac{1}{\lambda}\left\{\prod_{i=1}^{n}[1+\lambda\cdot\mu(E_i)]-1\right\}, & \text{as } \lambda \neq 0 \\ \sum_{i=1}^{n}\mu(E_i), & \text{as } \lambda = 0 \end{cases}$$

for any finite disjoint class $\{E_1, \dots, E_n\}$ of sets in **C** whose union is also in **C**; μ satisfies the *σ-λ-rule* (on **C**) iff there exists the above-mentioned λ, such that

$$\mu\left(\bigcup_{i=1}^{\infty} E_i\right) = \begin{cases} \frac{1}{\lambda}\left\{\prod_{i=1}^{\infty}[1+\lambda\cdot\mu(E_i)]-1\right\}, & \text{as } \lambda \neq 0, \\ \sum_{i=1}^{\infty}\mu(E_i), & \text{as } \lambda = 0, \end{cases}$$

for any disjoint sequence $\{E_n\}$ of sets in **C** whose union is also in **C**.

When $\lambda = 0$, the λ-rule, the finite λ-rule, or the σ-λ-rule is just the additivity, the finite additivity, or the σ-additivity, respectively.

Theorem 4.2. *If* **C** = **R** *is a ring and μ satisfies the λ-rule, then μ satisfies the finite λ-rule.*

Proof. The conclusion is obvious when $\lambda = 0$. Let $\lambda \neq 0$ and $\{E_1, \dots, E_n\}$ be a disjoint class of sets in **R**. We use the mathematical induction to prove

$$\mu\left(\bigcup_{i=1}^{n} E_i\right) = \frac{1}{\lambda}\left\{\prod_{i=1}^{n}[1+\lambda\cdot\mu(E_i)]-1\right\}. \tag{4.7}$$

From the definition we know directly that (4.7) is true when $n = 2$. Now, suppose that (4.7) is true for $n = k - 1$. We have

$$\begin{aligned}
\mu\left(\bigcup_{i=1}^{k} E_i\right) &= \mu\left(\left(\bigcup_{i=1}^{k-1} E_i\right) \cup E_k\right) \\
&= \mu\left(\bigcup_{i=1}^{k-1} E_i\right)[1+\lambda\cdot\mu(E_k)] + \mu(E_k) \\
&= \frac{1}{\lambda}\left\{\prod_{i=1}^{k-1}[1+\lambda\cdot\mu(E_i)]-1\right\}\cdot[1+\lambda\cdot\mu(E_k)] + \mu(E_k) \\
&= \frac{1}{\lambda}\left\{\prod_{i=1}^{k}[1+\lambda\cdot\mu(E_i)]-[1+\lambda\cdot\mu(E_k)]\right\} + \mu(E_k) \\
&= \frac{1}{\lambda}\left\{\prod_{i=1}^{k}[1+\lambda\cdot\mu(E_i)]-[1+\lambda\cdot\mu(E_k)]+\lambda\cdot\mu(E_k)\right\} \\
&= \frac{1}{\lambda}\left\{\prod_{i=1}^{k}[1+\lambda\cdot\mu(E_i)]-1\right\}.
\end{aligned}$$

That is, (4.7) is true for $n = k$. The proof is complete. □

In fact, Theorem 4.2 holds also when **C** is only a semiring. This is shown in Section 4.4, after introducing a new concept called *quasi-additivity*.

Example 4.2. Let $X = \{a, b\}$ and $\mathbf{C} = \mathbf{P}(X)$. If

$$\mu(E) = \begin{cases} 0 & E = \emptyset \\ 0.2 & E = \{a\} \\ 0.4 & E = \{b\} \\ 1 & E = X, \end{cases}$$

then μ satisfies the λ-rule, where $\lambda = 5$. Since **C** is a finite ring, μ satisfies the finite λ-rule and also the σ-λ-rule.

Definition 4.4. μ is called a *λ-measure* on **C** iff it satisfies the σ-λ-rule on **C** and there exists at least one set $E \in C$ such that $\mu(E) < \infty$.

Usually the λ-measure is denoted by g_λ. When **C** is a σ-algebra and $g_\lambda(X) = 1$, the λ-measure g_λ is also called a *Sugeno measure*. The set function given in Example 4.2 is a Sugeno measure.

Example 4.3. Let $X = \{x_1, x_2, \ldots\}$ be a countable set, **C** be the semiring consisting of all singletons of X and the empty set $\emptyset$, and $\{a_i\}$ be a sequence of nonnegative real numbers. Define $\mu(\{x_i\}) = a_i, i = 1, 2, \ldots,$ and $\mu(\emptyset) = 0$. Then μ is a λ-measure for any $\lambda \in (-1/\sup \mu, \infty) \cup \{0\}$, where $\sup \mu = \sup(\{a_i | i = 1, 2, \ldots\})$.

Theorem 4.3. *If g_λ is a λ-measure on a class* **C** *containing the empty set $\emptyset$, then $g_\lambda(\emptyset) = 0$, and g_λ satisfies the finite λ-rule.*

Proof. From Definition 4.4, there exists $E \in \mathbf{C}$ such that $g_\lambda(E) < \infty$. When $\lambda = 0, g_\lambda$ is a classical measure and therefore $g_\lambda(\emptyset) = 0$. Otherwise, $\lambda \neq 0$. Since $\{E, E_2, E_3, \ldots\}$, where $E_2 = E_3 = \cdots = \emptyset$ is a disjoint sequence of sets in **C** whose union is E, we have

$$g_\lambda(E) = \frac{1}{\lambda}\left\{\prod_{i=2}^{\infty}[1 + \lambda \cdot g_\lambda(E_i)] \cdot [1 + \lambda \cdot g_\lambda(E)] - 1\right\},$$

where $E_i = \emptyset$, and $i = 2, 3, \ldots$ That is,

$$1 + \lambda \cdot g_\lambda(E) = [1 + \lambda \cdot g_\lambda(E)] \cdot \left\{\prod_{i=2}^{\infty}[1 + \lambda \cdot g_\lambda(E_i)]\right\}.$$

Noting the fact that $\lambda \in (-1/\sup g_\lambda, \infty)$ and $g_\lambda(E) < \infty$, we know that

$$0 < 1 + \lambda \cdot g_\lambda(E) < \infty.$$

Thus, we have

$$\prod_{i=2}^{\infty}[1+\lambda \cdot g_\lambda(E_i)] = 1$$

and therefore,

$$1+\lambda g_\lambda(\emptyset) = 1.$$

Consequently, we have

$$g_\lambda(\emptyset) = 0.$$

By using this result, the second conclusion is clear. □

Theorem 4.4. *If g_λ is a λ-measure on a semiring* **S**, *then g_λ is monotone.*

Proof. When $\lambda = 0$ we refer the monotonicity of classical measures (Section 2.2). Now, let $\lambda \neq 0$ and let $E \in \mathbf{S}$, $F \in \mathbf{S}$, and $E \subset F$. Since **S** is a semiring, $F - E = \cup_{i=1}^{n} D_i$, where $\{D_i\}$ is a finite disjoint class of sets in **S**, and we have

$$\frac{1}{\lambda}\left\{\prod_{i=1}^{n}[1+\lambda \cdot g_\lambda(D_1) - 1\right\} \geq 0$$

in both cases where $\lambda > 0$ and $\lambda < 0$. By using Theorem 4.3, g_λ satisfies the finite λ-rule. So, we have

$$\begin{aligned} g_\lambda(F) &= g_\lambda(E \cup D_1 \cup \cdots \cup D_n) \\ &= \frac{1}{\lambda}\left\{\prod_{i=1}^{n}[1+\lambda \cdot g_\lambda(D_1)][1+\lambda \cdot g_\lambda(E)] - 1\right\} \\ &= g_\lambda(E) + \frac{1}{\lambda}\left\{\prod_{i=1}^{n}[1+\lambda \cdot g_\lambda(D_1)] - 1\right\}[1+\lambda \cdot g_\lambda(E)] \\ &\geq g_\lambda(E). \end{aligned}$$

□

Though we can prove directly that any λ-measure on a semiring possesses the continuity now, it seems more convenient to show this fact after introducing a new concept called a *quasi-measure*. However, from Theorem 4.3, Theorem 4.4, and the fact that λ-measures are continuous, we know that any λ-measure on a semiring is a monotone measure.

Theorem 4.5. *Let g_λ be a λ-measure on a semiring* **S**. *Then, it is subadditive when $\lambda < 0$; it is superadditive when $\lambda > 0$; and it is additive when $\lambda = 0$.*

Proof. From Theorems 4.3 and 4.4, we know that μ satisfies the λ-rule and is monotone. The conclusion of this theorem can be obtained directly from Definition 4.3. □

By selecting the parameter λ appropriately, we can use a λ-measure to fit a given monotone measure approximately.

Theorem 4.6. *Let g_λ be a λ-measure on a ring* **R**. *Then, for any $E \in \mathbf{R}$ and $F \in \mathbf{R}$,*

(1) $g_\lambda(E - F) = \dfrac{g_\lambda(E) - g_\lambda(E \cap F)}{1 + \lambda \cdot g_\lambda(E \cap F)}$,

(2) $g_\lambda(E \cup F) = \dfrac{g_\lambda(E) + g_\lambda(F) - g_\lambda(E \cap F) + \lambda \cdot g_\lambda(E) \cdot g_\lambda(F)}{1 + \lambda \cdot g_\lambda(E \cap F)}$.

Furthermore, if **R** *is an algebra and g_λ is normalized, then*

(3) $g_\lambda(\bar{E}) = \dfrac{1 - g_\lambda(E)}{1 + \lambda \cdot g_\lambda(E)}$.

Proof. From

$$\begin{aligned} g_\lambda(E) &= g_\lambda((E \cap F) \cup (E - F)) \\ &= g_\lambda(E \cap F) + g_\lambda(E - F)[1 + \lambda \cdot g_\lambda(E \cap F)] \end{aligned}$$

we obtain (1). As to (2), we have

$$\begin{aligned} g_\lambda(E \cup F) &= g_\lambda(E \cup [F - (E \cap F]) \\ &= g_\lambda(E) + g_\lambda(F - (E \cap F)) \cdot [1 + \lambda \cdot g_\lambda(E)] \\ &= g_\lambda(E) + \frac{g_\lambda(F) - g_\lambda(E \cap F)}{1 + \lambda \cdot g_\lambda(E \cap F)} \cdot [1 + \lambda \cdot g_\lambda(E)] \\ &= \frac{g_\lambda(E) + g_\lambda(F) - g_\lambda(E \cap F) + \lambda \cdot g_\lambda(E) \cdot g_\lambda(F)}{1 + \lambda \cdot g_\lambda(E \cap F)} \end{aligned}$$

Formula (3) is a direct result of (1) and the normalization of g_λ. □

How to construct a λ-measure on a semiring (or ring, algebra, σ-ring, σ-algebra, respectively) is a significant and interesting problem. If $X = \{x_1, \ldots, x_n\}$ is a finite set, **C** consists of X and all singletons of X, μ is defined on **C** such that $\mu(\{x_i\}) < \mu(X) < \infty$ for $i = 1, 2, \ldots, n$, and there are at least two points, x_{i_1} and x_{i_2}, satisfying $\mu(\{x_{i_j}\}) > 0$, $j = 1, 2$, then such a set function μ is always a λ-measure on **C** for some parameter λ. When $\mu(X) = \Sigma_{i=1}^n \mu(\{x_i\})$, $\lambda = 0$; otherwise, λ can be determined by the equation

$$\mu(X) = \frac{1}{\lambda}\left[\prod_{i=1}^{n}(1 + \lambda \cdot \mu(\{x_i\})) - 1\right]. \tag{4.8}$$

In fact, we have the following theorem.

Theorem 4.7. *Under the condition mentioned above, the equation*

$$1 + \lambda \cdot \mu(X) = \prod_{i=1}^{n} [1 + \lambda \cdot \mu(\{x_i\})]$$

determines the parameter λ uniquely:

(1) $\lambda > 0$ when $\sum_{i=1}^{n} \mu(\{x_i\}) < \mu(X)$;

(2) $\lambda = 0$ when $\sum_{i=1}^{n} \mu(\{x_i\}) = \mu(X)$;

(3) $-\frac{1}{\mu(X)} < \lambda < 0$ when $\sum_{i=1}^{n} \mu(\{x_i\}) > \mu(X)$.

Proof. Denote $\mu(X) = a, \mu(\{x_i\}) = a_i$ for $i = 1, 2, \ldots, n$, and $f_k(\lambda) = \prod_{i=1}^{k} (1 + a_i\lambda)$ for $k = 2, \ldots, n$. There is no loss of generality in assuming $a_1 > 0$ and $a_2 > 0$. From the given condition we know that $(1 + a_k\lambda) > 0$ for $k = 1, \ldots, n$ and any $\lambda \in (-1/a, \infty)$. Since

$$f_k(\lambda) = (1 + a_k\lambda) f_{k-1}(\lambda),$$

we have

$$f'_k(\lambda) = a_k \cdot f_{k-1}(\lambda) + (1 + a_k\lambda) f'_{k-1}(\lambda),$$

and

$$f''_k(\lambda) = 2a_k \cdot f'_{k-1}(\lambda) + (1 + a_k\lambda) f''_{k-1}(\lambda).$$

It is easy to see that, for any $k = 2, \ldots, n$ and any $\lambda \in (-1/a, \infty)$, if $f'_{k-1}(\lambda) > 0$ and $f''_{k-1} > 0$, then so are $f'_k(\lambda)$ and $f''_k(\lambda)$. Now, since

$$f'_2(\lambda) = a_1(1 + a_2\lambda) + a_2(1 + a_1\lambda) > 0$$

and

$$f''_2(\lambda) = 2a_1a_2 > 0,$$

we know that $f''_k(\lambda) > 0$. This means that the function $f_n(\lambda)$ is concave in $(-1/a, \infty)$. From the derivative of $f_n(\lambda)$,

$$f'_n(0) = \sum_{i=1}^{n} a_i.$$

Noting $\lim_{\lambda \to \infty} f_n(\lambda) = \infty$, we know that, if $\sum_{i=1}^{n} a_i < a$, the curve of $f_n(\lambda)$ has a unique intersection point with the line $f(\lambda) = 1 + a \cdot \lambda$ (illustrated in Fig. 4.2a) on some $\lambda > 0$. If $\sum_{i=1}^{n} a_i = a$, then the line $f(\lambda) = 1 + a \cdot \lambda$ is just a tangent of

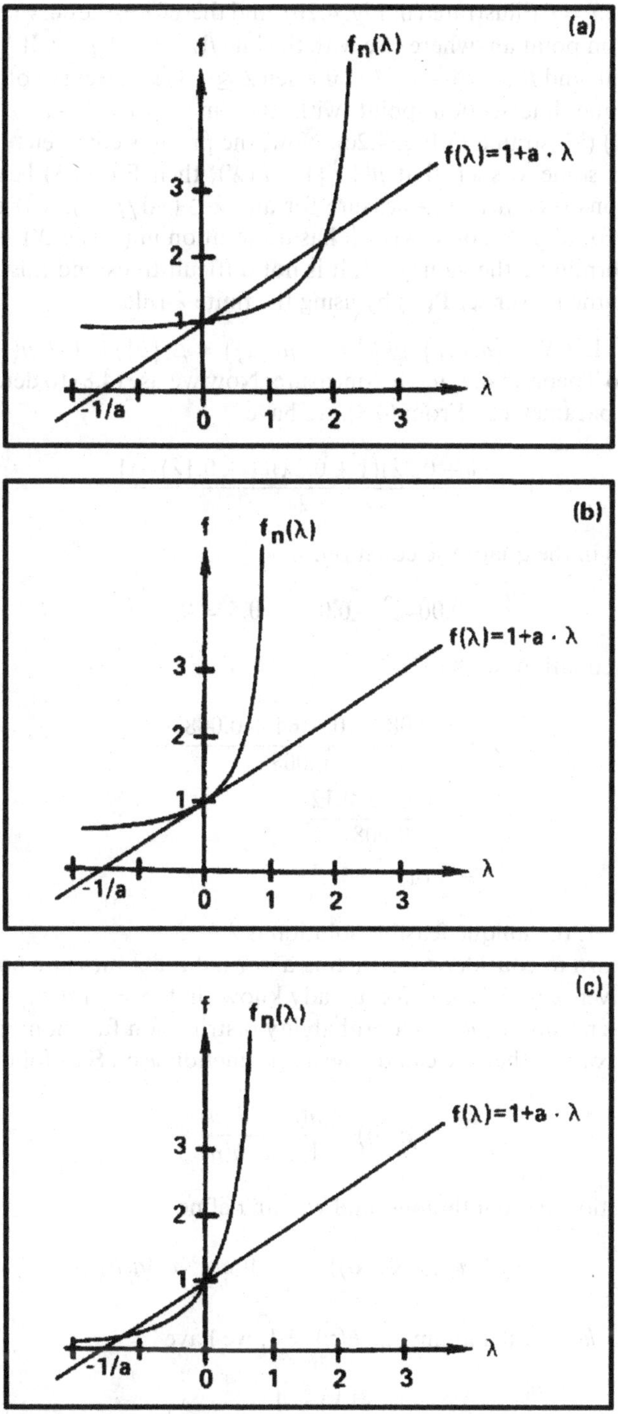

Fig. 4.2 The uniqueness of parameter λ

$f_n(\lambda)$ at point $\lambda = 0$ (illustrated in Fig. 4.2b), and therefore, the curve of $f_n(\lambda)$ has no intersection point anywhere else with the line $f(\lambda) = 1 + a \cdot \lambda$. If $\sum_{i=1}^{n} a_i > a$, since $f'_n(\lambda) > 0$, and $f(\lambda) = 1 + a \cdot \lambda \leq 0$ when $\lambda \leq -1/a$, the curve of $f_n(\lambda)$ must have a unique intersection point with the line $f(\lambda) = 1 + a \cdot \lambda$ on some $\lambda \in (-1/a, 0)$ (illustrated in Fig. 4.2c). Now, the proof is complete. □

If there is some x_i such that $\mu(\{x_i\}) = \mu(X)$, then Eq. (4.8) has infinitely many solutions (i.e., μ is a λ-measure for any $\lambda \in (-1/\mu(X), \infty)$) only when $\mu(\{x_j\}) = 0$ for all $j \neq i$; otherwise, it has no solution in $(-1/\mu(X), \infty)$.

After determining the value of λ, it is not difficult to extend this λ-measure from **C** onto the power set **P**(X) by using the finite λ-rule.

Example 4.4. Let $X = \{a, b, c\}, \mu(X) = 1, \mu(\{a\}) = \mu(\{b\}) = 0.2, \mu(\{c\}) = 0.1$. According to Theorem 4.7, μ is a λ-measure. Now we use (4.8) to determine the value of the parameter λ. From (4.8), we have

$$1 = \frac{(1 + 0.2\lambda)(1 + 0.2\lambda)(1 + 0.1\lambda) - 1}{\lambda},$$

which results in the quadratic equation,

$$0.004\lambda^2 + 0.08\lambda - 0.5 = 0.$$

Solving this equation, we have

$$\begin{aligned} \lambda &= \frac{-0.08 \pm (0.0064 + 0.008)^{1/2}}{0.008} \\ &= \frac{-0.08 \pm 0.12}{0.008} \\ &= 5 \text{ or } -25 \end{aligned}.$$

Since $-25 < -1$, the unique feasible solution is $\lambda = 5$.

Now we turn to consider constructing a normalized λ-measure on the Borel field for a given $\lambda \in (-1, \infty)$. We already know that $\mathbf{S} = \{[a, b) | -\infty < a \leq b < \infty\}$ is a semiring. If $h(x)$ is a probability distribution function (left continuous) on $(-\infty, \infty)$, then we can define a set function ψ on **S** as follows:

$$\psi([a, b)) = \frac{h(b) - h(a)}{1 + \lambda \cdot h(a)}.$$

This set function ψ is continuous, and we can define

$$\psi(X) = \psi((-\infty, \infty)) = \lim_{a \to -\infty, b \to \infty} \psi([a, b)).$$

Since $\lim_{x \to -\infty} h(x) = 0$ and $\lim_{x \to \infty} h(x) = 1$, we have

$$\psi(X) = 1.$$

Moreover, we can verify that such a set function ψ satisfies the λ-rule on **S**. In fact, for any $[a,b) \in \mathbf{S}$ and $[b,c) \in \mathbf{S}$, $[a,b) \cup [b,c) = [a,c) \in \mathbf{S}$ and

$$\begin{aligned}
&\psi([a,b)) + \psi([b,c)) + \lambda \cdot \psi([a,b)) \cdot \psi([b,c)) \\
&\qquad = \psi([a,b)) + \psi([b,c)) \cdot [1 + \lambda \cdot \psi(a,b)] \\
&\qquad = \frac{h(b)-h(a)}{1+\lambda \cdot h(a)} + \frac{h(c)-h(b)}{1+\lambda \cdot h(b)} \cdot \left[1 + \lambda \frac{h(b)-h(a)}{1+\lambda \cdot h(a)}\right] \\
&\qquad = \frac{h(b)-h(a)}{1+\lambda \cdot h(a)} + \frac{[h(c)-h(b)] \cdot [1+\lambda \cdot h(b)]}{[1+\lambda \cdot h(b)] \cdot [1+\lambda \cdot h(a)]} \\
&\qquad = \frac{h(c)-h(a)}{1+\lambda \cdot h(a)} \\
&\qquad = \psi([a,c)).
\end{aligned}$$

It is possible, but rather difficult to verify that such a set function ψ satisfies the σ-λ-rule on **S** and to extend ψ onto the Borel field in a way similar to that used for classical measures. However, if we use the aid of the concept of a quasi-measure, which is introduced in the next section, this problem becomes quite easy to solve.

4.4 Quasi-Measures

Definition 4.5. Let $a \in (0, \infty]$. An extended real function $\theta : [0, a] \to [0, \infty]$ is called a *T-function* iff it is continuous, strictly increasing, and such that $\theta(0) = 0$ and $\theta^{-1}(\{\infty\}) = \emptyset$ or $\{\infty\}$, according to a being finite or not.

Definition 4.6. μ is called *quasi-additive* iff there exists a T-function θ, whose domain of definition contains the range of μ, such that the set function $\theta \circ \mu$ defined on **C** by

$$(\theta \circ \mu)(E) = \theta(\mu(E)), \text{ for any } E \in \mathbf{C} \, ,$$

is additive; μ is called a *quasi-measure* iff there exists a T-function θ such that $\theta \circ \mu$ is a classical measure on **C**. The T-function θ is called the *proper T-function* of μ.

A normalized quasi-measure is called a *quasi-probability*.

Clearly, any classical measure is a quasi-measure with the identity function as its proper T-function.

Example 4.5. The monotone measure given in Example 3.4 is a quasi-measure. Its proper T-function is $\theta(y) = \sqrt{y}$, $y \in [0, 1]$.

Theorem 4.8. *Any quasi-measure on a semiring is a quasi-additive monotone measure.*

Proof. Let μ be a quasi-measure on a semiring **S** and θ be its proper T-function. Since any classical measure on a semiring is additive, μ is quasi-additive.

Furthermore, θ^{-1} exists, and it is continuous, strictly increasing, and $\theta^{-1}(0) = 0$. So, $\mu = \theta^{-1} \circ (\theta \circ \mu)$ is continuous, monotone, and $\mu(\emptyset) = 0$. That is, μ is a monotone measure. □

Theorem 4.9. *If μ is a classical measure, then, for any T-function θ whose range contains the range of μ, $\theta^{-1} \circ \mu$ is a quasi-measure with θ as its proper T-function.*

Proof. Since $\theta \circ (\theta^{-1} \circ \mu) = \mu$, the conclusion of this theorem is clear. □

Theorem 4.10. *Let μ be quasi-additive on a ring* **R** *with $\mu(\emptyset) = 0$. If μ is either continuous from below on* **R**, *or continuous from above at Ø and finite, then μ is a quasi-measure on* **R**.

Proof. Since μ is quasi-additive, there exists a T-function θ such that $\theta \circ \mu$ is additive on **R**. The composition $\theta \circ \mu$ is either continuous from below on **R**, or continuous from above at Ø and finite. So $\theta \circ \mu$ is a measure on **R** (Section 2.2, Theorem 2.32). That is, μ is a quasi-measure on **R**. □

Corollary 4.1. *Any quasi-additive monotone measure on a ring is a quasi-measure.*

Now, we return to solve the problems that are raised in Section 4.3.

Theorem 4.11. *Let $\lambda \neq 0$. Any λ-measure g_λ is a quasi-measure with*

$$\theta_\lambda(y) = \frac{\ln(1+\lambda y)}{k\lambda}, y \in [0, \sup g_\lambda],$$

as its proper T-function, where k is an arbitrary finite positive real number. Conversely, if μ is a classical measure, then $\theta_\lambda^{-1} \circ \mu$ is a λ-measure, where

$$\theta_\lambda^{-1}(x) = \frac{e^{k\lambda x} - 1}{\lambda}, \quad x \in [0, \infty],$$

and k is an arbitrary finite positive real number.

Proof. θ_λ is a T-function. Let $\{E_n\}$ be a disjoint sequence of sets in **C** whose union $\bigcup_{n=1}^{\infty} E_n$ is also in **C**. If g_λ is a λ-measure on **C** then it satisfies the σ-λ-rule and there exists $E_0 \in$ **C** such that $g_\lambda(E_0) < \infty$. Therefore, we have

$$\begin{aligned}
(\theta_\lambda \circ g_\lambda)\Big(\bigcup_{n=1}^{\infty} E_n\Big) &= \frac{1}{k \cdot \lambda} \cdot \ln\left[1 + \lambda \cdot g_\lambda\Big(\bigcup_{n=1}^{\infty} E_n\Big)\right] \\
&= \frac{1}{k \cdot \lambda} \cdot \ln\left(1 + \left[\prod_{n=1}^{\infty}[1 + \lambda \cdot g_\lambda(E_n)]\right] - 1\right) \\
&= \frac{1}{k \cdot \lambda} \cdot \sum_{n=1}^{\infty} \ln[1 + \lambda \cdot g_\lambda(E_n)] \\
&= \sum_{n=1}^{\infty} \frac{\ln[1 + \lambda \cdot g_\lambda(E_n)]}{k \cdot \lambda} \\
&= \sum_{n=1}^{\infty} (\theta_\lambda \circ g_\lambda)(E_n),
\end{aligned}$$

and $(\theta_\lambda \circ g_\lambda)(E_0) = \theta_\lambda(g_\lambda(E_0)) < \infty$. So $\theta_\lambda \circ g_\lambda$ is a classical measure on **C**. Conversely, if μ is a classical measure on **C**, then it is σ-additive, and there exists $E_0 \in \mathbf{C}$ such that $\mu(E_0) < \infty$. Therefore, we have

$$
\begin{aligned}
(\theta_\lambda^{-1} \circ \mu)\left(\bigcup_{n=1}^{\infty} E_n\right) &= \theta_\lambda^{-1}\left[\sum_{n=1}^{\infty} \mu(E_n)\right] \\
&= \frac{\exp\left[k\lambda \sum_{n=1}^{\infty} \mu(E_n)\right] - 1}{\lambda} \\
&= \frac{\prod_{n=1}^{\infty} e^{k\lambda \cdot \mu(E_n)} - 1}{\lambda} \\
&= (1/\lambda)\left\{\prod_{n=1}^{\infty}[1 + \lambda \cdot \theta_\lambda^{-1}(\mu(E_n))] - 1\right\} \\
&= (1/\lambda)\left\{\prod_{n=1}^{\infty}[1 + \lambda \cdot (\theta_\lambda^{-1} \circ \mu)(E_n)] - 1\right\};
\end{aligned}
$$

that is, $\theta_\lambda^{-1} \circ \mu$ satisfies the σ-λ–rule. Noting that $(\theta_\lambda^{-1} \circ \mu)(E_0) = \theta_\lambda^{-1}(\mu(E_0)) < \infty$, we conclude that $\theta_\lambda^{-1} \circ \mu$ is a λ-measure on **C**. □

Example 4.6. Let $X = \{a, b\}, \mathbf{F} = \mathbf{P}(X), g_\lambda$ be defined by

$$
g_\lambda(E) = \begin{cases} 0 & \text{if } E = \emptyset \\ 0.2 & \text{if } E = \{a\} \\ 0.4 & \text{if } E = \{b\} \\ 1 & \text{if } E = X. \end{cases}
$$

Then g_λ is a λ-measure with a parameter $\lambda = 5$. If we take

$$
\theta_\lambda(y) = \frac{\ln(1 + \lambda y)}{\ln(1 + \lambda)} = \frac{\ln(1 + 5y)}{\ln 6},
$$

then we have

$$
(\theta_\lambda \circ g_\lambda)(E) = \begin{cases} 0 & \text{if } E = \emptyset \\ 0.387 & \text{if } E = \{a\} \\ 0.613 & \text{if } E = \{b\} \\ 1 & \text{if } E = X. \end{cases}
$$

$\theta_\lambda \circ g_\lambda$ is a probability measure.

Example 4.7. Let $X = \{a, b\}$, $\mathbf{F} = \mathbf{P}(X)$, and let g_λ be a λ-measure defined by

$$g_\lambda(E) = \begin{cases} 0 & \text{if } E = \emptyset \\ 0.5 & \text{if } E = \{a\} \\ 0.8 & \text{if } E = \{b\} \\ 1 & \text{if } E = X. \end{cases}$$

with $\lambda = -0.75$. If we take

$$\theta_\lambda(y) = \frac{\ln(1 - 0.75y)}{\ln 0.25},$$

then

$$(\theta_\lambda \circ g_\lambda)(E) = \begin{cases} 0 & \text{if } E = \emptyset \\ 0.34 & \text{if } E = \{a\} \\ 0.66 & \text{if } E = \{b\} \\ 1 & \text{if } E = X, \end{cases}$$

which is a probability measure.

In a similar way, we know that under the mapping θ_λ the λ-rule and the finite λ-rule become the additivity and the finite additivity, respectively. Conversely, under the mapping θ_λ^{-1} the additivity and the finite additivity become the λ-rule and the finite λ-rule, respectively. Recalling some relevant knowledge in classical measure theory, we have the following corollaries.

Corollary 4.2. *On a semiring, the λ-rule is equivalent to the finite λ-rule.*

Corollary 4.3. *Any λ-measure on a semiring is continuous.*

Corollary 4.4. *On a ring, the λ-rule together with continuity are equivalent to the σ-λ-rule. Thus, on a ring, any monotone measure that satisfies the λ-rule is a λ-measure.*

Similarly as in classical measure theory, a monotone measure on a semiring that satisfies the λ-rule (or, is quasi-additive) may not satisfy the σ-λ-rule (or, may not be a quasi-measure).

Corollary 4.5. *If g_λ is a normalized λ-measure on an algebra $\mathbf{R}$, then its dual measure μ, which is defined by*

$$\mu(E) = 1 - g_\lambda(\bar{E}) \textit{ for any } E \in \mathbf{R},$$

is also a normalized λ-measure on $\mathbf{R}$, and the corresponding parameter is $\lambda' = -\lambda/(\lambda + 1)$.

Proof. Let $E \in \mathbf{R}, F \in \mathbf{R},$ and $E \cap F = \emptyset$. By using Theorem 4.6, we have

$$\begin{aligned}
&\mu(E) + \mu(F) - \frac{\lambda}{\lambda+1}\mu(E)\mu(F) \\
&= 1 - g_\lambda(\bar{E}) + 1 - g_\lambda(\bar{F}) - \frac{\lambda}{\lambda+1}[1 - g_\lambda(\bar{E})][1 - g_\lambda(\bar{F})] \\
&= \frac{(\lambda+1)g_\lambda(E)}{1+\lambda g_\lambda(E)} + \frac{(\lambda+1)g_\lambda(F)}{1+\lambda g_\lambda(F)} - \lambda\frac{(\lambda+1)g_\lambda(E)g_\lambda(F)}{[1+\lambda g_\lambda(E)][1+\lambda g_\lambda(F)]} \\
&= \frac{(\lambda+1)[g_\lambda(E) + g_\lambda(F) + \lambda g_\lambda(E)g_\lambda(F)]}{[1+\lambda g_\lambda(E)][1+\lambda g_\lambda(F)]} \\
&= \frac{(\lambda+1)g_\lambda(E \cup F)}{1+\lambda g_\lambda(E \cup F)} \\
&= 1 - g_\lambda(\overline{E \cup F}) \\
&= \mu(E \cup F).
\end{aligned}$$

Since μ is continuous, by Corollary 3.4, μ satisfies the σ-λ-rule with a parameter $\lambda' = -\lambda/(\lambda+1)$. So, noting that $\mu(X) = 1 - g_\lambda(\emptyset) = 1$, we know that μ is a normalized λ-measure on **R** with a parameter $\lambda' = -\lambda/(\lambda+1)$. □

As to the problem of constructing a λ-measure on the Borel field, we deal with it in Chapter 6.

4.5 Belief Measures and Plausibility Measures

In Section 4.4, a nonadditive measure is induced from a classical measure by a transformation of the range of the latter. In this section we attempt to construct a nonadditive measure in another way.

Definition 4.7. Let **P(P(**X**))** be the power set of **P(**X**)**. If p is a discrete probability measure on (**P(**X**)**, **P(P(**X**))**) with $p(\{\emptyset\}) = 0$, then the set function m: **P(**X**)** → [0, 1] determined by

$$m(E) = p(\{E\})\text{for any } E \in \mathbf{P}(X)$$

is called a *basic probability assignment* on **P(**X**)**.

Theorem 4.12. *A set function m:* **P(**X**)** → [0, 1] *is a basic probability assignment if and only if*

(1) $m(\emptyset) = 0$;

(2) $\sum_{E \in \mathbf{P}(X)} m(E) = 1$.

Proof. The necessity of these two conditions follows directly from Definition 4.7. As for their sufficiency, if we write

$$\mathbf{D}_n = \left\{E\middle|\tfrac{1}{n+1} < m(E) \leq \tfrac{1}{n}\right\}, n = 1, 2, \ldots,$$

then every D_n is a finite class,

$$\mathbf{D} = \bigcup_{n=1}^{\infty} \mathbf{D}_n = \{E|m(E) > 0\}$$

is a countable class, and $\hat{\mathbf{S}} = \{\{E|E \in \mathbf{P}(X)\} \cup \{\emptyset\}$ is a semiring. Define

$$p(\{E\}) = \begin{cases} m(E) & \text{if } E \in \mathbf{D} \\ 0 & \text{otherwise} \end{cases}$$

for any $E \in \mathbf{P}(X)$ and $p(\{\emptyset\}) = 0$. Then, p is a probability measure on $\hat{\mathbf{S}}$ with $p(\{\emptyset\}) = 0$, which can be extended uniquely to a discrete probability measure on $(\mathbf{P}(X), \mathbf{P}(\mathbf{P}(X)))$ by the formula

$$p(\mathbf{E}) = \sum_{E \in \mathbf{E}} p(\{E\}).$$

for any $\mathbf{E} \in \mathbf{P}(\mathbf{P}(X))$. □

Definition 4.8. If m is a basic probability assignment on $\mathbf{P}(X)$, then the set function Bel: $\mathbf{P}(X) \to [0, 1]$ determined by the formula

$$\mathrm{Bel}(E) = \sum_{F \subset E} m(F) \ \forall E \in \mathbf{P}(X) \tag{4.9}$$

is called a *belief measure* on $(X, \mathbf{P}(X))$, or, more specifically, a belief measure induced from m.

Lemma 4.1. *If E is a nonempty finite set, then*

$$\sum_{F \subset E} (-1)^{|F|} = 0.$$

Proof. Let $E = \{x_1, \ldots, x_n\}$Then, we have

$$\{|F||F \subset E\} = \{0, 1, \ldots, n\}$$

and

$$|\{F||F| = i\}| = \binom{n}{i}, \ i = 0, 1, \ \ldots, n.$$

So, we have

$$\sum_{F \subset E} (-1)^{|F|} = \sum_{i=0}^{n} (-1)^i \binom{n}{i} = (1-1)^n = 0. \qquad \square$$

Lemma 4.2. *If E is a finite set, $F \subset E$ and $F \neq E$, then*

$$\sum_{G | F \subset G \subset E} (-1)^{|G|} = 0.$$

Proof. $E - F$ is a nonempty finite set. Using Lemma 4.1, we have

$$\sum_{G | F \subset G \subset E} (-1)^{|G|} = \sum_{D \subset E-F} (-1)^{|F \cup D|} = (-1)^{|F|} \sum_{D \subset E-F} (-1)^{|D|} = 0. \qquad \square$$

Lemma 4.3. *Let X be finite, and λ and ν be finite set functions defined on* $\mathbf{P}(X)$. *Then we have*

$$\lambda(E) = \sum_{F \subset E} \nu(F) \quad \forall E \in \mathbf{P}(X) \tag{4.10}$$

if and only if

$$\nu(E) = \sum_{F \subset E} (-1)^{|E-F|} \lambda(F) \quad \forall E \in \mathbf{P}(X). \tag{4.11}$$

Proof. If (4.10) is true, then

$$\begin{aligned}
\sum_{F \subset E} (-1)^{|E-F|} \lambda(F) &= (-1)^{|E|} \sum_{F \subset E} (-1)^{|F|} \lambda(F) \\
&= (-1)^{|E|} \sum_{F \subset E} \left[(-1)^{|F|} \sum_{G \subset F} \nu(G) \right] \\
&= (-1)^{|E|} \sum_{G \subset E} \left[\nu(G) \sum_{F | G \subset F \subset E} (-1)^{|F|} \right] \\
&= (-1)^{|E|} \nu(E) (-1)^{|E|} \\
&= \nu(E).
\end{aligned}$$

Conversely, if (4.11) is true, then we have

$$\begin{aligned}\sum_{F\subset E}\nu(F) &= \sum_{F\subset E}\sum_{G\subset F}(-1)^{|F-G|}\lambda(G)\\&= \sum_{G\subset E}\left[(-1)^{|G|}\lambda(G)\sum_{F|G\subset F\subset E}(-1)^{|F|}\right]\\&= (-1)^{|E|}\lambda(E)(-1)^{|E|}\\&= \lambda(E).\end{aligned}$$

□

Theorem 4.13. *If Bel is a belief measure on* $(X, \mathbf{P}(X))$*, then*

(BM1) $\mathrm{Bel}(\emptyset) = 0$;

(BM2) $\mathrm{Bel}(X) = 1$;

(BM3) $\mathrm{Bel}\left(\bigcup_{i=1}^{n} E_i\right) \geq \sum_{I\subset\{1,\ldots,n\},I\neq\emptyset}(-1)^{|I|+1}\,\mathrm{Bel}\left(\bigcap_{i\in I} E_i\right)$,

where $\{E_1, \ldots, E_n\}$ *is any finite subclass of* $\mathbf{P}(X)$*;*

(BM4) Bel *is continuous from above.*

Proof. From Theorem 4.12 and Definition 4.8, it is easy to see that (BM1) and (BM2) are true. To show that (BM3) holds, let us consider an arbitrary finite subclass $\{E_1, \ldots, E_n\}$, and set $I(F) = \{i|1 \leq i \leq n, F \subset E_i\}$, for any $F \in \mathbf{P}(X)$. Using Lemma 4.1, we have

$$\begin{aligned}\sum_{I\subset\{1,\ldots,n\},I\neq\emptyset}(-1)^{|I|+1}\,\mathrm{Bel}\left(\bigcap_{i\in I}E_i\right) &= \sum_{I\subset\{1,\ldots,n\},I\neq\emptyset}\left[(-1)^{|I|+1}\sum_{F\subset\bigcap_{i\in I}E_i}m(F)\right]\\&= \sum_{F|I(F)\neq\emptyset}\left[m(F)\sum_{I\subset I(F),I\neq\emptyset}(-1)^{|I|+1}\right]\\&= \sum_{F|I(F)\neq\emptyset}\left[m(F)\left(1-\sum_{I\subset I(F)}(-1)^{|I|}\right)\right]\\&= \sum_{F|I(F)\neq\emptyset}m(F)\\&= \sum_{F\subset F_i \text{ for some } i}m(F)\\&\leq \sum_{F\subset\bigcup_{i=1}^{n}E_i}m(F)\\&= \mathrm{Bel}\left(\bigcup_{i=1}^{n}E_i\right)\end{aligned}$$

As to (BM4), let E_i be a decreasing sequence of sets in $\mathbf{P}(X)$, and $\bigcap_{i=1}^{\infty} E_i = E$. From Theorem 4.12, we know there exists a countable class $\{D_n\} \subset \mathbf{P}(X)$, such that $m(F) = 0$ whenever $F \notin \{D_n\}$, and for any $\epsilon > 0$ there exists n_0 such that $\sum_{n>n_0} m(D_n) < \varepsilon$. Then, for each D_n, where $n \leq n_0$, if $D_n \not\subset E$ (that is, $D_n - E \neq \emptyset$), there exists $i(n)$, such that $D_n \not\subset E_{i(n)}$. Let $i_0 = \max(i(1), \ldots, i(n_0))$. Then, if $D_n \not\subset E$, we have $D_n \not\subset E_{i0}$ for any $n \leq n_0$. Hence,

$$
\begin{aligned}
\mathrm{Bel}(E) &= \sum_{F \subset E} m(F) \\
&= \sum_{D_n \subset E} m(D_n) \\
&\geq \sum_{D_n \subset E, n \leq n_0} m(D_n) \\
&\geq \sum_{D_n \subset E_{i_0}, n \leq n_0} m(D_n) \\
&\geq \sum_{D_n \subset E_{i_0}} m(D_n) - \sum_{n > n_0} m(D_n) \\
&> \sum_{F \subset E_{i_0}} m(F) - \varepsilon \\
&= \mathrm{Bel}(E_{i_0}) - \varepsilon.
\end{aligned}
$$

Noting that $\mathrm{Bel}(E) \leq \mathrm{Bel}\{E_i\}$ for $i = 1, 2, \ldots$, and $\{\mathrm{Bel}(E_i)\}$ is decreasing with respect to i, we have $\mathrm{Bel}(E) = \lim_i \mathrm{Bel}(E_i)$. □

Observe that due to property (BM3), established for belief measures by Theorem 4.13, belief measures are ∞-monotone measures introduced in Section 4.2.

Theorem 4.14. *Any belief measure is monotone and superadditive.*

Proof. Let $E_1 \subset X, E_2 \subset X$, and $E_1 \cap E_2 = \emptyset$. We have

$$
\begin{aligned}
\mathrm{Bel}(E_1 \cup E_2) &\geq \mathrm{Bel}(E_1) + \mathrm{Bel}(E_2) - \mathrm{Bel}(E_1 \cap E_2) \\
&= \mathrm{Bel}(E_1) + \mathrm{Bel}(E_2) \geq \max\{\mathrm{Bel}(E_1), \mathrm{Bel}(E_2)\}.
\end{aligned}
$$

From this inequality, it is easy to see that Bel is monotone and superadditive. □

From Theorems 4.13 and 4.14, we know that the belief measure is an upper semicontinuous monotone measure.

On a finite space, we can express a basic probability assignment by the belief measure induced from it.

Theorem 4.15. *Let X be finite. If a set function* μ*:* $\mathbf{P}(X) \to [0, 1]$ *satisfies the conditions*

(1) $\mu(\emptyset) = 1$;
(2) $\mu(X) = 1$;
(3) $\mu\left(\bigcap_{i=1}^{n} E_i\right) \geq \sum_{I \subset \{1,\ldots,n\}, I \neq \emptyset} (-1)^{|I|+1} \mu\left(\bigcap_{i \in I} E_i\right)$,

where $\{E_1, \ldots, E_n\}$ *is any finite subclass of* $\mathbf{P}(X)$*, then the set function m determined by*

$$m(E) = \sum_{F \subset E} (-1)^{|E-F|} \mu(F) \ \forall E \in \mathbf{P}(X), \tag{4.12}$$

is a basic probability assignment, and μ *is the belief measure induced from m. That is,*

$$\mu(E) = \mathrm{Bel}(E) = \sum_{F \subset E} m(F).$$

Proof. First, $m(\emptyset) = \sum_{F \subset \emptyset} (-1)^{|\emptyset - F|} \mu(F) = \mu(\emptyset) = 0$. Next, from (4.12) and Lemma 4.3, we have

$$\sum_{E \subset X} m(E) = \mu(X) = 1.$$

To prove that m is a basic probability assignment, we should show that $m(E) \geq 0$ for any $E \subset X$. Indeed, since X is finite, E is also finite, and we can write $E = \{x_1, \ldots, x_n\}$. If we denote $E_i = E - \{x_i\}$, then $E = \bigcup_{i=1}^{n} E_i$ and

$$\begin{aligned}
m(E) &= \sum_{F \subset E} (-1)^{|E-F|} \mu(F) \\
&= \mu(E) - \sum_{I \subset \{1,\ldots,n\}, I \neq \emptyset} (-1)^{|I|+1} \mu\left(\bigcap_{i \in I} E_i\right) \\
&= \mu\left(\bigcup_{i=1}^{n} E_i\right) - \sum_{I \subset \{1,\ldots,n\}, I \neq \emptyset} (-1)^{|I|+1} \mu\left(\bigcap_{i \in I} E_i\right) \\
&\geq 0.
\end{aligned}$$

The last conclusion in this theorem is a direct result of Lemma 4.3. □

Definition 4.9. If m is a basic probability assignment on $\mathbf{P}(X)$, then the set function Pl: $\mathbf{P}(X) \to [0, 1]$ determined by

$$\mathrm{Pl}(E) = \sum_{F \cap E \neq \emptyset} m(F) \ \forall\, E \in \mathbf{P}(X) \tag{4.13}$$

is called a *plausibility measure* on $(X, \mathbf{P}(X))$, or, more exactly, a plausibility measure induced from m.

Theorem 4.16. *If* Bel *and* Pl *are the belief measure and plausibility measure, respectively, induced from the same basic probability assignment then*

$$\mathrm{Bel}(E) = 1 - \mathrm{Pl}(\bar{E}) \tag{4.14}$$

and

$$\mathrm{Bel}(E) \leq \mathrm{Pl}(E)$$

for any $E \subset X$.

Proof.

$$\begin{aligned}\mathrm{Bel}(E) &= \sum_{F \subset E} m(F) \\ &= \sum_{F \subset X} m(F) - \sum_{F \not\subset E} m(F) \\ &= 1 - \sum_{F \cap \bar{E} \neq \emptyset} m(F) \\ &= 1 - \mathrm{Pl}(\bar{E}).\end{aligned}$$

The second conclusion can be obtained directly from Definitions 4.8 and 4.9.□

Theorem 4.17. *If* Pl *is a plausibility measure on* $(X, \mathbf{P}(X))$, *then*

(PMI) $\mathrm{Pl}(\emptyset) = 0;$

(PM2) $\mathrm{Pl}(X) = 1;$

(PM3) $$\mathrm{Pl}\left(\bigcap_{i=1}^{n} E_i\right) \leq \sum_{I \subset \{1,\ldots,n\}, I \neq \emptyset} (-1)^{|I|+1}\, \mathrm{Pl}\left(\bigcup_{i \in I} E_i\right),$$

where $\{E_1, \ldots, E_n\}$ *is any finite subclass of* $\mathbf{P}(X)$.

(PM4) Pl *is continuous from below.*

Proof. From Theorem 4.13 and Theorem 4.16, we can directly obtain (PM1), (PM2), and (PM4). As to (PM3), by using Lemma 4.1, we have

$$\begin{aligned}
\mathrm{Pl}\left(\bigcap_{i=1}^{n} E_i\right) &= 1 - \mathrm{Bel}\left(\overline{\bigcap_{i=1}^{n} E_i}\right) \\
&= 1 - \mathrm{Bel}\left(\bigcup_{i=1}^{n} \bar{E}_i\right) \\
&\leq 1 - \sum_{I\subset\{1,\ldots,n\},\ I\neq\emptyset} (-1)^{|I|+1}\mathrm{Bel}\left(\bigcap_{i\in I} \bar{E}_i\right) \\
&= \sum_{I\subset\{1,\ldots,n\},\ I\neq\emptyset} (-1)^{|I|+1}\left[1 - \mathrm{Bel}\left(\bigcap_{i\in I} \bar{E}_i\right)\right] \\
&= \sum_{I\subset\{1,\ldots,n\},\ I\neq\emptyset} (-1)^{|I|+1}\left[1 - \mathrm{Bel}\left(\overline{\bigcup_{i\in I} E_i}\right)\right] \\
&= \sum_{I\subset\{1,\ldots,n\},\ I\neq\emptyset} (-1)^{|I|+1}\mathrm{Pl}\left(\bigcup_{i\in I} E_i\right).
\end{aligned}$$

□

Due to the property (PM3), which is established for plausibility measures by Theorem 4.17, plausibility measures are ∞-alternating measures introduced in Section 4.2.

Theorem 4.18. *Any plausibility measure is monotone and subadditive.*

Proof. $E \subset F \subset X$, then $\bar{F} \subset \bar{E} \subset X$. From Theorem 4.14 and Theorem 4.16, we have

$$\mathrm{Pl}(E) = 1 - \mathrm{Bel}(\bar{E}) \leq 1 - \mathrm{Bel}(\bar{F}) = \mathrm{Pl}(F)$$

As to subadditivity, if $E_1 \subset X$ and $E_2 \subset X$, then

$$\begin{aligned}
0 &\leq \mathrm{Pl}(E_1 \cap E_2) \\
&\leq \mathrm{Pl}(E_1) + \mathrm{Pl}(E_2) - \mathrm{Pl}(E_1 \cup E_2).
\end{aligned}$$

So $\mathrm{Pl}(E_1 \cup E_2) \leq \mathrm{Pl}(E_1) + \mathrm{Pl}(E_2)$. □

From Theorem 4.17 and Theorem 4.18, we know that the plausibility measure is a lower semicontinuous monotone measure.

Theorem 4.19. *Any discrete probability measure p on $(X, \mathbf{P}(X))$ is both a belief measure and a plausibility measure. The corresponding basic probability assignment focuses on the singletons of $\mathbf{P}(X)$. Conversely, if m is a basic probability* *assignment focusing on the singletons of* **P**(X)*, then the belief measure and the plausibility measure induced from m coincide, resulting in a discrete probability measure on* $(X, \mathbf{P}(X))$.

Proof. Since p is a discrete probability measure, there exists a countable set $\{x_1, x_2, \cdots\} \subset X$, such that

$$\sum_{i=1}^{\infty} p(\{x_i\}) = 1.$$

Let

$$m(E) = \begin{cases} p(E) & \text{if } E = \{x_i\} \text{ for some } i \\ 0 & \text{otherwise} \end{cases}$$

for any $E \in \mathbf{P}(X)$. Then, m is a basic probability assignment, and

$$p(E) = \sum_{x_i \in E} p(\{x_i\}) = \sum_{F \subset E} m(F) = \sum_{F \cap E \neq \emptyset} m(F)$$

for any $E \in \mathbf{P}(X)$. That is, p is both a belief measure and a plausibility measure. Conversely, if a basic probability assignment m focuses only on the singletons of $\mathbf{P}(X)$, then, for any $E \in \mathbf{P}(X)$,

$$\mathrm{Bel}(E) = \sum_{F \subset E} m(F) = \sum_{x \in E} m(\{x\}) = \sum_{F \cap E \neq \emptyset} m(F) = \mathrm{P1}(E).$$

So, Bel and Pl coincide, and it is easy to verify that they are σ-additive. Consequently, they are discrete probability measures on $(X, \mathbf{P}(X))$. □

Theorem 4.20. *Let* Bel *and* Pl *be the belief measure and the plausibility measure, respectively, induced from a basic probability assignment m. If* Bel *coincides with* Pl*, then m focuses only on singletons.*

Proof. If there exists $E \in \mathbf{P}(X)$ that is not a singleton of $\mathbf{P}(X)$ such that $m(E) > 0$, then, for any $x \in E$,

$$\mathrm{Bel}(\{x\}) = m(\{x\}) < m(\{x\}) + m(E) \leq \sum_{F \cap \{x\} \neq \emptyset} m(F) = \mathrm{P1}(\{x\}).$$

This contradicts the coincidence of Bel and Pl. □

The Sugeno measures defined on the power set $\mathbf{P}(X)$ are special examples of belief measures and plausibility measures when X is countable.

Theorem 4.21. *Let X be countable, and $g_\lambda (\lambda \neq 0)$ be a Sugeno measure on $(X, \mathbf{P}(X))$. Then g_λ is a belief measure when $\lambda > 0$, and is a plausibility measure when $\lambda < 0$.*

Proof. Let $X = \{x_1, x_2, \ldots\}$. When $\lambda > 0$, we define m: $\mathbf{P}(X) \to [0, 1]$ by

$$m(E) = \begin{cases} \lambda^{|E|-1} \prod_{x_i \in E} g_\lambda(\{x_i\}) & \text{if } E \neq \emptyset \\ 0 & \text{if } E = \emptyset \end{cases}$$

for any $E \in$ P(X). Obviously, $m(E) \geq 0$ for any $E \in \mathbf{P}(X)$. From Definition 4.3, we have

$$\begin{aligned} g_\lambda(E) &= \frac{1}{\lambda}\left[\prod_{x_i \in E} (1 + \lambda \cdot g_\lambda(\{x_i\})) - 1\right] \\ &= \frac{1}{\lambda} \sum_{F \subset E, F \neq \emptyset} \left[\lambda^{|F|} \cdot \prod_{x_i \in F} g_\lambda(\{x_i\})\right] \\ &= \sum_{F \subset E, F \neq \emptyset} \left[\lambda^{|F|-1} \cdot \prod_{x_i \in F} g_\lambda(\{x_i\})\right] \\ &= \sum_{F \subset E} m(F). \end{aligned}$$

Since $g_\lambda(X) = 1$, we have

$$\sum_{F \subset X} m(F) = 1.$$

Therefore, m is a basic probability assignment, and thus, g_λ is the belief measure induced from m. When $\lambda < 0$, we have $\lambda' = -\lambda/(\lambda + 1) > 0$. By using Corollary 4.5 and Theorem 4.16, we know that g_λ is a plausibility measure. □

4.6 Possibility Measures and Necessity Measures

Definition 4.10. A monotone measure μ is called *maxitive* on **C** iff

$$\mu\left(\bigcup_{t \in T} E_t\right) = \sup_{t \in T} \mu(E_t) \tag{4.15}$$

for any subclass $\{E_t | t \in T\}$ of **C** whose union is in **C**, where T is an arbitrary index set.

If **C** is a finite class, then the maxitivity of μ on **C** is equivalent to the simpler requirement that

$$\mu(E_i \cup E_2) = \mu(E_1) \vee \mu(E_2) \tag{4.16}$$

whenever $E_i \in \mathbf{C}$, $E_2 \in \mathbf{C}$, and $E_1 \cup E_2 \in \mathbf{C}$. Symbol $\vee$ denotes the maximum of $\mu(E_1)$ and $\mu(E_2)$.

Definition 4.11. A monotone measure μ is called a *generalized possibility measure* on **C** iff it is maxitive on **C** and there exists $E \in \mathbf{C}$ such that $\mu(E) < \infty$.

Usually, a generalized possibility measure is denoted by π.

Definition 4.12. If π is a generalized possibility measure defined on **P**(X), then the function f defined on X by

$$f(x) = \pi(\{x\}) \text{ for any } x \in X$$

is called its *possibility profile*.

Theorem 4.22. *Any generalized possibility measure π (on* **C***) is a lower semicontinuous monotone measure (on* **C***).*

Proof. According to the convention, when $T = \emptyset$ we have $\cup_{t\in T} E_t = \emptyset$ and $\sup_{t\in T} \mu(E_t) = 0$. So, if $\emptyset \in \mathbf{C}$, then $\pi(\emptyset) = 0$. Furthermore, if $E \in \mathbf{C}$, $F \in \mathbf{C}$, and $E \subset F$, then, by using maxitivity, we have

$$\pi(F) = \pi(E \cup F) = \pi(E) \vee \pi(F) \geq \pi(E).$$

At last, π is continuous from below. In fact, if $\{E_n\}$ is an increasing sequence of sets in **C** whose union E is also in **C**, from the definition of the supremum, for any $\varepsilon > 0$, there exists n_0 such that

$$\pi(E_{n_0}) \geq \sup_n \pi(E_n) - \varepsilon = \pi(E) - \varepsilon.$$

Noting that π is monotone, we know that

$$\lim_n \pi(E_n) = \pi(E).$$

□

Definition 4.13. When a generalized possibility measure π defined on **P**(X) is normalized, it is called a *possibility measure*.

The following example shows that a possibility measure is not necessarily continuous from above.

Example 4.8. Let $X = (-\infty, \infty)$. A set function $\pi : \mathbf{P}(X) \to [0, 1]$ is defined by

$$\pi(E) = \begin{cases} 1 & \text{if } E \neq \emptyset, \\ 0 & \text{if } E = \emptyset \end{cases}$$

for any $E \in \mathbf{P}(X)$. Clearly, π is maxitive and $\pi(X) = 1$; therefore it is a possibility measure on **P**(X). But it is not continuous from above. In fact, if we take $E = (0,1/n)$, then $\{E_n\}$ is decreasing, and $\bigcap_{n=1}^{\infty} E_n = \emptyset$. We have $\pi(E_n) = 1$ for all $n = 1, 2, \ldots$, but $\pi(\emptyset) = 0$. So $\lim_n \pi(E_n) \neq \pi(\bigcap_{n=1}^{\infty} E_n)$.

Theorem 4.23. *If f is the possibility profile of a possibility measure π, then*

$$\sup_{x \in X} f(x) = 1. \tag{4.17}$$

Conversely, if a function $f: X \to [0, 1]$ satisfies (4.17), then f can determine a possibility measure π uniquely, and f is the possibility profile of π.

Proof. From (4.15), we have

$$\begin{aligned} \sup_{x \in X} f(x) &= \sup_{x \in X} \pi(\{\pi\}) \\ &= \pi(\bigcup_{x \in X} \{x\}) \\ &= \pi(X) \\ &= 1. \end{aligned}$$

Conversely, let

$$\pi(E) = \sup_{x \in E} f(x)$$

for any $E \in \mathbf{P}(X)$, then π is a possibility measure, and

$$\pi(\{x\}) = \sup_{x \in \{x\}} f(x) = f(x).$$

□

A similar result can be easily obtained for generalized possibility measures: Any function $f: X \to [0, \infty)$ uniquely determines a generalized possibility measure π on $\mathbf{P}(X)$ by

$$\pi(E) = \sup_{x \in E} f(x) \text{ for any } E \in \mathbf{P}(X).$$

Definition 4.14. A basic probability assignment is called *consonant* iff it focuses on a *nest* (that is, a class fully ordered by the inclusion relation of sets).

Theorem 4.24. *Let X be finite. Then any possibility measure is a plausibility measure, and the corresponding basic probability assignment is consonant. Conversely, the plausibility measure induced by a consonant basic probability assignment is a possibility measure.*

Proof. Let $X = \{x_1, \ldots, x_n\}$ and π be a possibility measure. There is no loss of generality in assuming

$$1 = \pi(\{x_1\}) \geq \pi(\{x_2\}) \geq \cdots \geq \pi(\{x_n\}).$$

Define a set function m on $\mathbf{P}(X)$ by

$$m(E) = \begin{cases} \pi(\{x_i\}) - \pi(\{x_{i+1}\}) & \text{if } E = F_i, i = 1, \ldots, n-1 \\ \pi(\{x_n\}) & \text{if } E = F_n \\ 0 & \text{otherwise,} \end{cases}$$

where $F_i = \{x_1, \ldots, x_i\}, i = 1, \ldots, n$. Then m is a basic probability assignment focusing on $\{F_1, \ldots, F_n\}$, which is a nest. The plausibility measure induced from this basic probability assignment m is just π. Conversely, let m be a basic probability assignment focusing on a nest $\{F_1, \ldots, F_k\}$ that satisfies $F_1 \subset F_2 \subset \cdots \subset F_k$ and Pl be the plausibility measure induced from m. For any $E_1 \in \mathbf{P}(X), E_2 \in \mathbf{P}(X)$, denote

$$j_0 = \min\{j | F_j \cap (E_1 \cup E_2) \neq \emptyset\},$$

and

$$j_{0i} = \min\{j | F_j \cap E_i \neq \emptyset\}, i = 1, 2.$$

Then we have

$$\begin{aligned} \mathrm{Pl}(E_1 \cup E_2) &= \sum_{F_j \cap (E_1 \cup E_2) \neq \emptyset} m(F_j) \\ &= \sum_{j \geq j_0} m(F_j) \\ &= \left[\sum_{j \geq j_{01}} m(F_j)\right] \vee \left[\sum_{j \geq j_{02}} m(F_j)\right] \\ &= \left[\sum_{F_j \cap E_1 \neq \emptyset} m(F_j)\right] \vee \left[\sum_{F_j \cap E_2 \neq \emptyset} m(F_j)\right] \\ &= \mathrm{Pl}(E_1) \vee \mathrm{Pl}(E_2). \end{aligned}$$

That is, Pl satisfies (4.16) on $\mathbf{P}(X)$. So, Pl is a possibility measure. □

Example 4.9. Let $X = \{x_1, x_2, x_3, x_4, x_5\}$, π be a possibility measure on $(X, \mathbf{P}(X))$ with a possibility profile $f(x) = \pi(\{x\}), x = x_1, \ldots, x_5$, as follows:

$$f(x_1) = 1,\ f(x_2) = 0.9,\ f(x_3) = 0.5,\ f(x_4) = 0.5,\ f(x_5) = 0.3.$$

The corresponding basic probability assignment m focuses on four subsets of $X: F_1 = \{x_1\}, F_2 = \{x_1, x_2\}, F_4 = \{x_1, x_2, x_3, x_4\}$, and $F_5 = X$, with $m(F_1) = 0.1, m(F_2) = 0.4, m(F_4) = 0.2$, and $m(F_5) = 0.3$. This is illustrated in Fig. 4.3. $\{F_1, F_2, F_4, F_5\}$ forms a nest. In this example, $m(F_3) = m(\{x_1, x_2, x_3\}) = 0$.

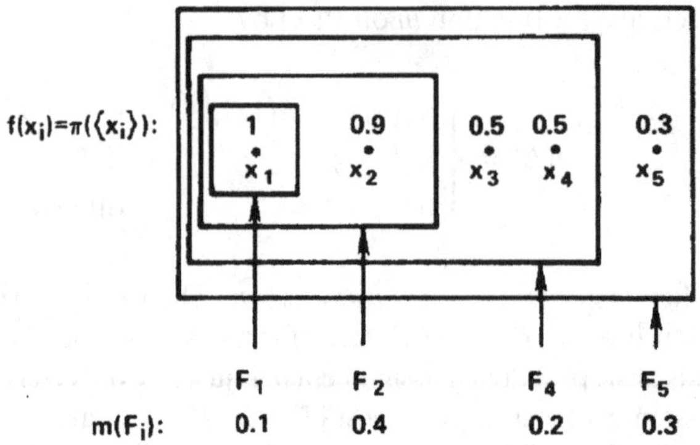

Fig. 4.3 A possibility profile on a finite space and the corresponding basic probability assignment

When X is not finite, a possibility measure on $\mathbf{P}(X)$ may not be a plausibility measure even when X is countable.

Example 4.10. Let X be the set of all rational numbers in [0, 1] and $f(x) = x, \forall x \in X$. X is a countable set. Define a set function π on $\mathbf{P}(X)$ as follows:

$$\pi(E) = \sup_{x \in E} f(x), \forall E \in \mathbf{P}(X).$$

Then, π is a possibility measure on $\mathbf{P}(X)$, but it is not a plausibility measure.

Definition 4.15. If π is a possibility measure on $\mathbf{P}(X)$, then its dual set function ν, which is defined by

$$\nu(E) = 1 - \pi(\bar{E}) \text{ for any } E \in \mathbf{P}(X)$$

is called a *necessity measure* (or *consonant belief measure*) on $\mathbf{P}(X)$.

Theorem 4.25. *A set function $\nu : \mathbf{P}(X) \to [0, 1]$ is a necessity measure if and only if it satisfies*

$$\nu\left(\bigcap_{t \in T} E_t\right) = \inf_{t \in T} \nu(E_t),$$

for any subclass $\{E_t | t \in T\}$ of $\mathbf{P}(X)$, where T is an index set, and $\nu(\emptyset) = 0$.

Proof. From Definitions 4.13 and 4.15, the conclusion is easy to obtain. □

Theorem 4.26. *Any necessity measure is an upper semicontinuous monotone measure. Moreover, if X is finite, then any necessity measure is a special example of belief measure and the corresponding basic probability assignment is consonant.*

Proof. The conclusion follows directly from Definition 4.15, Theorem 4.16, and Theorem 4.24. □

4.7 Properties of Finite Monotone Measures

In this section, we take a σ-ring **F** as the class **C**.

Theorem 4.27. *If μ is a finite monotone measure, then we have*

$$\lim_n \mu(E_n) = \mu(\lim_n E_n)$$

for any sequence $\{E_n\} \subset \mathbf{F}$ whose limit exists.

Proof. Let $\{E_n\}$ be a sequence of sets in **F** whose limit exists. Write $E = \lim_n E_n = \limsup_n E_n = \liminf_n E_n$. By applying the finiteness of μ, we have

$$\begin{aligned}\mu(E) = \ & \mu(\limsup_n E_n) = \lim_n \mu(\bigcup_{i=n}^{\infty} E_i) = \limsup_n \mu(\bigcup_{i=n}^{\infty} E_i) \\ & \geq \limsup_n \mu(E_n) \geq \liminf_n \mu(E_n) \\ & \geq \liminf_n \mu\left(\bigcap_{i=n}^{\infty} E_i\right) = \mu(\liminf_n E_n) = \mu(E)\end{aligned}$$

Therefore, $\lim_n \mu(E_n)$ exists and

$$\lim_n \mu(E_n) = \mu(E)$$

□

Definition 4.16. μ is *exhaustive* iff

$$\lim_n \mu(E_n) = 0$$

for any disjoint sequence $\{E_n\}$ of sets in **F**.

Theorem 4.28. *If μ is a finite upper semicontinuous monotone measure, then it is exhaustive.*

Proof. Let $\{E_n\}$ be a disjoint sequence of sets in **F**. If we write $F_n = \bigcup_{i=n}^{\infty} E_i$, then $\{F_n\}$ is a decreasing sequence of sets in **F**, and

$$\lim_n F_n = \bigcap_{n=1}^{\infty} F_n = \limsup_n E_n = \emptyset.$$

Since μ is a finite upper semicontinuous monotone measure, by using the finiteness and the continuity from above of μ, we have

$$\lim_n \mu(F_n) = \mu(\mathit{lim}_n F_n) = \mu(\emptyset) = 0.$$

Noting that

$$0 \leq \mu(E_n) \leq \mu(F_n),$$

we obtain

$$\lim_n \mu(E_n) = 0.$$

So, μ is exhaustive. □

Corollary 4.6. *Any finite monotone measure on a measurable space is exhaustive.*

Notes

4.1. The special nonadditive measures that are now called Choquet capacities were introduced by Gustave Choquet in the historical context outlined in Chapter 1. After their introduction [Choquet, 1953–54], they were virtually ignored for almost twenty years. They became a subject of interest of a small group of researchers in the early 1970s, primarily in the context of statistics. Among them, Peter Huber played an important role by recognizing that Choquet capacities are useful in developing *robust statistics* [Huber, 1972, 1973, 1981, Huber and Strassen, 1973]. Another researcher in this group, Anger [1971, 1977], focused more on further study of mathematical properties of Choquet capacities. It seems that the interest of these researchers in Choquet capacities was stimulated by an important earlier work of Dempster on upper and lower probabilities [Dempster, 1967a,b, 1968a,b]. Although Dempster was apparently not aware of Choquet capacities (at least he does not refer to the seminal paper by Choquet in his papers), the mathematical structure he developed for dealing with upper and lower probabilities is closely connected with Choquet capacities. It is well documented that Dempster's work on upper and lower probabilities also stimulated in the 1970s the development of *evidence theory*, which is based on ∞-monotone and ∞-alternating measures (Note 4.5). References to Choquet capacities in the literature have increased significantly since the late 1980s, primarily within the emerging areas of imprecise probabilities [Kyburg, 1987, Chateauneuf and Jaffray, 1989, De Campos and Bolanos, 1989, Wasserman and Kadane, 1990, 1992, Grabisch et al., 1995, Kadane and Wasserman, 1996, Walley, 1991].

4.2. The class of λ-measures was introduced and investigated by Sugeno [1974, 1977]. The fact that any λ-measure can be induced from a classical measure was shown by Wang [1981]. λ-measures were also investigated by Kruse [1980, 1982ab, 1983], Banon [1981], and Wierzchon [1982, 1983].

4.3. The concept of *quasi-measures* (often referred to in the literature as *pseudo-additive measures*) was introduced and investigated by Wang [1981]. Important examples of quasi-measures are special monotone measures that are called *decomposable measures*. These are normalized monotone measures, $\mu_\perp$, on measurable space $(X, \mathbf{C})$ that are semicontinuous from below and satisfy the property $\mu_\perp(A \cup B) = \perp[\mu_\perp(A), \mu_\perp(B)]$ for all $A, B, A \cup B \in \mathbf{C}$ such that $A \cap B = \emptyset$. Symbol $\perp$ denotes here a function from $[0,1]^2$ to [0,1] that qualifies as a triangular conorm (or t-conorm) [Klement, et al., 2000] and **C** is usually a σ-algebra. Since decomposable measures are not covered in this book, the following are some useful references for their study: [Dubois and Prade, 1982, Weber, 1984, Chateauneuf, 1996, Pap, 1997a,b, 1999, 2002b, Grabisch, 1997d].

4.4. In an early paper, Banon [1981] presents a comprehensive overview of the various types of monotone measures (defined on finite spaces) and discusses their classification. Lamata and Moral [1989] continue this discussion by introducing a classification of pairs of dual monotone measures. This classification is particularly significant in the area of imprecise probabilities, where one of the dual measures represents the lower probability and the other one the upper probability.

4.5 A theory based upon belief and plausibility measures was originated and developed by Shafer [1976]. Its emergence was motivated by previous work on lower and upper probabilities by Dempster [1967a,b, 1968a,b], as well as by Shafer's historical reflection upon the concept of probability [Shafer, 1978] and his critical examination of the Bayesian treatment of evidence [Shafer, 1981]. The theory is now usually referred to as the *Dempster–Shafer theory of evidence* (or just *evidence theory*). Although the seminal book by Shafer [1976] is still the best introduction to the theory (even though it is restricted to finite sets), several other books devoted to the theory, which are more up-to-date, are now available: [Guan and Bell, 1991–92, Kohlas and Monney, 1995, Kramosil, 2001, Yager et al., 1994]. There are too many articles dealing with the theory and its applications to be listed here, but most of them can be found in reference lists of the mentioned books and in two special journal issues devoted to the theory: *Intern. J. of Approximate Reasoning,* **31**(1–2), 2002, pp. 1–154, and *Intern. J. of Intelligent Systems*, **18**(1), 2003, pp. 1–148. The theory is well covered from different points of view in articles by Shafer [1979, 1981, 1982, 1990], Höhle [1982], Dubois and Prade [1985, 1986a], Walley [1987], Smets [1988, 1992, 2002], and Smets and Kennes [1994]. Possible ways of fuzzifying the theory are suggested by Höhle [1984], Dubois and Prade [1985], and Yen [1990]. Axiomatic characterizations of comparative belief structures, which are generalizations of comparative probability structures [Walley and Fine, 1979], were formulated by Wong, Yao, and Bollmann [1992].

4.6. A mathematical theory that is closely connected with Dempster-Shafer theory, but which is beyond the scope of this book, is the *theory of random sets*. Random sets were originally conceived in connection with stochastic geometry. They were proposed in the 1970s independently by two authors, Kendall [1973, 1974] and Matheron [1975]. The connection of random sets with belief measures is examined by Nguyen [1978] and Smets [1992], and it is also the subject of several articles in a book edited by Goutsias et al. [1997]. A recent book by Molchanov [2005] is currently the most comprehensive and up-to-date reference for the theory and applications of random sets. A good introduction to random sets was written by Nguyen [2006].

4.7. Possibility measures were introduced in several different contexts. In the late 1940s the British economist George Shackle introduced possibility measures indirectly, via monotone decreasing set functions that he called measures of *potential surprise* [Shackle, 1949]. He argued that these functions are essential in dealing with uncertainty in economics [Shackle 1955, 1961]. As shown by Klir [2002], measures of potential surprise can be reformulated in terms of monotone increasing measures—possibility measures. In the late 1970s possibility measures were introduced in two very different contexts: the context of fuzzy sets [Zadeh, 1978] and the context of plausibility measures [Shafer, 1976, 1987]. The literature on the theory based on possibility measures (and their dual necessity measures) is now very extensive. An early book by Dubois and Prade [1988] is a classic in this area. More recent developments in the theory are covered in a text by Kruse et al. [1994] and in monographs by Wolkenhauer [1998] and Borgelt and Kruse [2002]. Important sources are also edited books by De Cooman et al. [1995] and Yager [1982]. A sequence of three papers by De Cooman [1997] is perhaps the most comprehensive and general treatment of possibility theory. Thorough surveys of possibility theory with extensive bibliographies were written by Dubois et al. [1998, 2000].

4.8. An interesting connection between modal logic [Chellas, 1980; Hughes and Cresswell, 1996] and the various nonadditive measures is suggested in papers by Resconi et al. [1992, 1993]. Modal logic interpretation of belief and plausibility measures on finite sets is studied in detail by Harmanec et al. [1994] and Tsiporkova et al. [1999], and on infinite sets by Harmanec et al. [1996]. A modal logic interpretation of possibility theory is established in a paper by Klir and Harmanec [1994].

Exercises

4.1. Consider the monotone measures μ_i $(i = 1, 2, \ldots, 9)$ on $(X, \mathbf{P}(X))$, where $X = \{a, b, c\}$, which are defined in Table 4.2. Determine for each of these measures the following:

Table 4.2 Monotone measures in Exercises 4.1. and 4.2

	a b c	$\mu_1(A)$	$\mu_2(A)$	$\mu_3(A)$	$\mu_4(A)$	$\mu_5(A)$	$\mu_6(A)$	$\mu_7(A)$	$\mu_8(A)$	$\mu_9(A)$
A:	0 0 0	0.0	0.0	0.0	0.0	0.0	0.0	0.0	0.0	0.0
	1 0 0	0.0	0.2	0.4	0.2	0.0	1.0	0.2	0.3	0.2
	0 1 0	0.0	0.2	0.2	0.3	0.0	1.0	0.0	0.1	0.3
	0 0 1	0.0	0.2	0.0	0.4	0.0	1.0	0.0	0.3	0.4
	1 1 0	0.7	0.6	0.5	0.6	1.0	1.0	0.5	0.3	0.6
	1 0 1	0.8	0.6	0.6	0.6	1.0	1.0	0.2	0.6	0.7
	0 1 1	0.9	0.4	0.5	0.7	1.0	1.0	0.0	1.0	0.8
	1 1 1	1.0	1.0	1.0	1.0	1.0	1.0	1.0	1.0	1.0

(a) Is the measure superadditive or subadditive?
(b) Is the measure 2-monotone or 2-alternating?
(c) Is the measure a belief measure or a plausibility measure?
(d) Is the measure a possibility measure or a necessity measure?

4.2. Determine the dual measures for each of the measures in Exercise 4.1, and answer for each of them the questions stated in Exercise 4.1.

4.3. Check for each of the following set functions whether it is a λ-measure. If the answer is affirmative, determine the parameter λ.

(a) $X = \{a, b\}, \mathbf{F} = \mathbf{P}(X)$, and μ is given by $\mu(\emptyset) = 0, \mu(\{a\}) = 1/2, \mu(\{b\}) = 3/4, \mu(X) = 1$.
(b) $X = \{a, b\}, \mathbf{F} = \mathbf{P}(X)$, and μ is given by $\mu(\emptyset) = 0, \mu(\{a\}) = 1/2, \mu(\{b\}) = 1/3, \mu(X) = 1$.
(c) $X = \{a, b, c\}, \mathbf{F} = \mathbf{P}(X)$, and μ is given by

$$\mu(E) = \begin{cases} 1 & \text{if } E = X \\ 0 & \text{if } E = \emptyset \\ 1/2 & \text{otherwise} \end{cases}$$

for any $E \in \mathbf{F}$
(d) $X = \{a, b, c\}, \mathbf{F} = \mathbf{P}(X)$, and μ is given by

$$\mu(E) = \begin{cases} 1 & \text{if } E = X \\ 0 & \text{otherwise} \end{cases}$$

for any $E \in \mathbf{F}$.

4.4. Is any of the set functions defined in Exercise 4.3 a normalized λ-measure? For each that is a normalized λ-measure, determine the dual λ-measure as well as the value of the corresponding parameter λ.

4.5. Prove that the σ-λ-rule is equivalent to the continuity and the λ-rule for a nonnegative set function defined on a ring. Give an example to show that a similar conclusion need not be true on a semiring.

4.6. Let $X = \{x_1, x_2, x_3, x_4\}$, and $a_1 = 0.1, a_2 = 0.2, a_3 = 0.3, a_4 = 0.4$. Find the λ-measure, g_λ, defined on (X, $\mathbf{P}(X)$) and subject to $g_\lambda(\{x_i\}) = a_i$, $i = 1, 2, 3, 4$, for each of the following values of parameter λ:

(a) $\lambda = 5$; (b) $\lambda = 2$; (c) $\lambda = 1$; (d) $\lambda = 0$; (e) $\lambda = -1$; (f) $\lambda = -2$; (g) $\lambda = -2.4$.

Can you use $\lambda = -2.5$ or $\lambda = -5$ to find a λ-measure satisfying the above-mentioned requirement? Justify your answer.

4.7. Prove the following: If μ is a Dirac measure on (X, $\mathbf{F}$), then μ is a Sugeno measure for any $\lambda \in (-1/\sup \mu, \infty) \cup \{0\}$; conversely, if X is countable, $\mathbf{F} = \mathbf{P}(X)$, and μ is a Sugeno measure on (X, $\mathbf{F}$) for two different parameters λ and λ', then μ is a Dirac measure.

4.8. Let $X = \{a, b, c\}$ and $\mu(\{a\}) = 0.25, \mu(\{b\}) = \mu(\{c\}) = 0.625, \mu(X) = 1$. Viewing μ as a λ-measure, determine the value of the associated parameter λ.

4.9. Let $X = \{a, b\}, \mathbf{F} = \mathbf{P}(X)$, and let m be a measure on $(X, \mathbf{F})$ defined by

$$m(E) = \begin{cases} 1 & \text{if } E = X \\ 3/4 & \text{if } E = \{b\} \\ 1/4 & \text{if } E = \{a\} \\ 0 & \text{if } E = \emptyset. \end{cases}$$

Find a quasi-measure μ by using $\theta(y) = \sqrt{y}, y \in [0, 1]$, as its proper T-function. Is there any other T-function (say θ') such that $\mu = \theta' \circ m$? If you find any such T-functions, what can you conclude from them?

4.10. Let $X = \{a_1, a_2\}$ and μ be a nonnegative set function of $\mathbf{P}(X)$. Show that if $0 = \mu(\emptyset) < \mu(a_i) < \mu(X) < \infty, i = 1, 2, \ldots$, then μ is a quasi-measure.

4.11. Let $X = \{a, b, c, d\}$ and let $m(\{a\}) = 0.4, m(\{b, c\}) = 0.1, m(\{a, c, d\}) = 0.3, m(X) = 0.2$ be a basic probability assignment. Determine the corresponding belief measure and plausibility measure.

4.12. Repeat Exercise 4.11 for each of the basic probability assignments given in Table 4.3, where subsets of X are defined by their characteristic functions.

4.13. Determine which basic probability assignments given in Table 4.3 are consonant.

4.14. Determine which basic probability assignments given in Table 4.3 induce a discrete probability measure on $(X, \mathbf{P}(X))$.

4.15. Given $X = \{a, b, c, d\}$,
$\text{Bel}(\emptyset) = \text{Bel}(\{b\}) = \text{Bel}(\{c\}) = \text{Bel}(\{d\}) = \text{Bel}(\{b, d\}) = \text{Bel}(\{c, d\}) = 0$,
$\text{Bel}(\{a\}) = \text{Bel}(\{a,b\}) = \text{Bel}(\{a,c\}) = \text{Bel}(\{a,d\}) = \text{Bel}(\{a,b,d\}) = 0.1$,

Table 4.3 Basic probability assignments employed in Exercises 4.12–4.14

a	b	c	d	m_1	m_2	m_3	m_4	m_5	m_6	m_7	m_8	m_9	m_{10}
0	0	0	0	0	0	0	0	0	0	0	0	0	0
0	0	0	1	0.2	0	0	0.2	0.2	0	0.05	0	0	0
0	0	1	0	0	0.4	0	0	0.2	0	0.05	0	0	0
0	0	1	1	0	0	0	0.1	0	0	0.05	0	0	0
0	1	0	0	0	0.5	0	0	0.3	1	0.05	0.2	0	0.9
0	1	0	1	0	0	0	0	0	0	0.05	0	0	0
0	1	1	0	0.3	0	0	0	0	0	0.05	0	0	0
0	1	1	1	0	0	0	0	0	0	0.05	0.5	0	0
1	0	0	0	0.1	0.1	0.2	0	0.3	0	0.05	0	0	0.1
1	0	0	1	0	0	0	0	0	0	0.05	0	0	0
1	0	1	0	0.1	0	0.3	0	0	0	0.05	0	0	0
1	0	1	1	0	0	0	0	0	0	0.1	0	0	0
1	1	0	0	0	0	0	0	0	0	0.1	0	1	0
1	1	0	1	0.2	0	0	0	0	0	0.1	0	0	0
1	1	1	0	0.1	0	0.4	0	0	0	0.1	0	0	0
1	1	1	1	0	0	0.1	0.7	0	0	0.1	0.3	0	0

Table 4.4 Possibility profiles employed in Exercises 4.16 and 4.17

	f_1	f_2	f_3	f_4	f_5	f_6
a	1	1	0	0.9	1	1
b	0.8	1	1	0	1	1
c	0.4	0.2	0.3	0	1	0
d	0.1	0.6	0.3	1	1	0

$\text{Bel}(\{b,c\}) = \text{Bel}(\{b,c,d\}) = 0.2$, $\text{Bel}(\{a,b,c\}) = 0.3$, $\text{Bel}(\{a,c,d\}) = 0.4$, $\text{Bel}(X) = 1$, determine the corresponding basic probability assignment.

4.16. Let $X = \{a,b,c,d\}$. Use each of the possibility profiles given in Table 4.4 to determine the corresponding possibility measures and basic probability assignments.

4.17. Determine the dual necessity measure for each possibility measure obtained in Exercise 4.16.

4.18. Find an example that illustrates that a possibility measure defined on an infinite space need not be a plausibility measure.

Chapter 5
Extensions

5.1 A General Discussion on Extensions

The extension of set functions is one of the important parts of the theory of set functions. It is also one of the ways for constructing nonadditive set functions.

Definition 5.1. Let **E** and **F** be classes of subsets of a nonempty set X such that $\mathbf{E} \subset \mathbf{F}$, and let μ and ν be real-valued set functions defined on **E** and **F**, respectively. then, set function ν is called an *extension* of μ from **E** to **F** iff $\nu\ (E) = \mu(E)$ for every $E \in \mathbf{E}$.

Given a monotone measure μ on **E**, considering how to obtain one of its extensions on **F** without any additional requirements is trivial. In fact, define

$$\nu(F) = \sup_{E \subset F} \mu(E),$$

for every $F \in \mathbf{F}$. It is easy to verify that ν is a monotone measure on **F**, and ν coincides with μ on **E**. That is, monotone measure ν is an extension of μ from **E** to **F**. Hence, we restrict our discussion of extensions in this chapter only to continuous monotone measures with some additional structural requirements.

Extension can be used to construct monotone measures on a σ-ring. However, not all continuous monotone measures defined on a ring **R** can be extended onto the σ-ring **F**(**R**) while keeping the continuity. The following is an example of continuous monotone measure for which a required extension does not exist.

Example 5.1. Let $X = \{1, 2, \ldots\}$, **R** be the class of all finite subsets of X. We know that

$$\mathbf{F}(\mathbf{R}) = \mathbf{P}(X).$$

Define a set function μ on **R** as follows:

$$\mu(E) = \begin{cases} 0 & \text{if } E = \emptyset \\ 1 & \text{otherwise} \end{cases} \quad \forall\ E \in \mathbf{R}.$$

Z. Wang, G.J. Klir, *Generalized Measure Theory*,
DOI: 10.1007/978-0-387-76852-6_5, © Springer Science+Business Media, LLC 2009

μ is a finite continuous monotone measure on **R**. If a nonnegative monotone set function μ' on **F(R)** is an extension of μ, then $\mu'(E) \geq 1$ for any infinite set E in **F(R)** due to the monotonicity of μ'. Take

$$E_n = \{n, n+1, \ldots\},\ n = 1, 2, \ldots.$$

If there exists some E_{n_0} such that $\mu'(E_{n_0}) = \infty$, then we take

$$F_i = \{n_0, n_0 + 1, \ldots, n_0 + i - 1\},\ i = 1, 2, \ldots,$$

and we have $F_i \nearrow E_{n_0}$ and $\mu'(F_i) = \mu(F_i) = 1$ for any $i = 1, 2, \ldots$. This shows that μ' is not continuous from below. Otherwise, $\mu'(E_n) < \infty$ for any $n = 1, 2, \ldots$. From $E_n \searrow \emptyset$, and $\mu'(E_n) \geq 1$ for every $n = 1, 2, \ldots$, we know that μ' is not continuous from above at $\emptyset$. Consequently, μ' is not a continuous monotone measure on **F(R)**.

From this example we know that it is impossible to establish a unified extension theorem for continuous monotone measures corresponding to the extension theorem in classical measure theory. We can only give extension theorems for some special classes of continuous monotone measures and semi-continuous monotone measures.

5.2 Extension of Quasi-Measures and λ-Measures

In this section an extension theorem of quasi-measures is established, and thereby the problem that remained in Chapter 4 of constructing a λ-measure on the Borel field from a given probability distribution function is solved satisfactorily.

Theorem 5.1. *Let μ be a σ-finite quasi-measure on a semiring* **S** *with a proper T-function θ. Then μ can be uniquely extended to a quasi-measure on* **F(S)** *with the same T-function θ.*

Proof. $\theta \circ \mu$ is a classical measure on **S**. Since $\theta^{-1}(\{\infty\}) = \emptyset$ or $\{\infty\}$, that is, $\theta(x) < \infty$ if and only if $x < \infty$, we know that $\theta \circ \mu$ is σ-finite as is μ. So, $\theta \circ \mu$ can be uniquely extended to a classical measure ν on **F(S)**. By Theorem 4.8, $\mu' = \theta^{-1} \circ \nu$ is a quasi-measure on **F(S)** with the proper T-function θ. On **S**, since $\nu = \theta \circ \mu$, μ' coincides with μ, that is, μ' is an extension of μ. Uniqueness can be obtained from the uniqueness of the extension of a classical measure. □

We should note that the restriction "with the same proper T-function θ" in the above theorem is necessary. Otherwise, the extension of μ may not be unique. The following example shows that, without such a restriction, the extensions of a quasi-measure on **S**, even only to **R(S)**, may not be unique.

Example 5.2. Let $X = \{a, b, c\}$. $\mathbf{S} = \{\emptyset, \{a\}, \{b\}, \{c\}\}$ is a semiring. Define

$$\mu(E) = \begin{cases} 0 & \text{if } E = \emptyset \\ 1/3 & \text{otherwise} \end{cases}$$

for every $E \in \mathbf{S}$. Then μ is a classical measure on $\mathbf{S}$. Of course, it is also a quasi-measure with proper T-function $\theta(y) = y$. The ring generated by $\mathbf{S}$ is

$$\mathbf{R}(\mathbf{S}) = \{\emptyset, \{a\}, \{b\}, \{c\}, \{a, b\}, \{b, c\}\{a, c\}, X\}.$$

If we define

$$\mu'(E) = \begin{cases} \mu(E) & \text{if } E \in \mathbf{S} \\ 1 & \text{if } E = X \\ 2/3 & \text{otherwise} \end{cases}$$

for any $E \in \mathbf{R}(\mathbf{S})$, the μ' is a quasi-measure on $\mathbf{R}(\mathbf{S})$ with the proper T-function $\theta'(y) = \theta(y) = y$, and it is an extension of μ from $\mathbf{S}$ onto $\mathbf{R}(\mathbf{S})$. If we also define

$$\mu''(E) = \begin{cases} \mu(E) & \text{if } E = \mathbf{S} \\ 1 & \text{if } E = X \\ 1/2 & \text{otherwise} \end{cases}$$

for any $E \in \mathbf{R}(\mathbf{S})$, then μ'' is an extension of μ from $\mathbf{S}$ onto $\mathbf{R}(\mathbf{S})$, and is also a quasi-measure on $\mathbf{R}(\mathbf{S})$, but with another proper T-function

$$\theta''(y) = \begin{cases} y & \text{if } 0 \le y \le 1/3 \\ \frac{1+[1-6(\frac{1}{2}-y)]^{1/2}}{3} & \text{if } y > 1/3. \end{cases}$$

μ' and μ'' are even both normalized, but they are different.

However, we can prove that, for a finite quasi-measure defined on an algebra $\mathbf{R}$, even without the restriction "with the same proper T-function," the extension from $\mathbf{R}$ onto $\mathbf{F}(\mathbf{R})$ is unique.

Theorem 5.2. *Let* $\mathbf{R}$ *be an algebra. If* $\mu : \mathbf{R} \to [0, \infty)$ *is a* σ*-finite quasi-measure on* $\mathbf{R}$*, then* μ *can be uniquely extended to a quasi-measure on* $\mathbf{F}(\mathbf{R})$.

Proof. We only need to prove uniqueness. By using a method that is common in classical measure theory, we can reduce the σ-finite case into a finite case for μ.

Suppose that, on **F**(**R**), there are two extensions μ' and μ'' of μ; then μ' and μ'' as well as μ are finite. If we write

$$\mathbf{M} = \{E|E \in \mathbf{F}(\mathbf{R}), \mu'(E) = \mu''(E)\},$$

then, relying on the finiteness of μ' and μ'', for any monotone sequence $\{E_1\}$ of sets in **M**, we have

$$\mu'(\lim_i E_i) = \lim_i \mu'(E_i) = \lim_i \mu''(E_i) = \mu''(\lim_i E_i);$$

that is,

$$\lim_i E_i \in \mathbf{M}.$$

So, **M** is a monotone class. Since $\mu'(E) = \mu''(E)$ for any $E \in \mathbf{R}$, then we have

$$\mathbf{M} \supset \mathbf{R},$$

and by Corollary 2.1, we have

$$\mathbf{M} \supset \mathbf{F}(\mathbf{R}).$$

Consequently,

$$\mu'(E) = \mu''(E)$$

for any $E \in \mathbf{F}(\mathbf{R})$. The proof is complete. □

The conclusion in Theorem 5.2 is stronger than the result of the uniqueness of the extension in classical measure theory. Here we do not give a restriction "with the same proper T-function," but in classical measure theory there is actually a restriction "with the same proper T-function—identity function." This shows that the structure of a quasi-measure on **F**(**R**) is fully determined by its structure on **R**.

The following example gives a general method to construct a λ-measure on a discrete space (finite or infinitely countable).

Example 5.3. The set function μ given in Example 4.3 is a finite λ-measure for any $\lambda \in (-1/\sup_i a_i, \infty) \cup \{0\}$ on the semiring consisting of all singletons of

$X = \{x_i, x_2, \ldots\}$ and the empty set Ø. By Theorems 4.11 and 5.1, μ can be uniquely extended to a λ-measure on $\mathbf{F}(\mathbf{S}) = \mathbf{P}(X)$ with the same parameter λ. In fact, according to the σ-λ-rule, we can define

$$\mu(E) = \begin{cases} (1/\lambda)\left[\prod_{i|x_i \in E} (1 + \lambda a_i) - 1\right] & \text{if } \lambda \neq 0 \\ \sum_{i|x_i \in E} a_i & \text{if } \lambda = 0 \end{cases}$$

for any $E \in \mathbf{P}(X)$. It is easy to show that this set function μ is the unique λ-measure on $\mathbf{P}(X)$ that satisfies

$$\mu(\{x_i\}) = a_i, \; i = 1, 2, \ldots$$

Now we return to the construction of Sugeno measures on the Borel field. Let $h(x)$ be a probability distribution function (continuous from the left). From $h(x)$, we can get another probability distribution function $h'(x)$, by

$$h' = \theta_\lambda \circ h,$$

where θ_λ is just the T-function given in Theorem 4.11 with $k = \ln(1 + \lambda)/\lambda$:

$$\theta_\lambda(y) = \frac{\ln(1 + \lambda y)}{\ln(1 + \lambda)}, \qquad y \in [0, 1].$$

$h'(x)$ determines a probability measure p on the semiring $\mathbf{S} = \{[a, b)| -\infty < a \leq b < \infty\}$:

$$P([a, b)) = h'(b) - h'(a), -\infty < a \leq b < \infty,$$

and p can be uniquely extended onto the Borel field with $p((-\infty, \infty)) = 1$. By using the second conclusion in Theorem 4.11, we get a Sugeno measure $g_\lambda = \theta_\lambda^{-1} \circ p$ with the proper T-function θ_λ, where

$$\theta_\lambda^{-1}(y) = \frac{e^{y \ln(1+\lambda)} - 1}{\lambda} = \frac{(1 + \lambda)^y - 1}{\lambda}.$$

For such a Sugeno measure, on the semiring $\mathbf{S}$, we have

$$\begin{aligned} g_\lambda([a, b)) &= (\theta_\lambda^{-1} \circ p)([a, b)) = \theta_\lambda^{-1}[h'(b) - h'(a)] \\ &= \theta_\lambda^{-1}\left(\frac{\ln[1 + \lambda \cdot h(b)] - \ln\,[1 + \lambda \cdot h(a)]}{\ln\,(1 + \lambda)}\right) \end{aligned}$$

$$= \frac{\exp\left[\ln \dfrac{1+\lambda\cdot h(b)}{1+\lambda\cdot h(a)}\right] - 1}{\lambda} = \frac{\dfrac{1+\lambda\cdot h(b)}{1+\lambda\cdot h(a)} - 1}{\lambda} = \frac{h(b)-h(a)}{1+\lambda\cdot h(a)}$$

This coincides with the set function ψ defined in Section 4.2. Noting

$$(\theta_\lambda \circ \psi)([a,b)) = \frac{\ln\left(1+\lambda\dfrac{h(b)-h(a)}{1+\lambda\cdot h(a)}\right)}{\ln\,(1+\lambda)} = \frac{\ln\dfrac{1+\lambda\cdot h(b)}{1+\lambda\cdot h(a)}}{\ln(1+\lambda)} = h'(b) - h'(a),$$

that is, $\theta_\lambda \circ \psi$ is a probability measure on **S** and, therefore, we know that ψ is a quasi-measure on **S**. So, by Theorem 5.1, g_λ is the unique extension of ψ from **S** onto the Borel field **F(S)**.

Conversely, if g_λ is a Sugeno measure on the Borel field, then

$$h(x) = g_\lambda((-\infty, x))$$

is a probability distribution function (continuous from the left), and, by using Theorem 4.6(1), we have

$$g_\lambda([a,b])) = g_\lambda((-\infty,b) - (-\infty,a)) = \frac{h(b)-h(a)}{1+\lambda\cdot h(a)}.$$

$h(x)$ is called the *distribution function* of g_λ.

Summing up these discussions, we have actually proved the following theorem:

Theorem 5.3. *Fixed* $\lambda \in (-1,\infty)$, *the relation*

$$h(\cdot) = g_\lambda((-\infty,\cdot))$$

establishes a one-to-one correspondence between probability distribution functions and Sugeno measures on the Borel field.

This theorem is just a generalization of the relevant results in classical measure theory.

5.3 Extension of Semicontinuous Monotone Measures

Let **R** be an algebra of sets in **P**(X). The class of all those sets that can be expressed by the limit of an increasing sequence of sets in **R** is denoted by $\mathbf{R}_\sigma$. Similarly, the notation $\mathbf{R}_\delta$ is used to denote the class of all those sets that can be expressed by the limit of a decreasing sequence of sets in **R**. Obviously, we have

$$\mathbf{R}_\delta = \{E|\overline{E} \in \mathbf{R}_\sigma\}$$

and vice versa.

For the sake of simplicity, in this section we assume that the set functions we discuss are finite.

Definition 5.2. A nondecreasing set function $\mu : \mathbf{C} \to [0, \infty)$ is *lower* (or *upper*) *consistent* on **C** iff for any $F \in \mathbf{C}$ and any $\{E_n\} \subset \mathbf{C}$,

$$E_n \nearrow \bigcup_{n=1}^{\infty} E_n \supset F$$

implies

$$\lim_n \mu(E_n) \geq \mu(F)$$

(or

$$E_n \searrow \bigcup_{n=1}^{\infty} E_n \subset F$$

implies

$$\lim_n \mu(E_n) \leq \mu(F),$$

respectively).

Lemma 5.1. *Let* $\mu : \mathbf{C} \to [0, \infty)$ *be nondecreasing. If* **C** *is closed under the formation of finite intersections (or finite unions), then, for* μ *on* **C**, *lower (or upper) consistency is equivalent to continuity from below (or from above, respectively).*

Proof. Assume μ is continuous from below on **C**. For any $F \in \mathbf{C}$ and any $\{E_n\} \subset \mathbf{C}$ if

$$E_n \nearrow \bigcup_{n=1}^{\infty} E_n \supset F,$$

then we have

$$E_n \cap F \nearrow F.$$

By using the monotonicity and the continuity from below of μ on **C**, we have

$$\lim_n \mu(E_n) \geq \lim_n \mu(E_n \cap F) = \mu(F);$$

that is, μ is lower consistent. The converse implication is obvious.

The proof of the upper consistency is similar. □

Theorem 5.4. *If μ is a lower semicontinuous monotone measure on* **R** *then μ may be uniquely extended to a lower semicontinuous monotone measure on* $\mathbf{R}_\sigma$.

Proof. For any $E \in \mathbf{R}_\sigma$, define

$$\mu'(E) = \lim_n \mu(E_n)$$

when $\{E_n\} \subset \mathbf{R}$ and $E_n \nearrow E$. This definition is unambiguous. In fact, if there exist two sequences $\{E_n\}$ and $\{E'_n\}$ in **R** such that both $E_n \nearrow E$ and $E'_n \nearrow E$, then, for any positive integer $n_0, E_n \nearrow E \supset E'_{n_0}$, and by using Lemma 5.1, we have

$$\lim_n \mu(E_n) \geq \mu(E'_{n_0}).$$

Therefore, we have

$$\lim_n \mu(E_n) \geq \lim_n \mu(E'_n).$$

The converse inequality also holds. So we have

$$\lim_n \mu(E_n) = \lim_n \mu(E'_n).$$

Now we prove the monotonicity of μ' on $\mathbf{R}_\sigma$. Suppose $E \in \mathbf{R}_\sigma, F \in \mathbf{R}_\sigma$, and $E \subset F$. Then, there exist $\{E_n\} \subset \mathbf{R}$ and $\{F_n\} \subset \mathbf{R}$, such that $E_n \nearrow E$ and $F_n \nearrow F$. For any positive integer n_0, since

$$F_n \nearrow F \supset E \supset E_{n_0},$$

we have

$$\lim_n \mu(F_n) \geq \mu(E_{n_0}),$$

and therefore,

$$\mu'(F) = \lim_n \mu(F_n) \geq \lim_n \mu(E_n) = \mu'(E).$$

The continuity from below of μ' may be proved as follows. Suppose $\{E_n|\ n = 0, 1, 2, \ldots\} \subset R_\sigma$, and $E_n \nearrow E_0$. By the construction of $\mathbf{R}_\sigma$, for every $n = 0, 1, 2, \ldots$, there exists $\{E_{ni}|\ i = 1, 2, \ldots\} \subset \mathbf{R}$ such that $E_{ni} \nearrow E_n$. According to the zigzag diagonal method, write $F_1 = E_{11}, F_2 = E_{12},\ F_3 = E_{21}$, $F_4 = E_{13}, F_5 = E_{22}, F_6 = E_{31}, F_7 = E_{14}, \ldots,$ and denoting $F'_n = \cup_{i=1}^n F_i \in \mathbf{R}$, then $F'_n \nearrow \cup_{i=1}^\infty E_n = E_0$, and therefore,

$$\mu'(E_0) = \lim_n \mu(F'_n).$$

Observing the fact that, for any positive integer n_0, there exists $j = j(n_0)$, such that $F'_{n_0} \subset E_j$, we have, by the monotonicity of μ',

$$\mu(F'_{n_0}) = \mu'(F'_{n_0}) \leq \mu'(E_j).$$

Consequently, we have

$$\mu'(E_0) \leq \lim_j \mu'(E_j).$$

The converse inequality is assured by the monotonicity of μ'

Clearly, μ' is an extension of μ, because they coincide on $\mathbf{R}$. The uniqueness of the extension is obvious. □

Theorem 5.5. *If μ is an upper semicontinuous monotone measure on $\mathbf{R}$, then μ may be uniquely extended to an upper semicontinuous monotone measure on $\mathbf{R}_\delta$.*

Proof. If we define a set function ν on $\mathbf{R}$ by

$$\nu(E) = \mu(X) - \mu(\overline{E})$$

for every $E \in \mathbf{R}$, then ν is a lower semicontinuous monotone measure, and $\nu(X) = \mu(X)$. By Theorem 5.4, ν can be extended to a lower semicontinuous monotone measure ν' on $\mathbf{R}_\sigma$. It is easy to verify that μ' defined on $\mathbf{R}_\delta$ by

$$\mu'(E) = \mu(X) - \nu'(\overline{E})$$

is an extension of μ. The uniqueness of the extension is guaranteed by the uniqueness in Theorem 5.4. □

5.4 Absolute Continuity and Extension of Monotone Measures

We assume that **R** is an algebra, **F** is a σ-algebra containing **R**, and all set functions that we discuss in this section are finite.

Definition 5.3. Let μ and ν be two continuous monotone measures on **C**. We say that μ is *absolutely continuous* with respect to ν, denoted as $\mu \ll \nu$, iff for any $\varepsilon > 0$ there exists $\delta > 0$ such that $\mu(F) - \mu(E) < \varepsilon$ whenever $E \in \mathbf{C}$, $F \in \mathbf{C}, E \subset F$, and $\nu(F) - \nu(E) < \delta$.

The concept of absolute continuity given in the above definition is a generalization of the one in classical measure theory.

Theorem 5.6. *Let μ be a continuous monotone measure on* **R**. *μ can be extended onto $\mathbf{R}_\sigma$ if there exists a continuous monotone measure ν on $\mathbf{R}_\sigma$, such that $\mu \ll \nu$ on* **R**. *The extension is unique, and it preserves the absolute continuity with respect to ν.*

Proof. We only need to prove the continuity from above of μ' given in the proof of Theorem 5.4. Suppose $\{E_n\} \subset \mathbf{R}_\sigma$ and $E_n \searrow E_0 \in \mathbf{R}_\sigma$. Take set sequence $\{E_{ni} | i = 1, 2, \ldots\} \subset$ R , which satisfy $E_{ni} \nearrow E_n$ for every $n = 0, 1, 2, \ldots$. Since $E_0 \subset E_n$ for any $n = 1, 2, \ldots$, we may assume that $E_{0i} \subset E_{ni}$ for any $n = 1, 2, \ldots$, and $i = 1, 2, \ldots$, without any loss of generality. As $\mu \ll \nu$ on **R**, for any $\varepsilon > 0$ there exists $\delta > 0$ such that $\mu(F) < \mu(E) + \varepsilon/2$ whenever $E \in \mathbf{R}$, $F \in \mathbf{R}$, $E \subset F$ and $\nu(F) < (E) + \delta$. By using the continuity of ν and the definition of μ' on $\mathbf{R}_\sigma$, there exist N and N', such that

$$\nu(E_N) < \nu(E_0) + \delta/2,$$

$$\nu(E_0) < \nu(E_{0N'}) + \delta/2,$$

and

$$\mu'(E_N) < \mu(E_{NN'}) + \varepsilon/2.$$

Thus, we have

$$\nu(E_{NN'}) \leq \nu(E_N) < \nu(E_{0N'}) + \delta,$$

and therefore,

$$\mu(E_{NN'}) < \mu(E_{0N'}) + \varepsilon/2.$$

Consequently, we have

$$\mu'(E_N) < \mu(E_{0N'}) + \varepsilon \leq \mu'(E_0) + \varepsilon.$$

Observing the monotonicity of μ', we obtain

$$\lim_n \mu'(E_n) = \mu'(E_0).$$

Now, we turn to prove that $\mu' \ll \nu$ on $\mathbf{R}_\sigma$.

For any given $\varepsilon > 0$, since $\mu \ll \nu$ on $\mathbf{R}$, we know that there exists $\delta > 0$, such that $\mu(F_0) < \mu(E_0) + \varepsilon/2$ whenever $E_0 \in \mathbf{R}$, $F_0 \in \mathbf{R}$, $E_0 \subset F_0$, and $\nu(F_0) < \nu(E_0) + 2\delta$. Now, for any given $E \in \mathbf{R}_\sigma, F \in \mathbf{R}_\sigma$, satisfying $E \subset F$ and $\nu(F) < \nu(E) + \delta$, we take two set sequences $\{E_n\} \subset \mathbf{R}$ and $\{F_n\} \subset \mathbf{R}$ such that $E_n \nearrow E$ and $F_n \nearrow F$. There is no loss of generality in assuming $E_n \subset F_n$ for any $n = 1, 2, \ldots$ (otherwise, we can take $E_n \cap F_n$ instead of E_n). By using the continuity of μ' and ν on $\mathbf{R}_\sigma$, there exists a positive integer n_0 such that

$$\mu(F_{n_0}) > \mu'(F) - \varepsilon/2$$

and

$$\nu(E_{n_0}) > \nu(E) - \delta.$$

Since $\mu \ll \nu$ on $\mathbf{R}$, and

$$\nu(F_{n_0}) \leq \nu(F) < \nu(E) + \delta < \nu(E_{n_0}) + 2\delta,$$

we have

$$\mu(F_{n_0}) < \mu(E_{n_0}) + \varepsilon/2,$$

and therefore,

$$\mu'(F) < \mu'(F_{n_0}) + \varepsilon/2 = \mu(F_{n_0}) + \varepsilon/2 < \mu(E_{n_0}) + \varepsilon = \mu'(E_{n_0}) + \varepsilon \leq \mu'(E) + \varepsilon.$$

This means $\mu' \ll \nu$ on $\mathbf{R}_\sigma$. The uniqueness of the extension has been shown in Theorem 5.4. □

Since a continuous monotone measure is both continuous from below and continuous from above, regarding it as a lower semicontinuous monotone measure, we can obtain an extension from Theorem 5.4, and regarding it as an upper semicontinuous monotone measure, we can also obtain another extension from Theorem 5.5. Because of Theorem 5.6, we know that these two extensions coincide under the condition given in Theorem 5.6.

To extend a continuous monotone measure from an algebra onto a σ-algebra containing this algebra, we need a new concept, which is called $\mathbf{R}_\sigma$-approachability of a monotone measure on a σ-algebra containing $\mathbf{R}_\sigma$.

Definition 5.4. A continuous monotone measure μ on $\mathbf{F}$ is $\mathbf{R}_\sigma$-*approachable* iff for any set $E \in \mathbf{F}$ and $\varepsilon > 0$, there exists $F \in \mathbf{R}_\sigma$, such that $F \supset E$ and $\mu(F) \leq \mu(E) + \varepsilon$.

Theorem 5.7. *A continuous monotone measure μ on $\mathbf{R}$ may be extended to an $\mathbf{R}_\sigma$-approachable monotone measure on $\mathbf{F}$ if there exists an $\mathbf{R}_\sigma$-approachable continuous monotone measure ν on $\mathbf{F}$ such that $\mu \ll \nu$ on $\mathbf{R}$. The extension is unique and preserves the absolute continuity with respect to ν.*

Proof. It is clear by Theorem 5.6 that μ may be uniquely extended to a continuous monotone measure μ' on $\mathbf{R}_\sigma$, and $\mu' \ll \nu$ on $\mathbf{R}_\sigma$. If we define

$$\mu''(E) = \inf\{\mu'(F) | E \subset F \in \mathbf{R}_\sigma\}$$

for any $E \in \mathbf{F}$, then μ'' is nondecreasing, and it coincides with μ' on $\mathbf{R}_\sigma$

To prove the continuity from above of μ'' on $\mathbf{F}$, we suppose $\{E_n\} \subset \mathbf{F}$ and $E_n \searrow E_0 \in \mathbf{F}$. Since $\mu' \ll \nu$ on $\mathbf{R}_\sigma$, for any $\varepsilon > 0$, there exists $\delta > 0$ such that

$$\mu'(F) < \mu'(E) + \varepsilon/2$$

whenever $E \in \mathbf{R}_\sigma, F \in \mathbf{R}_\sigma, E \subset F$, and $\nu(F) < \nu(E) + \delta$. By using the continuity of ν on $\mathbf{F}$, there exists N such that

$$\nu(E_N) < \nu(E_0) + \delta/2.$$

Noting that $\mathbf{R}_\sigma$ is closed under the formation of finite intersections, by using the $\mathbf{R}_\sigma$-approachability of ν on $\mathbf{F}$ and the definition of μ'', we may take $F_0 \in \mathbf{R}_\sigma, F_N \in \mathbf{R}_\sigma$, such that $E_0 \subset F_0, E_N \subset F_N, F_0 \subset F_N$, and

$$\mu'(F_0) < \mu''(E_0) + \varepsilon/2,$$

$$\nu(F_N) < \nu(E_N) + \delta/2.$$

Thus, we have

$$\nu(F_N) < \nu(E_0) + \delta \leq \nu(F_0) + \delta$$

and, therefore,

$$\mu'(F_N) < \mu'(F_0) + \varepsilon/2.$$

Consequently, we have

$$\mu''(E_N) \leq \mu'(F_N) < \mu'(F_0) + (\varepsilon/2) < \mu''(E_0) + \varepsilon.$$

That is, μ'' is continuous from above on **F**. The continuity from below of μ'' on **F** can be proved by a completely analogous method. So μ'' is a continuous monotone measure on **F**.

Obviously, μ'' is $\mathbf{R}_\sigma$-approachable, and it is the unique extension possessing $\mathbf{R}_\sigma$-approachability. We can also prove the absolute continuity of μ'' with respect to ν in a similar way. □

Since a classical measure on **F**(**R**) is $\mathbf{R}_\sigma$-approachable, we have the following corollary.

Corollary 5.1. *A continuous monotone measure μ on* **R** *may be uniquely extended onto* **F**(**R**) *if there exists a finite measure ν on* **R** *such that $\mu \ll \nu$ on* **R**.

Noting that any T-function is continuous, we know that any quasi-measure is absolutely continuous with respect to a certain classical measure. So, from Corollary 5.1, we have the following result again, which was obtained in Section 5.2.

Corollary 5.2. *Any quasi-measure on* **R** *can be uniquely extended onto* **F**(**R**).

This shows that the result in Theorem 5.7 is a generalization of the result in Section 5.2.

5.5 Extension of Possibility Measures and Necessity Measures

In this section we restrict the range of set functions that we discuss to [0, 1]. However, there is no essential difficulty for generalizing most results in this section to cases where the range of set functions is $[0, \infty]$.

Definition 5.5. A set function $\mu : \mathbf{C} \to [0, 1]$ is *P-consistent* on **C** iff, for any $\{E_t | t \in T\} \subset \mathbf{C}$ and any $E \in \mathbf{C}$,

$$E \subset \bigcup_{t \in T} E_t$$

implies

$$\mu(E) \leq \sup_{t \in T} \mu(E_t),$$

where T is an arbitrary index set.

From Definition 5.5, we know that if μ: $\mathbf{C} \to [0, 1]$ is P-consistent and $\emptyset \in \mathbf{C}$, then $\mu(\emptyset) = 0$. In fact, since $\emptyset \subset \emptyset = \bigcup_{t \in \emptyset} E_t$, we should have $\mu(\emptyset) \leq \sup_{t \in \emptyset} \mu(E_t) = 0$.

Theorem 5.8. *If $\mu : \mathbf{C} \to [0, 1]$ is P-consistent, then it is monotone and maxitive on* **C**.

Proof. If we take a singleton as the index set T, then the monotonicity of μ can be immediately obtained from the definition of P-consistency. Furthermore, when $E = \bigcup_{t \in T} E_t$, then on the one hand, from the definition of P-consistency, we have

$$\mu(E) \le \sup_{t \in T} \mu(E_t);$$

and on the other hand, since $E \supset E_t$ for any $t \in T$, by monotonicity, we have $\mu(E) \ge \mu(E_t)$ for any $t \in T$, so

$$\mu(E) \ge \sup_{t \in T} \mu(E_t).$$

Consequently, we have

$$\mu(E) = \sup_{t \in T} \mu(E_t).$$

That is, μ is maxitive on **C**. □

In general, P-consistency is not equivalent to maxitivity on an arbitrary class **C**, as is illustrated in the following example.

Example 5.4. $X = \{a, b, c, \}, \mathbf{C} = \{\{a\}, \{a, b, \}, \{b, c\}\}$. If μ is a set function on **C** with

$$\mu(\{a\}) = 0.5, \mu(\{a, b\}) = 0.7, \mu(\{b, c\}) = 0.6,$$

then μ is maxitive on **C**, but it is not P-consistent. In fact, $\{a\} \cup \{b, c\} \supset \{a, b\}$, but $\mu(\{a\}) \vee \mu(\{b, c\}) = 0.6 < 0.7 = \mu(\{a, b\})$.

However, if μ is a nonnegative monotone set function defined on the class **C** that is closed under the formation of arbitrary unions, then P-consistency and maxitivity are equivalent for μ.

Theorem 5.9. *A set function μ: $\mathbf{C} \to [0, 1]$ can be extended to a generalized possibility measure π on $\mathbf{P}(X)$ if and only if μ is P-consistent on* **C**.

Proof. Necessity: Let μ be extendable to a generalized possibility measure π on $\mathbf{P}(X)$. Noting that π is nondecreasing, we know for any $\{E_t | t \in T\} \subset \mathbf{C}$ and any $E \in \mathbf{C}$, if $E \subset \bigcup_{t \in T} E_t$ then

$$\mu(E) = \pi(E) \le \pi(\bigcup_{t \in T} E_t) = \sup_{t \in T} \pi(E_t) = \sup_{t \in T} \mu(E_t).$$

That is, μ is P-consistent on **C**.

Sufficiency: Let μ be P-consistent on **C**. We define a set function

$$\begin{aligned} \pi : \mathbf{P}(X) &\to [0,\ 1] \\ F &\mapsto \sup_{x \in F} \inf_{E | x \in E \in \mathbf{C}} \mu(E). \end{aligned} \tag{5.1}$$

π is a generalized possibility measure on $\mathbf{P}(X)$. In fact, if we write

$$f(x) = \inf_{E|x \in E \in \mathbf{C}} \mu(E),$$

for any $x \in X$, then, similarly to the result in Theorem 4.22, $f(x)$ can uniquely determine a generalized possibility measure. It is just the set function π defined above. The following shows that this set function π is an extension of μ on $\mathbf{C}$, i.e., for any $F \in \mathbf{C}, \pi(F) = \mu(F)$. Take $F \in \mathbf{C}$ arbitrarily. On the one hand, from (5.1), since $F \in \{E|x \in E \in \mathbf{C}\}$ when $x \in F$, we have

$$\pi(F) \leq \sup_{x \in F} \mu(F) = \mu(F).$$

On the other hand, arbitrarily given $\varepsilon > 0$, for any $x \in F$ there exists $E_x \in \mathbf{C}$ such that $x \in E_x$ and

$$\inf_{E|x \in E \in \mathbf{C}} \mu(E) \geq \mu(E_x) - \varepsilon.$$

Since $\cup_{x \in F} E_x \supset F$, by using the P-consistency of μ on $\mathbf{C}$ we have

$$\sup_{x \in F} \mu(E_x) \geq \mu(F),$$

and therefore,

$$\pi(F) = \sup_{x \in F} \inf_{E|x \in E \in \mathbf{C}} \mu(E) \geq \sup_{x \in F}[\mu(E_x) - \varepsilon] \geq \mu(F) - \varepsilon.$$

Because ε may be arbitrarily close to zero, we obtain

$$\pi(F) \geq \mu(F).$$

Consequently, we have

$$\pi(F) = \mu(F).$$

□

Example 5.5. X and $\mathbf{C}$ are given in Example 5.4. If μ is a set function on $\mathbf{C}$ with

$$\mu(\{a\}) = 0.5, \mu(\{a, b\}) = 0.7, \mu(\{b, c\}) = 0.7,$$

then μ is P-consistent on $\mathbf{C}$, and therefore, it can be extended to a generalized possibility measure π on $\mathbf{P}(X)$ with

$$\pi(\{a\}) = 0.5, \pi(\{b\}) = 0.7, \pi(\{c\}) = 0.7.$$

In general, the above-mentioned extension may not be unique. For instance, in Example 5.5 the generalized possibility measure π':

$$\pi'(\{a\}) = 0.5, \pi'(\{b\}) = 0.7, \pi'(\{c\}) = 0.6, \pi'(\{a,b\}) = 0.7,$$
$$\pi'(\{b,c\}) = 0.7, \pi'(\{a,c\}) = 0.6, \pi'(\emptyset) = 0, \pi'(X) = 0.7,$$

is an extension of μ too.

Denote all of generalized possibility measure extensions of the set function $\mu : \mathbf{C} \to [0,1]$ by $\mathbf{E}_\pi(\mu)$; then $\mathbf{E}_\pi(\mu)$ is nonempty if μ is P-consistent on $\mathbf{C}$.

Given two arbitrary set functions $\mu_1 : \mathbf{P}(X) \to [0,1]$ and $\mu_2 : \mathbf{P}(X) \to [0,1]$, if we define a relation "$\leq$" as follows:

$$\mu_1 \leq \mu_2 \text{ iff } \mu_1(E) \leq \mu_2(E) \text{ for every } E \in \mathbf{P}(X),$$

then the relation "$\leq$" is a partial ordering on $\mathbf{E}_\pi(\mu)$. Furthermore, if we denote $\bar{\mu} = \sup\{\mu_1, \mu_2\}$, then

$$\bar{\mu}(E) = \mu_1(E) \vee \mu_2(E)$$

for any $E \in \mathbf{P}(X)$.

Theorem 5.10. $(\mathbf{E}_\pi(\mu), \leq)$ *is an upper semilattice, and the extension given by* (5.1) *is the greatest element of* $(\mathbf{E}_\pi(\mu), \leq)$.

Proof. If π_1 and π_2 are generalized possibility measures then so is their supremum. Furthermore, if both π_1 and π_2 belong to $\mathbf{E}_\pi(\mu)$ then so does their supremum. Therefore, $(\mathbf{E}_\pi(\mu), \leq)$ is an upper semilattice. Now we turn to show the second conclusion of the theorem. Let π be the generalized possibility measure extension of μ which has the expression (5.1). Given an arbitrary $\pi' \in \mathbf{E}_\pi(\mu)$, since

$$\pi'(\{x\}) \leq \pi'(E) = \mu(E)$$

for any $E \in \mathbf{C}$ and any singleton $\{x\}$ satisfying $x \in E \in \mathbf{C}$, we have

$$\pi'(\{x\}) \leq \inf_{E|x \in E \in \mathbf{C}} \mu(E).$$

Therefore, for any $F \in \mathbf{P}(X)$,

$$\pi(F) = \sup_{x \in F} \inf_{E|x \in E \in \mathbf{C}} \mu(E) \geq \sup_{x \in F} \pi'(\{x\}) = \pi'(F).$$

That is, π is the greatest element of $(\mathbf{E}_\pi(\mu), \leq)$. □

If $\pi_1 \in \mathbf{E}_\pi(\mu), \pi_2 \in \mathbf{E}_\pi(\mu)$, and $\pi_1 \leq \pi_2$, then any generalized possibility measure π on $\mathbf{P}(X)$ that satisfies $\pi_1 \leq \pi \leq \pi_2$ is an extension of μ. So, if $\mathbf{E}_\pi(\mu)$ possesses two or more different elements, then it has a potency not less than the continuum.

$\mathbf{E}_\pi(\mu)$ may be also obtained by solving a certain fuzzy relation equation.

From Theorem 5.9, it is easy to determine whether a set function μ: $\mathbf{C} \rightarrow [0, 1]$ can be extended to a possibility measure on $\mathbf{P}(X)$. In this problem, there are three cases:

1. If $X \in \mathbf{C}$, and $\mu(X) \neq 1$, then μ cannot be extended to any possibility measure on $\mathbf{P}(X)$.
2. If $X \in \mathbf{C}$, and $\mu(X) = 1$, then μ can be extended to a possibility measure on $\mathbf{P}(X)$ when μ is *P*-consistent on $\mathbf{C}$.
3. If $X \notin \mathbf{C}$, let $\mathbf{C}' = \mathbf{C} \cup \{X\}$, and define μ' on $\mathbf{C}'$ by

$$\mu'(E) = \begin{cases} \mu(E) & \text{if } E \in \mathbf{C} \\ 1 & \text{if } E = X, \end{cases}$$

then μ can be extended to a possibility measure on $\mathbf{P}(X)$ when μ' is *P*-consistent on $\mathbf{C}'$.

The discussion regarding extensions of generalized possibility measures (Theorem 5.10) is equally applicable to cases 2 and 3.

It is natural to ask under what conditions the above-mentioned extension is unique. To answer this question, we need the concepts of atom and plump class defined in Chapter 2.

Lemma 5.2. *Let* $\mathbf{C}$ *be an AU-class. If a set function* μ: $\mathbf{C} \rightarrow [0, 1]$ *is nondecreasing, then it is a generalized possibility measure. Furthermore, if* $\mathbf{C}$ *is just the class of all atoms of some class* $\mathbf{C}'$, *that is,* $\mathbf{C} = \mathbf{A}[\mathbf{C}']$, *then any nondecreasing set function* μ: $\mathbf{C} \rightarrow [0, 1]$ *is P-consistent on* $\mathbf{C}$.

Proof. If $\{E_t | t \in T\} \subset \mathbf{C}, \bigcup_{t \in T} E_t \in \mathbf{C}$, where T is an arbitrary index set, then, since $\mathbf{C}$ is an AU-class, there exists $t_0 \in T$ such that $E_{t_0} = \bigcup_{t \in T} E_t$. So

$$\mu(\bigcup_{t \in T} E_t) = \mu(E_{t_0}) \leq \sup_{t \in T} \mu(E_t).$$

Noting that μ is nondecreasing, we have the converse inequality:

$$\mu(\bigcup_{t \in T} E_t) \geq \sup_{t \in T} \mu(E_t).$$

Consequently, we have

$$\mu(\bigcup_{t \in T} E_t) = \sup_{t \in T} \mu(E_t).$$

That is, μ is maxitive.

Furthermore, let $\mathbf{C} = \mathbf{A}[\mathbf{C}']$. If $\mathbf{A} = \mathbf{A}(x) \in \mathbf{A}[\mathbf{C}']$,

$$\{A_t | t \in T\} = \{A(x_t) | t \in T\} \subset \mathbf{A}[\mathbf{C}'],$$

and $A \subset \cup_{t \in T} A_t$, where T is an arbitrary index set, then there exists $t_0 \in T$ such that $x \in A_{t_0}$, and, therefore, by Theorem 2.12, $A \subset A_{t_0}$. So we have

$$\mu(A) \leq \mu(A_{t_0}) \leq \sup_{t \in T} \mu(A_t).$$

That is, μ is P-consistent on $\mathbf{A}$ $[\mathbf{C}']$. □

Theorem 5.11. *Any nondecreasing set function $\mu : \mathbf{A}[\mathbf{C}] \to [0, 1]$ is a generalized possibility measure on $\mathbf{A}[\mathbf{C}]$, and it can be uniquely extended to a generalized possibility measure π on $\mathbf{F}_p(\mathbf{C})$.*

Proof. By Lemma 5.2 and Theorem 5.8 we know that μ is a generalized possibility measure on $\mathbf{A}[\mathbf{C}]$, and by Theorem 5.9 μ can be extended to a generalized possibility measure π on $\mathbf{P}(X)$ containing $\mathbf{F}_p(\mathbf{C})$. So, we only need to prove that on $\mathbf{F}_p(\mathbf{C})$ the extension is unique. For any $E \in \mathbf{F}_p(\mathbf{C})$, through Theorem 3.17, E can be expressed by

$$E = \bigcup_{t \in T} A_t,$$

where $A_t \in \mathbf{A}[\mathbf{C}]$ and T is an index set. Since π is a generalized possibility measure, it should hold that

$$\pi(E) = \sup_{t \in T} \mu(A_t).$$

But, the expression of E may not be unique. If there exists another expression

$$E = \bigcup_{s \in S} A'_s,$$

where $A'_s \in \mathbf{A}[\mathbf{C}]$ and S is an index set, we must prove that

$$\sup_{t \in T} \mu(A_t) = \sup_{s \in S} \mu(A'_s)$$

to show the uniqueness of the extension. In fact, for any A_t there exists $x_t \in A_t$ such that $A_t = A(x_t)$. From $x_t \in E = \cup_{s \in S} A'_s$ we know that there exists $s_t \in S$ such that $x_t \in A'_{s_t}$. Therefore, by Theorem 3.12, $A_t \subset A'_{s_t}$. Using the monotonicity of μ, we have

$$\mu(A_t) \leq \mu(A'_{s_t}) \leq \sup_{s \in S} \mu(A'_s).$$

This inequality holds for any $t \in T$. So, we have

$$\sup_{t \in T} \mu(A_t) \leq \sup_{s \in S} \mu(A'_s).$$

Analogously, the converse inequality holds. Consequently, we have

$$\sup_{t \in T} \mu(A_t) = \sup_{s \in S} \mu(A'_s).$$

□

As to the extension of necessity measures, we have a similar discussion.

Definition 5.6. A set function $\mu : \mathbf{C} \to [0, 1]$ is *N-consistent* on $\mathbf{C}$ iff for any $\{E_t | t \in T\} \subset \mathbf{C}$ and any $E \in \mathbf{C}$,

$$E \supset \bigcap_{t \in T} E_t$$

implies

$$\mu(E) \geq \inf_{t \in T} \mu(E_t),$$

where T is an arbitrary index set.

Theorem 5.12. *Let* $\emptyset \in \mathbf{C}$. *A set function* $\mu : \mathbf{C} \to [0, 1]$ *with* $\mu(\emptyset) = 0$ *can be extended to a necessity measure* ν *on* $\mathbf{P}(X)$ *if and only if* μ *is N-consistent on* $\mathbf{C}$.

This extension may be not unique. Denoting all necessity measure extensions of the set function μ given in the above theorem by $\mathbf{E}_\nu(\mu)$, then we have the following theorem.

Theorem 5.13. $(\mathbf{E}_\nu(\mu), \leq)$ *is a lower semilattice, and the set function* ν *given by*

$$\nu : \mathbf{P}(X) \to [0, 1]$$

$$F \mapsto \inf_{x \notin F} \sup_{E | x \notin F \in \mathbf{C}} \mu(E)$$

is the smallest element of $(\mathbf{E}_\nu(\mu), \leq)$.

Using the concept of the hole, we can address the uniqueness of the necessity measure extension.

Notes

5.1. The issue of extensions of possibility and necessity measures was first addressed by Wang [1985b, 1986, 1987]. It was shown by Wang [1986b] that these extensions can be obtained by solving appropriate fuzzy relation equations. It follows from this result that, in general, there are several (possibly even infinitely many) smallest extensions on **P**(X) for a given possibility measure on **C**. It was shown by Qiao [1989] that the extensions of possibility and necessity measures can be generalized to monotone sets.
5.2. The work on extensions of quasi-measures (including λ-measures as a special case) was initiated by Wang [1981].
5.3. Extensions of semicontinuous monotone measures as well as some other kinds of monotone measures were studied by Wang [1990a].

Exercises

5.1. Let $X = \{x_1, x_2, x_3\}$. Using the concepts of the quasi-measure and its proper T-function, determine a λ-measure on μ on **P**(X) with a parameter $\lambda = 1$ and constrained by $\mu(\{x_1\}) = 1, \mu(\{x_2\}) = 2, \mu(\{x_3\}) = 3$.
5.2. Given a probability distribution function

$$h(x) = \begin{cases} 0 & \text{if } x \le -1 \\ 1/4 & \text{if } -1 < x \le 1 \\ 1 & \text{if } x > 1, \end{cases}$$

determine the corresponding Sugeno measure g_λ on the Borel field **B** with a parameter $\lambda = 2$. In particular, list the values of $g_\lambda(\{1\}), g_\lambda(\{0\})$, and $g_\lambda([-2, 0))$.
5.3. Repeat Exercise 5.2 for each of the following probability distribution functions:

(a) $$h(x) = \begin{cases} 0 & \text{if } x \le -1 \\ 1/4 & \text{if } -1 < x \le 0 \\ 3/4 & \text{if } 0 < x \le 1 \\ 1 & \text{if } x > 1; \end{cases}$$

(b) $$h(x) = \begin{cases} 0 & \text{if } x \le -1 \\ (1+x)/2 & \text{if } -1 < x \le 1 \\ 1 & \text{if } x > 1; \end{cases}$$

(c) $$h(x) = \begin{cases} 0 & \text{if } x \le 0 \\ 1 & \text{if } x > 0. \end{cases}$$

5.4. Let X be a nonempty set and **R** be an algebra of subsets of X. Prove that $\mathbf{R}_\sigma$ is closed under the formation of countable unions. Assuming that we call the class that is closed under the formation of countable unions a σ-class, show that $\mathbf{R}_\sigma$ is the smallest σ-class containing $\mathbf{R}$ (we can also refer to it as *σ-class generated by* $\mathbf{R}$). Similarly, assuming that we call the class that is closed under the formation of countable intersections a δ-class and denote it by $\mathbf{R}_\delta$, show that $\mathbf{R}_\delta$ is the smallest δ-class containing $\mathbf{R}$ (*δ-class generated by* $\mathbf{R}$).

5.5. In analogy to the concept of $\mathbf{R}_\sigma$-approachability define an appropriate concept of *$\mathbf{R}_\delta$-approachability*. What relevant result can be obtained for this new concept?

5.6. Determine $\mathbf{R}_\sigma$ and $\mathbf{R}_\delta$ for each $\mathbf{R}$ specified as follows:

(a) $X = \{a, b, c, d\}, \mathbf{R} = \{\emptyset, \{a, b\}, \{c, d\}, X\}$;
(b) $X\{1, 2, \ldots\}, \mathbf{R}$ is the class of all finite sets and their complements in X;
(c) $X = [0, 1], \mathbf{R}$ is the class of all finite sets and their complements in X.

5.7. Let $X = \{1, 2, \ldots\}, \mathbf{R}$ be the class of all finite sets in X, and μ be a set function defined on $\mathbf{R}$ as follows:

$$\mu(E) = \begin{cases} 0 & \text{if } E = \emptyset \\ 1 & \text{otherwise} \end{cases}$$

Extend μ as a lower semicontinuous monotone measure on a class, as large as possible, that contains $\mathbf{R}$.

5.8. Repeat Exercise 5.7 for the following set function:

$$\mu(E) = 1 - \frac{1}{|E| + 1}$$

5.9. Let $\boldsymbol{M}$ be the class of all monotone measures defined on a measurable space $(X, \mathbf{F})$. Prove that "$\ll$" is a transitive relation on $\boldsymbol{M}$.

5.10. Let $X = \{a, b, c, d, e\}$. Determine for each of the following set functions μ defined on given classes $\mathbf{C}$ whether it is P-consistent on $\mathbf{C}$:

(a) $\mathbf{C} = \mathbf{P}(X), \mu(E) = 1$ for any $E \in \mathbf{P}(X)$;
(b) $\mathbf{C} = \mathbf{P}(X)$,

$$\mu(E) = \begin{cases} 0 & \text{if } E = \emptyset \\ 1 & \text{otherwise;} \end{cases}$$

(c) $\mathbf{C}$ is the class consisting of all subsets of X that contain at most two points, and

$$\mu(E) = |E|/5$$

for any $E \in \mathbf{C}$;
(d) $\mathbf{C} = \{\{a\}, \{b\}, \{a, c\}, \{b, d, e\}, \{a, c, d, e\}\}, \mu(\{a\}) = 0.1, \mu(\{b\}) = 0.8, \mu(\{a, c\}) = 0.5, \mu(\{b, d, e\}) = 0.8, \mu(\{a, c, d, e\}) = 0.6$;

(e) $\mathbf{C} = \{\{a\}, \{b\}, \{a, c\}, \{b, d, e\}, \{a, c, d, e\}\}, \mu(\{a\}) = 0.1, \mu(\{b\}) = 0.5, \mu(\{a, e\}) = 0.5, \mu(\{b, d, e\}) = 0.8, \mu(\{a, c, d, e\}) = 0.6$;
(f) $\mathbf{C} = \{\{a,b\},\{b,c\},\{a,b,c\},\{a,d,e\},\{b,c,d\}\}, \mu(\{a,b\}) = 0.2, \mu(\{b,c\}) = 0.5, \mu(\{a,b,c\}) = 0.5, \mu(\{a,d,e\}) = 1, \mu(\{b,c,d\}) = 0.9$;
(g) $\mathbf{C} = \{\{a, b), \{a, d, e\}, \{b, d\}, \{a, b, e\}\}, \mu(\{a, b\}) = 0.5, \mu(\{a, d, e\}) = 1, \mu(\{b, d\}) = 1, \mu(\{a, b, e\}) = 0.8$;
(h) $\mathbf{C} = \{\{a,b\},\{a,b,d\},\{c,d,e\},\{a,b,c,e\},\{d,e\}\}, \mu(\{a,b\}) = 0.3, \mu(\{a,b,d\}) = 0.4, \mu(\{c,d,e\}) = 1, \mu(\{a,b,c,e\}) = 1, \mu(\{d,e\}) = 0.6$;
(i) $\mathbf{C} = \{\{a, b, c\}, \{b\}, \{c\}, \{b, c, d\}, \{b, e\}\}, \mu(\{a, b, c\}) = 0.5, \mu(\{b\}) = 0.1, \mu(\{c\}) = 0.2, \mu(\{b, c, d\}) = 1, \mu(\{b, e\}) = 0.6$.

5.11. Extend each P-consistent set function μ given in Exercise 5.10 onto $\mathbf{P}(X)$ as a generalized possibility measure. Determine whether the extension is unique. If it is not unique, find the greatest extension. Can you also show that there are extensions other than the greatest one?

5.12. In Exercise 5.10, determine which classes are AU-classes.

5.13. Using the set functions listed in Exercise 5.10 and the classes on which the respective set functions are defined, confirm the conclusions given in Section 5.4.

Chapter 6
Structural Characteristics for Set Functions

6.1 Null-Additivity

Up to now, we have used some structural characteristics such as nonnegativity, monotonicity, additivity, subadditivity, λ-rule, maxitivity, continuity, and so on to describe the features of a set function. Since the monotone measures in general lose additivity, they appear much looser than the classical measures. Thus, it is quite difficult to develop a general theory of monotone measures without any additional condition. Before 1981 it was thought that monotone measures additionally possessed subadditivity (even maxitivity), or satisfied the λ-rule. But these conditions are so strong that the essence of the problem is concealed in most propositions. Since 1981, many new concepts on structural characteristics, which monotone measures may possess (e.g., null-additivity, autocontinuity, and uniform autocontinuity), have been introduced. As we shall see later, they are substantially weaker than subadditivity or the λ-rule, but can effectively depict most important monotone measures and are powerful enough to guarantee that many important theorems presented in the following chapters will be justified. In several theorems, they go so far as to be a necessary and sufficient condition. In generalized measure theory, these new concepts replace additivity and thus play important roles.

We assume that $\mathbf{F}$ is a σ-algebra of sets in $\mathbf{P}(X)$, and we define these new concepts in a wider scope: Set functions $\mu : \mathbf{F} \to [-\infty, \infty]$ are considered to be extended real-valued.

Definition 6.1. $\mu : \mathbf{F} \to [-\infty, \infty]$ is *null-additive* iff

$$\mu(E \cup F) = \mu(E)$$

whenever $E \in \mathbf{F}$, $F \in \mathbf{F}$, $E \cap F = \emptyset$, and $\mu(F) = 0$.

Theorem 6.1. *If for any nonempty set $F \in \mathbf{F}$, $\mu(F) \neq 0$, then μ is null-additive.*

Proof. If there exists some set $F \in \mathbf{F}$ such that $\mu(F) = 0$, then $F = \emptyset$. Therefore, for any $E \in \mathbf{F}$, we have

$$\mu(E \cup F) = \mu(E)$$ □

Z. Wang, G.J. Klir, *Generalized Measure Theory*,
DOI: 10.1007/978-0-387-76852-6_6,

Theorem 6.2. *If μ: $\mathbf{F} \rightarrow [0, \infty]$ is a nondecreasing set function, then the following statements are equivalent:*

(1) *μ is null-additive;*
(2) *$\mu(E \cap F) = \mu(E)$ whenever $E \in \mathbf{F}$, $F \in \mathbf{F}$, and $\mu(F) = 0$;*
(3) *$\mu(E - F) = \mu(E)$ whenever $E \in \mathbf{F}$, $F \in \mathbf{F}$, $F \subset E$, and $\mu(F) = 0$;*
(4) *$\mu(E - F) = \mu(E)$ whenever $E \in \mathbf{F}$, $F \in \mathbf{F}$, and $\mu(F) = 0$;*
(5) *$\mu(E \Delta F) = \mu(E)$ whenever $E \in \mathbf{F}$, $F \in \mathbf{F}$, and $\mu(F) = 0$.*

Proof. (1) $\Rightarrow$ (2): If $\mu(F) = 0$, noting $0 \le \mu(F - E) \le \mu(F) = 0$, $(F - E) \cap E = \emptyset$, we have $\mu(E \cup F) = \mu(E \cup (F - E)) = \mu(E)$;
(2) $\Rightarrow$ (1): Evident;
(1) $\Rightarrow$ (3): It is only necessary to note $\mu(E) = \mu((E - F) \cup F)$; (3) $\Rightarrow$ (4): Since $\mu(E - F) = \mu(E - (F \cap E))$, and $F \cap E \subset E$, noting $0 \le \mu(F \cap E) \le \mu(F) = 0$, we have the conclusion;
(4) $\Rightarrow$ (1): The conclusion follows from $\mu(E) = \mu((E \cup F) - F)$ when $E \cap F = \emptyset$;
(2) and (4) $\Rightarrow$ (5): The conclusion follows from the inequality

$$\mu(E - F) \le \mu(E \Delta F) \le \mu(E \cup F);$$

(5) $\Rightarrow$ (1): We only need to point out that $E \Delta F = E \cup F$ when $E \cap F = \emptyset$. □

One of the simplest monotone measures that is not null-additive is given as follows.

Example 6.1. $X = \{a, b\}$, $\mathbf{F} = \mathbf{P}(X)$, and

$$\mu(E) = \begin{cases} 1 & \text{if } E = X \\ 0 & \text{if } E \neq X. \end{cases}$$

Theorem 6.3. *Let μ be a null-additive, continuous, monotone measure, and $E \in \mathbf{F}$. Then, we have*

$$\lim_n \mu(E \cup F_n) = \mu(E)$$

for any decreasing set sequence $\{F_n\} \subset \mathbf{F}$ for which $\lim_n \mu(F_n) = 0$ and there exists at least one positive integer n_0 such that $\mu(E \cup F_{n_0}) < \infty$ as $\mu(E) < \infty$.

Proof. It is sufficient to prove this theorem for $\mu(E) < \infty$. If we write $F = \cap_{n=1}^{\infty} F_n$, we have $\mu(F) = \lim_n \mu(F_n) = 0$. Since $E \cup F_n \searrow E \cup F$, it follows from the continuity and the null-additivity of μ that

$$\lim_n \mu(E \cup F_n) = \mu(E \cup F) = \mu(E).$$ □

Theorem 6.4. *Let μ be a null-additive continuous monotone measure, and $E \in \mathbf{F}$. We have*

$$\lim_n \mu(E - F_n) = \mu(E)$$

for any decreasing set sequence $\{F_n\} \subset \mathbf{F}$ for which $\lim_n \mu(F_n) = 0$.

Proof. Since $E - F_n \nearrow E - (\cap_{n=1}^{\infty} F_n)$ and $\mu(\cap_{n=1}^{\infty} F_n) = 0$, by Theorem 6.2 and the continuity of μ it follows that

$$\lim_n \mu(E - F_n) = \mu\left(E - \left(\bigcap_{n=1}^{\infty} F_n\right)\right) = \mu(E). \qquad \square$$

The following example shows that the conclusion in Theorem 6.3 is not true when the finiteness condition is abandoned.

Example 6.2. Let $X = \{0, 1, 2, ...\}$, $\mathbf{F} = \mathbf{P}(X)$,

$$\mu(E) = \begin{cases} \sum_{i \in E} 2^{-(i+1)} & \text{if } 0 \notin E \\ \infty & \text{if } 0 \in E \text{ and } E - \{0\} \neq \emptyset \\ 1 & \text{if } E = \{0\}. \end{cases}$$

It is not too difficult to verify that μ is a continuous monotone measure. By Theorem 6.1 μ is null-additive. If we take $E = \{0\}$, $F_n = \{n, n + 1, ...\}$, $n = 1, 2, ...$, then $\{F_n\}$ is decreasing, and $\lim_n \mu(F_n) = 0$, but $\mu(E \cup F_n) = \infty$ for any $n = 1, 2, ...$, and $\mu(E) = 1$. So we have

$$\lim_n \mu(E \cup F_n) \neq \mu(E).$$

6.2 Autocontinuity

Definition 6.2. $\mu: \mathbf{F} \to [-\infty, \infty]$ is *autocontinuous from above* (or *from below*) iff

$$\lim_n \mu(E \cup F_n) = \mu(E) \qquad (\text{or } \lim_n \mu(E - F_n) = \mu(E))$$

whenever $E \in \mathbf{F}$, $F_n \in \mathbf{F}$, $E \cap F_n = \emptyset$ (or $F_n \subset E$, respectively), $n = 1, 2, ...$, and $\lim_n \mu(F_n) = 0$; μ is *autocontinuous* iff it is both autocontinuous from above and autocontinuous from below.

Theorem 6.5. *Let $\mu: \mathbf{F} \to [-\infty, \infty]$ be an extended real-valued set function. If there exists $\varepsilon > 0$ such that $|\mu(E)| \geq \varepsilon$ for any $E \in \mathbf{F}$, $E \neq \emptyset$, then μ is autocontinuous.*

Proof. Under the condition of this theorem, if $\{F_n\} \subset \mathbf{F}$ is such that $\lim_n \mu(F_n) = 0$, then there must be some n_0 such that $F_n = \emptyset$ whenever $n \geq n_0$, and, therefore,

$$\lim_n \mu(E \cup F_n) = \lim_n \mu(E - F_n) = \lim_n \mu(E) = \mu(E). \qquad \square$$

Theorem 6.6. *If* μ*:* $\mathbf{F} \to [-\infty, \infty]$ *is autocontinuous from above or autocontinuous from below then it is null-additive.*

Proof. For any $E \in \mathbf{F}$, $F \in \mathbf{F}$, $E \cap F = \emptyset$, and $\mu(F) = 0$, taking $F_n = F$, $n = 1, 2, \ldots$, we have $\lim_n \mu(F_n) = \mu(F) = 0$. If μ is autocontinuous from above, then

$$\mu(E \cup F) = \lim_n \mu(E \cup F_n) = \mu(E),$$

so that μ is null-additive; if μ is autocontinuous from below, then

$$\mu(E \cup F) = \lim_n \mu((E \cup F) - F_n) = \mu(E),$$

and μ is null-additive as well. □

Obviously, if μ: $\mathbf{F} \to [0, \infty]$ is nondecreasing, then the restrictions "$E \cap F_n = \emptyset$" and "$F_n \subset E$" in the statement of Definition 6.2 may be omitted.

Theorem 6.7. *Let* μ*:* $\mathbf{F} \to [0, \infty]$ *be nondecreasing.* μ *is autocontinuous if and only if*

$$\lim_n \mu(E \Delta F_n) = \mu(E)$$

whenever $E \in \mathbf{F}$*,* $\{F_n\} \subset \mathbf{F}$*, and* $\lim_n \mu(F_n) = 0$.

Proof. Necessity: For any $E \in \mathbf{F}$, $\{F_n\} \subset \mathbf{F}$, with $\lim_n \mu(F_n) = 0$, noting

$$E - F_n \subset E \Delta F_n \subset E \cup F_n,$$

by the monotonicity of μ we have

$$\mu(E - F_n) \le \mu(E \Delta F_n) \le \mu(E \cup F_n).$$

Since μ is both autocontinuous from above and autocontinuous from below, we have

$$\lim_n \mu(E \cup F_n) = \mu(E)$$

and

$$\lim_n \mu(E - F_n) = \mu(E).$$

Thus, we have

$$\lim_n \mu(E \Delta F_n) = \mu(E).$$

Sufficiency: For any $E \in \mathbf{F}$, $\{F_n\} \subset \mathbf{F}$, with $\lim_n \mu(F_n) = 0$, we have $F_n - E \in \mathbf{F}$ and $\mu(F_n - E) \le (F_n)$. So we have

$$\lim_n \mu(F_n - E) = 0,$$

and therefore, by the condition given in this theorem, we have

$$\lim_n \mu(E \cup F_n) = \lim_n \mu(E\Delta(F_n - E)) = \mu(E).$$

That is, μ is autocontinuous from above. Similarly, from

$$\lim_n \mu(F_n \cap E) = 0,$$

it follows that

$$\lim_n \mu(E - F_n) = \lim_n \mu(E\Delta(F_n \cap E)) = \mu(E).$$

That is, μ is autocontinuous from below. □

The following two theorems indicate the relation between the autocontinuity and the continuity of nonnegative set functions.

Theorem 6.8. *If* μ: $\mathbf{F} \rightarrow [0, \infty)$ *is continuous from above at* $\emptyset$ *and autocontinuous from above (or from below), then* μ *is continuous from above (or from below, respectively).*

Proof. If $\{E_n\}$ is a decreasing sequence of sets in $\mathbf{F}$, and $E = \cap_{n=1}^{\infty} E_n$, then $E_n - E \searrow \emptyset$. From the finiteness and the continuity from above at $\emptyset$ of μ, we know

$$\lim_n \mu(E_n - E) = 0$$

and, therefore, by using the autocontinuity from above of μ, we have

$$\lim_n \mu(E_n) = \lim_n \mu(E \cup (E_n - E)) = \mu(E).$$

That is, μ is continuous from above. The proof of the continuity from below is similar. □

Theorem 6.9. *If* μ: $\mathbf{F} \rightarrow [0, \infty]$ *is nondecreasing, continuous from above at* $\emptyset$*, and autocontinuous from above, then* μ *is continuous from above.*

Proof. If $\{E_n\}$ is a decreasing sequence of sets in $\mathbf{F}$ with $\mu(E_1) < \infty$, we know from the monotonicity of μ that

$$0 \le \mu\left(E_1 - \bigcap_{n=1}^{\infty} E_n\right) \le \mu(E_1) < \infty.$$

Thus, the proof is similar to the proof of Theorem 6.8. □

When the set function we consider is a continuous monotone measure (or semicontinuous monotone measure) there are some interesting results on the autocontinuity from above and the autocontinuity from below.

Lemma 6.1. *Let* μ*:* $\mathbf{F} \to [0, \infty]$ *be a lower semicontinuous monotone measure (or upper semicontinuous monotone measure),* $\{E_n\} \subset \mathbf{F}$ *with* $\lim_n \mu(E_n) = 0$*. If* μ *is autocontinuous from above (or from below and finite), then there exists some sequence* $\{E_k\}$ *of subsequences of* $\{E_n\}$*, where* $E_k = \{E_{n_i^{(k)}}\}$*,* $k = 1, 2, \ldots$*, such that*

$$\lim_k \mu\left(\bigcup_{i=1}^{\infty} E_{n_i^{(k)}}\right) = 0$$

(or $\lim_k \mu(A - \bigcup_{i=1}^{\infty} E_{n_i^{(k)}}) = \mu(A)$ *for any fixed* $A \in \mathbf{F}$*, respectively).*

Proof. Assume μ is autocontinuous from above. For arbitrarily given $\varepsilon > 0$, we take n_1 such that $\mu(E_{n1}) < \varepsilon/2$. Since μ is autocontinuous from above and $\lim_n \mu(E_n) = 0$, we can take $n_2 > n_1$ such that

$$\mu(E_{n_1} \cup E_{n_2}) < \mu(E_{n_1}) + \varepsilon/4 < 3\varepsilon/4.$$

Generally, for $\cup_{i=1}^{j} E_{n_i}$, we can take $n_{j+1} > n_j$ such that

$$\mu\left(\bigcup_{i=1}^{j+1} E_{n_i}\right) = \mu\left(\left(\bigcup_{i=1}^{j} E_{n_i}\right) \cup E_{n_{j+1}}\right) < (1 - 2^{-(j+1)})\varepsilon < \varepsilon,$$

$j = 1, 2, \ldots$. Consequently, by using the continuity from below of μ we get a subsequence $\{E_{n_i}\}$ of $\{E_n\}$ such that

$$\mu\left(\bigcup_{i=1}^{\infty} E_{n_i}\right) = \lim_j \mu\left(\bigcup_{i=1}^{j+1} E_{n_i}\right) \leq \varepsilon.$$

In a similar way, we can also prove the case when μ is autocontinuous from below. □

Theorem 6.10. *Let* μ*:* $\mathbf{F} \to [0, \infty]$ *be a lower semicontinuous monotone measure. If* μ *is autocontinuous from above, then for any* $\{E_n\} \subset \mathbf{F}$ *with* $\lim_n \mu(E_n) = 0$*, there exists some subsequence* $\{E_{n_i}\}$ *of* $\{E_n\}$ *such that*

$$\mu(\overline{\lim_i}\, E_{n_i}) = 0.$$

The inverse proposition is also true when μ *is a finite null-additive continuous monotone measure.*

Proof. Suppose μ is autocontinuous from above. From Lemma 6.1, there exists a subsequence $\{E_{n_i^{(1)}}\}$ such that

$$\mu\left(\bigcup_{i=1}^{\infty} E_{n_i^{(1)}}\right) < 1.$$

As $\lim_i \mu(E_{n_i^{(1)}}) = 0$, too, there exists a subsequence $\{E_{n_i^{(2)}}\}$ of $\{E_{n_i^{(1)}}\}$ such that

$$\mu\left(\bigcup_{i=1}^{\infty} E_{n_i^{(2)}}\right) < 1/2.$$

Generally, there exists a subsequence $\{E_{n_i^{(k-1)}}\}$ of $\{E_{n_i^{(k-1)}}\}$ such that

$$\mu\left(\bigcup_{i=1}^{\infty} E_{n_i^{(k)}}\right) < 1/k.$$

$k = 2, 3, \ldots.$ If we take $n_i = n_i^{(i)}$, then $\{E_{n_i}\}$ is a subsequence of $\{E_n\}$ and satisfies

$$\bigcup_{i=k}^{\infty} E_{n_i} \subset \bigcup_{i=1}^{\infty} E_{n_i^{(k)}}$$

for any $k = 1, 2, \ldots.$ Consequently, we have

$$\mu(\overline{\lim_i}\, E_{n_i}) = \mu\left(\bigcap_{k=1}^{\infty}\bigcup_{i=k}^{\infty} E_{n_i}\right) \leq \mu\left(\bigcup_{i=k}^{\infty} E_{n_i}\right) \leq \mu\left(\bigcup_{i=1}^{\infty} E_{n_i^{(k)}}\right) < 1/k$$

for any $k = 1, 2, \ldots$, and thus we have

$$\mu(\overline{\lim_i}\, E_{n_i}) = 0.$$

Conversely, for any $E \in \mathbf{F}$, $\{F_n\} \subset \mathbf{F}$ with $\lim_n \mu(F_n) = 0$, there exists a subsequence $\{F_{n_k}\}$ of $\{F_n\}$ *such that*

$$\overline{\lim_i}\, \mu(E \cup F_n) = \lim_k \mu(E \cup F_{n_k}).$$

Since $\lim_n \mu(F_{n_k}) = 0$, too, by the condition in the inverse proposition, there exists some subsequence $\{F_{n_{k_i}}\}$ such that

$$\mu(\overline{\lim_i}\, E_{n_{k_i}}) = 0.$$

Therefore, by the finiteness, monotonicity, and continuity from above of μ (see the proof of Theorem 4.27), we have

$$\overline{\lim_n}\, \mu(E \cup F_n) = \lim_i \mu(E \cup F_{n_{k_i}}) \leq \mu(\overline{\lim_i}\,(E \cup F_{n_{k_i}})) = \mu(E \cup \overline{\lim_i}\, F_{n_{k_i}}).$$

By applying the null-additivity of μ, we have

$$\overline{\lim_n}\, \mu(E \cup F_n) = \mu(E).$$

Noting

$$\mu(E) \leq \underline{\lim}_n \mu(E \cup F_n) \leq \overline{\lim}_n \mu(E \cup F_n),$$

we have

$$\lim_n \mu(E \cup F_n) = \mu(E).$$

That is, μ is autocontinuous from above. □

Theorem 6.11. *Let* μ*:* $\mathbf{F} \to [0, \infty)$ *be a finite continuous monotone measure.* μ *is autocontinuous from below if and only if it is null-additive, and for any* $A \in \mathbf{F}$ *and any* $\{E_n\}$ *with* $\lim_n \mu(E_n) = 0$ *there exists some subsequence* $\{E_{n_i}\}$ *of* $\{E_n\}$ *such that*

$$\mu(A - \overline{\lim_i} E_{n_i}) = \mu(A).$$

Proof. Similar to the proof of Theorem 6.10. □

In Theorem 6.10, from $\lim_n \mu(E_n) = 0$, we know that there exists a subsequence $\{E_{ni}\}$ of $\{E_n\}$ such that $\sum_{n=1}^{\infty} \mu(E_{n_i}) < \infty$. If μ is a classical measure, we obtain the same conclusion, $\mu(\overline{\lim}_i E_{n_i}) = 0$, by the Borel–Cantelli lemma. Hence, Theorem 6.11 may be regarded as a generalization of the Borel–Cantelli lemma onto monotone measure spaces.

Theorem 6.12. *Let* μ*:* $\mathbf{F} \to [0, \infty]$ *be a continuous monotone measure. If* μ *is autocontinuous from above, then it is autocontinuous from below. Furthermore, when* μ *is finite, the autocontinuity from below implies the autocontinuity from above and, therefore, the autocontinuity, the autocontinuity from above, and the autocontinuity from below are equivalent.*

Proof. Suppose μ is autocontinuous from above. For any $E \in \mathbf{F}$, $\{F_n\} \subset \mathbf{F}$ with $\lim_n \mu(F_n) = 0$, there exists some subsequence $\{F_{n_k}\}$ of $\{F_n\}$ such that

$$\underline{\lim}_n \mu(E - F_n) = \lim_k \mu(E - F_{n_k}).$$

Since μ is autocontinuous from above, by Theorem 6.10 we know that there exists some subsequence $\{F_{n_{k_i}}\}$ of $\{F_{n_k}\}$ such that

$$\mu(\overline{\lim_i} F_{n_{k_i}}) = 0.$$

Thus, by applying the null-additivity of μ (see Theorem 6.6), we have

$$\mu(E) \geq \overline{\lim_n} \mu(E - F_n) \geq \underline{\lim}_n \mu(E - F_n) = \lim_i \mu(E - F_{n_{k_i}}) \geq \mu(\underline{\lim}_i (E - F_{n_{k_i}}))$$
$$= \mu(E - \overline{\lim_i} F_{n_{k_i}}) = \mu(E).$$

Consequently, we have

$$\lim_n \mu(E - F_n) = \mu(E).$$

Therefore, μ is autocontinuous from below.

Now, we turn to the proof of the second part of the theorem. Assume μ is finite and autocontinuous from below. For any $E \in \mathbf{F}$, $\{F_n\} \subset \mathbf{F}$ with $\lim_n \mu(F_n) = 0$, there exists some subsequence $\{F_{n_k}\}$ of $\{F_n\}$ such that

$$\overline{\lim_n}\, \mu(E \cup F_n) = \lim_k \mu(E \cup F_{n_k}).$$

Since μ is autocontinuous from below, by Theorem 6.11, for the given E, there exists a subsequence $\{G_i^{(1)}\}$ of $\{F_{n_k}\}$ such that

$$\mu(E - \overline{\lim_i}\, G_i^{(1)}) = \mu(E);$$

since $\lim_i \mu(G_i^{(1)}) = 0$, for $E \cup G_1^{(1)}$, there exists a subsequence $\{G_i^{(2)}\}$ of $\{G_i^{(1)}\}$ such that

$$G_1^{(2)} = G_1^{(1)} \text{ and } \mu(E \cup G_1^{(1)} - \overline{\lim_i}\, G_i^{(2)}) = \mu(E \cup G_1^{(1)});$$

also, by a similar reasoning, for $E \cup G_2^{(2)}$ there exists a subsequence $\{G_i^{(3)}\}$ of $\{G_i^{(2)}\}$ such that $G_j^{(3)} = G_j^{(2)}, j = 1, 2$, and

$$\mu(E \cup G_2^{(2)} - \overline{\lim_i}\, G_i^{(3)}) = \mu(E \cup G_2^{(2)});$$

generally, for $E \cup G_n^{(n)}$, there exists a subsequence $\{G_i^{(n+1)}\}$ of $\{G_i^{(n)}\}$ such that $G_j^{(n+1)} = G_j^{(n)}, j = 1, 2, \ldots, n$, and

$$\mu(E \cup G_n^{(n)} - \overline{\lim_i}\, G_i^{(n+1)}) = \mu(E \cup G_n^{(n)}).$$

Continuing this process until infinity and writing $G_i = G_i^{(i)}$ for $i = 1, 2, \ldots$, we obtain a subsequence $\{G_i\}$ of $\{F_{n_k}\}$. Since $\{G_i\}$ is also a subsequence of $\{G_i^{(n)}\}$ for each $n = 1, 2, \ldots$, we have

$$\overline{\lim_i}\, G_i \subset \overline{\lim_i}\, G_i^{(n)}$$

for each $n = 1, 2, \ldots$. Hence, by using the monotonicity of μ,

$$\begin{aligned}\mu(E \cup G_i) &\geq \mu(E \cup G_i - \overline{\lim_i}\, G_i) = \mu(E \cup G_i^{(i)} - \overline{\lim_i}\, G_i)\\ &\geq \mu(E \cup G_i^{(i)} - \overline{\lim_j}\, G_j^{(i+1)}) = \mu(E \cup G_i^{(i)}) = \mu(E \cup G_i);\end{aligned}$$

that is,

$$\mu(E \cup G_i - \overline{\lim_i}\ G_i) = \mu(E \cup G_i)$$

for any $i = 1, 2, \ldots$. Finally, denoting G_i by $F_{n_{k_i}}$, $i = 1, 2, \ldots$, we obtain

$$\mu(E \cup F_{n_{k_i}} - \overline{\lim_i}\ F_{n_{k_i}}) = \mu(E \cup F_{n_{k_i}})$$

for any $i = 1, 2, \ldots$. Thus, noting $\lim_i \cup_{l=i}^{\infty}(F_{n_{k_i}} - \overline{\lim}_j F_{n_{k_j}}) = \emptyset$ and μ is finite, we have

$$\begin{aligned}
\mu(E) &\le \underline{\lim_n}\ \mu(E \cup F_n) \le \overline{\lim_n}\ \mu(E \cup F_n) \\
&= \lim_i \mu(E \cup F_{n_{k_i}}) = \lim_i \mu(E \cup F_{n_{k_i}} - \overline{\lim_j} F_{n_{k_j}}) \\
&\le \lim_i \mu\left(E \cup \left(\bigcup_{l=i}^{\infty}(F_{n_{k_l}} - \overline{\lim_j} F_{n_{k_j}})\right)\right) \\
&= \mu(E).
\end{aligned}$$

Consequently we have

$$\lim_n \mu(E \cup F_n) = \mu(E).$$

Therefore, μ is autocontinuous from above. □

However, when the continuous monotone measure μ is not finite the autocontinuity from below may not imply the autocontinuity from above.

Example 6.3. We use the continuous monotone measure given in Example 3.5. $X = \{1, 2, \ldots\}$, $\mathbf{F} = \mathbf{P}(X)$, $\mu(E) = |E| \cdot \sum_{i \in E} 2^{-i}$ for any $E \in \mathbf{F}$. For such a set function μ it is easy to see that a set E is finite if and only if $\mu(E) < \infty$. Now, let us show that μ is autocontinuous from below. For any $E \in \mathbf{F}$, $\{F_n\} \subset \mathbf{F}$ with $\lim_n \mu(F_n) = 0$, without any loss of generality we may suppose $F_n \neq \emptyset$ for any $n = 1, 2, \ldots$, and, therefore, we have

$$\lim_n\ (\inf\{i | i \in F_n\}) = \infty.$$

Thus, there are two cases: (1) If $\mu(E) < \infty$, then there exists n_0 such that $E \cap F_n = \emptyset$ (that is, $E - F_n = E$) for any $n \ge n_0$, and thus we have

$$\lim_n\ \mu(E - F_n) = \mu(E);$$

(2) If $\mu(E) = \infty$, then E is an infinite set. Since $\lim_n \mu(F_n) = 0$, there exists n_0 such that $\mu(F_n) < \infty$ (that is, F_n is a finite set) for any $n \ge n_0$. So $E - F_n$ is an infinite set. Consequently,

$$\lim_n \mu(E - F_n) = \infty = \mu(E).$$

Therefore, μ is autocontinuous from below. However, μ is not autocontinuous from above. In fact, if we take $E = \{1\}$, $F_n = \{n\}$, $n = 1, 2, \ldots$, then we have

$$\lim_n \mu(F_n) = \lim_n 2^{-n} = 0,$$

$$\lim_n \mu(E \cup F_n) = \lim_n [2 \cdot (2^{-1} + 2^{-n})] = 1,$$

but

$$\mu(E) = 2^{-1}.$$

This example also shows that, for monotone measures, the autocontinuity is really stronger than the null-additivity.

6.3 Uniform Autocontinuity

Definition 6.3. μ: $\mathbf{F} \to [-\infty, \infty]$ is *uniformly autocontinuous from above* (or *from below*) iff for any $\varepsilon > 0$, there exists $\delta = \delta(\varepsilon) > 0$ such that

$$\mu(E) - \varepsilon \leq \mu(E \cup F) \leq \mu(E) + \varepsilon \text{ (or } \mu(E) - \varepsilon \leq \mu(E - F) \leq \mu(E) + \varepsilon)$$

whenever $E \in \mathbf{F}$, $F \in \mathbf{F}$, $E \cap F = \emptyset$ (or $F \subset E$, respectively), and $|\mu(F)| \leq \delta$; μ is *uniformly autocontinuous* iff it is both uniformly autocontinuous from above and uniformly autocontinuous from below.

Theorem 6.13. *If μ: $\mathbf{F} \to [-\infty, \infty]$ is uniformly autocontinuous from above (or from below), then it is autocontinuous from above (or from below, respectively). Therefore, the uniform autocontinuity implies the autocontinuity.*

Proof. It is evident. □

Similar to the case in Section 6.2, if μ: $\mathbf{F} \to [0, \infty]$ is nondecreasing, then the restriction "$E \cap F = \emptyset$" and "$F \subset E$" in the statement of Definition 6.3 may be omitted.

Theorem 6.14. *If μ: $\mathbf{F} \to [0, \infty]$ is nondecreasing, then the following statements are equivalent:*

(1) *μ is uniformly autocontinuous;*
(2) *μ is uniformly autocontinuous from above;*
(3) *μ is uniformly autocontinuous from below;*
(4) *for any $\varepsilon > 0$ there exists $\delta = \delta(\varepsilon) > 0$ such that*

$$\mu(E) - \varepsilon \leq \mu(E \Delta F) \leq \mu(E) + \varepsilon$$

whenever $E \in \mathbf{F}$, $F \in \mathbf{F}$, and $\mu(F) \leq \delta$.

Proof. (1) $\Rightarrow$ (2): Obvious.
(2) $\Rightarrow$ (3): Since $\mu(E \cap F) \leq \mu(F) \leq \delta$, the desired conclusion follows from

$$\mu(E) = \mu((E - F) \cup (E \cap F)) \leq \mu(E - F) + \varepsilon$$

and the monotonicity of μ.
(3) $\Rightarrow$ (4): On the one hand, from $\mu(E \cap F) \leq \mu(F) \leq \delta$, we have

$$\mu(E \;\Delta\; F) = \mu((E \cup F) - (E \cap F)) \geq \mu(E \cup F) - \varepsilon \geq \mu(E) - \varepsilon.$$

On the other hand, since $\mu(F - E) \leq \mu(F) \leq \delta$, we have

$$\mu(E) \geq \mu(E - F) = \mu((E \Delta F) - (F - E)) \geq \mu(E \Delta F) - \varepsilon.$$

(4) $\Rightarrow$ (1): Obvious. □

The following example shows that not all autocontinuous monotone measures are uniformly autocontinuous.

Example 5.4. Let $X = X^{-} \cup X^{+}$, where $X^{-} = \{-1, -2, \ldots\}$, $X^{+} = \{1, 2, \ldots\}$, and let $\mathbf{F} = \mathbf{P}(X)$. A set function μ: $\mathbf{F} \to [0,\infty]$ is defined as

$$\mu(E) = |E^{*}| + \sum_{i \in E - E^{*}} 2^{i}$$

for any $E \in \mathbf{F}$, where

$$E^{*} = \{i | i \in E \cap X^{-}, |i| \leq \sup\{j | j \in E \cap X^{+}\}\} \cup \{E \cap X^{+}\}.$$

Obviously, $E - E^{*} \subset X^{-}$. It is not difficult to verify that such a set function μ is a continuous monotone measure and is autocontinuous. But μ is not uniformly autocontinuous. In fact, for any $\delta > 0$ and $0 < \varepsilon < 1$ there exist $i \in X^{-}$ and $j = -i \in X^{+}$ such that $\mu(\{i\}) = 2^{i} < \delta$ and

$$\mu(\{j, i\}) - \mu(\{j\}) = 2 - 1 = 1 > \varepsilon.$$

We should note in this example that the set function μ is not finite. When X is countable and μ is finite we have a heartening result on the uniform autocontinuity which will be given in Section 6.5.

6.4 Structural Characteristics of Monotone Set Functions

In this section, we summarize the relations among the structural characteristics when the set function μ: $\mathbf{F} \to [0, \infty]$ is nondecreasing.

Theorem 6.15. *If* μ: $\mathbf{F} \to [0, \infty]$ *is quasi-additive, then it is autocontinuous; furthermore, when* μ *is finite, the quasi-additivity implies the uniform autocontinuity.*

Proof. Let μ: **F** $\to [0, \infty]$ be quasi-additive with a proper T-function θ. For any $E \in \mathbf{F}$, $\{F_n\} \subset \mathbf{F}$ with $E \cap F_n = \emptyset$, $n = 1, 2, \ldots$, and $\lim_n \mu(F_n) = 0$, applying the additivity of $\theta \circ \mu$ and the continuity of θ and θ^{-1} we have

$$\begin{aligned}\lim_n \mu(E \cup F_n) &= \lim_n \theta^{-1}[\theta(\mu(E)) + \theta(\mu(F_n))] \\ &= \theta^{-1}[\theta(\mu(E)) + \theta(\lim_n \mu(F_n))] \\ &= \theta^{-1}[\theta(\mu(E))] \\ &= \mu(E).\end{aligned}$$

That is, μ is autocontinuous from above. Similarly, for any $E \in \mathbf{F}$, $\{F_n\} \subset \mathbf{F}$ with $F_n \subset E$, $n = 1, 2, \ldots$, and $\lim_n \mu(F_n) = 0$, without any loss of generality in assuming $\mu(F_n) < \infty$, $n = 1, 2, \ldots$, we have

$$\begin{aligned}\lim_n \mu(E - F_n) &= \lim_n \theta^{-1}[\theta(\mu(E)) - \theta(\mu(F_n))] \\ &= \theta^{-1}[\theta(\mu(E)) - \theta(\lim_n \mu(F_n))] \\ &= \theta^{-1}[\theta(\mu(E))] \\ &= \mu(E).\end{aligned}$$

That is, μ is autocontinuous from below. Consequently, μ is autocontinuous.

Furthermore, if $\mu(X) = a < \infty$ then θ is uniformly continuous on $[0, a]$, and so is θ^{-1} on $[0, \theta(a)]$. From

$$\mu(E \cup F) = \theta^{-1}[\theta(\mu(E)) + \theta(\mu(F))]$$

for any disjoint $E, F \in \mathbf{F}$, it is easy to see that μ is uniformly autocontinuous from above, and therefore, by Theorem 6.14 and the fact that the quasi-additivity implies the monotonicity, it is uniformly autocontinuous. □

Corollary 6.1. *If* μ: **F** $\to [0, \infty]$ *satisfies the* λ*-rule, then it is autocontinuous. Furthermore, it is uniformly autocontinuous when* $\mu(X) < \infty$.

Theorem 6.16. *If* μ: **F** $\to [0, \infty]$ *is nondecreasing and subadditive, then it is uniformly autocontinuous.*

Proof. From

$$\mu(E) \le \mu(E \cup F) \le \mu(E) + \mu(F)$$

for any $E \in \mathbf{F}$, $F \in \mathbf{F}$, it is easy to obtain the conclusion. □

Corollary 6.2. *If* μ: **F** $\to [0, \infty]$ *is additive, then it is uniformly autocontinuous.*

Corollary 6.3. *If* μ: **F** $\to [0, \infty]$ *is f-additive, then it is uniformly autocontinuous.*

The scheme shown in Fig. 6.1 illustrates the relations among these structural characteristics for nonnegative increasing set functions defined on a σ-algebra.

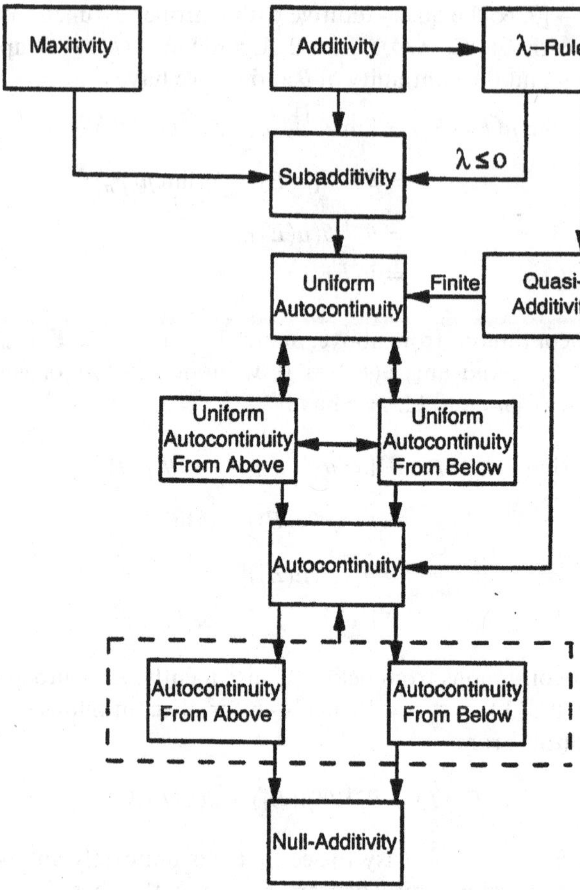

Fig. 6.1 The relation among structural characteristics of nonnegative set functions

The following result is useful in constructing examples of monotone measures (or continuous monotone measures) possessing the null-additivity, the autocontinuity, or the uniform autocontinuity, and in judging which structural characteristic a monotone measure possesses.

Theorem 6.17. *Let both* μ_1: $\mathbf{F} \to [0, \infty]$ *and* μ_2: $\mathbf{F} \to [0, \infty]$ *be null-additive (or autocontinuous, or uniformly autocontinuous). If we define* μ: $\mathbf{F} \to [0, \infty]$ *by*

$$\mu(E) = \mu_1(E) + \mu_2(E)$$

for any $E \in \mathbf{F}$, *then* μ *is null-additive (or autocontinuous, or uniformly autocontinuous, respectively), too.*

Proof. We only prove the conclusion for null-additivity. The remaining may be proved in a similar way.

Let μ_1 and μ_2 be null-additive. For any $E \in \mathbf{F}$, $F \in \mathbf{F}$ with $E \cap F = \emptyset$, if $\mu(F) = 0$, that is, $\mu_1(F) + \mu_2(F) = 0$, then it must hold that $\mu_1(F) = \mu_2(F) = 0$. So, we have

$$\mu(E \cup F) = \mu_1(E \cup F) + \mu_2(E \cup F) = \mu_1(E) + \mu_2(E) = \mu(E).$$

This shows μ is null-additive.

6.5 Monotone Measures on *S*-Compact Space

Theorem 6.18. *If $(X, \mathbf{F})$ is S-compact, and μ: $\mathbf{F} \to [0, \infty]$ is a finite continuous monotone measure, then the null-additivity, the autocontinuity, and the uniform autocontinuity are equivalent for μ.*

Proof. It is sufficient to prove that the null-additivity implies the uniform autocontinuity.

Assume μ is null-additive. If μ is not uniformly autocontinuous, then there exist $\{E_n\} \subset \mathbf{F}$, $\{F_n\} \subset \mathbf{F}$ and $\varepsilon > 0$ such that

$$\mu(E_n \cup F_n) > \mu(E_n) + \varepsilon,$$

$n = 1, 2, \ldots$, but

$$\lim_n \mu(F_n) = 0.$$

Since $(X, \mathbf{F})$ is S-compact, we can choose a subsequence $\{n_i\}$ from $\{n\}$ such that

$$\overline{\lim_i} E_{n_i} = \underline{\lim_i} E_{n_i} = \lim_i E_{n_i} \in \mathbf{F}$$

and

$$\overline{\lim_i} F_{n_i} = \underline{\lim_i} F_{n_i} = \lim_i F_{n_i} \in \mathbf{F}$$

Now, we denote $\lim_i E_{n_i} = E$ and $\lim_i F_{n_i} = F$. By Theorem 4.27,

$$\mu(F) = \lim_i \mu(F_{n_i}) = \lim_i \mu(F_{n_i}) = 0.$$

Noting the finiteness and the null-additivity of μ, we have, on one hand,

$$\overline{\lim_i} \mu(E_{n_i} \cup F_{n_i}) \le \mu(\overline{\lim_i}(E_{n_i} \cup F_{n_i})) = \mu(\overline{\lim_i} E_{n_i} \cup \overline{\lim_i} F_{n_i}) = \mu(E \cup F) = \mu(E);$$

on the other hand, we have

$$\mu(E_{n_i} \cup F_{n_i}) > \mu(E_{n_i}) + \varepsilon,$$

and, therefore, by Theorem 4.27 again, we have

$$\overline{\lim_i}\,\mu(E_{n_i} \cup F_{n_i}) \geq \lim_i \mu(E_{n_i}) + \varepsilon = \mu(E) + \varepsilon.$$

This is a contradiction. □

As a direct result of Theorem 2.22 and Theorem 6.18, we obtain the following proposition.

Corollary 6.4. *If X is a countable set and μ is a finite continuous monotone measure on $(X, \mathbf{F})$, then the null-additivity is equivalent to the uniform autocontinuity for μ.*

For a continuous monotone measure, the null-additivity is quite a light requirement and easy to verify, while the uniform autocontinuity is very powerful. Since most spaces we meet in praxis are countable, the result shown in this section is extremely important.

Notes

6.1. The use of subadditivity and maxitivity of continuous monotone measures (fuzzy measures) to develop a theory of Sugeno integral was initiated by Batle and Trillas [1979] and Ralescu and Adams [1980].

6.2. The concepts of null-additivity, autocontinuity, and uniform autocontinuity were introduced by Wang [1984]. In fact, these concepts and relevant results on Sugeno integrals (fuzzy integrals) were already reported in 1981, at the 12th European Fuzzy Mathematics Workshop in Hamburg, Germany. The concept of autocontinuity is further discussed by Sun and Wang [1988] and Wang [1992].

6.3. In addition to the structural characteristics defined in this chapter, Wang [1985a] introduced several other structural characteristics for monotone measures (fuzzy measures), such as converse null-additivity, pseudo-null-additivity, converse autocontinuity, and pseudo-autocontinuity, and discussed the relationship among them and their applicability. Pseudo-autocontinuity was further studied by Sun [1992].

6.4. The concept of S-compact spaces was introduced by Wang [1990b].

6.5. Some concepts and results presented in this chapter were generalized to continuous monotone measures (fuzzy measures) defined on a fuzzy measurable space by Qiao [1990].

Exercises

6.1. Let μ be a monotone measure on $(X, \mathbf{F})$. A set $E \in \mathbf{F}$ is called a *μ-null set* iff $\mu(E) = 0$; μ is called *weakly null-additive* iff $\mu(E \cup F) = 0$ whenever $E \in \mathbf{F}$, $F \in \mathbf{F}$, $\mu(E) = \mu(F) = 0$. Prove that the class of all μ-null sets is a σ-ring if and only if μ is weakly null-additive.

6.2. Function μ: $\mathbf{F} \to [-\infty, \infty]$ is said to be *null-subtractive* iff $\mu(E - F) = \mu(E)$ whenever $E \in \mathbf{F}$, $F \in \mathbf{F}$, $E \supset F$, and $\mu(F) = 0$. Prove that the null-subtractivity is equivalent to the null-additivity for μ if it is nonnegative and nondecreasing.

6.3. Is the statement "If μ: $\mathbf{F} \to [0, \infty]$ is nondecreasing, continuous from above at $\emptyset$, and autocontinuous from below, then μ is continuous from below" true? (Observe that this statement is similar to Theorem 6.9.)

6.4. Verify that the set function μ given in Example 6.4 is a continuous monotone measure and is autocontinuous.

6.5. Let μ_1 and μ_2 be monotone measures on $(X, \mathbf{F})$. Further, let μ be defined by $\mu(E) = \mu_1(E) + \mu_2(E)$ for any $E \in \mathbf{F}$ and let it be denoted by $\mu_1 + \mu_2$. Prove that:

(a) μ is a monotone measure on $(X, \mathbf{F})$;

(b) if both μ_1 and μ_2 are autocontinuous (uniformly autocontinuous) then so is μ.

6.6. In Exercise 6.5, if we replace $\mu(E) = \mu_1(E) + \mu_2(E)$ by $\mu(E) = \mu_1(E) \cdot \mu_2(E)$, shall we get a similar result? Justify your answer.

6.7. Let $(X, \mathbf{F}, \mu)$ be a monotone measure space and μ be finite, continuous, and autocontinuous. Prove that

$$\overline{\lim_n}\, \mu(A_n \cup B_n) = \overline{\lim_n}\, \mu(A_n)$$

whenever $\{A_n\} \subset \mathbf{F}$, $\{B_n\} \subset \mathbf{F}$, $\mu(B_n) \to 0$, and $\overline{\lim}_n\, \mu(A_n) = \mu(\overline{\lim}_n\, A_n)$.

6.8. Let $(X, \mathbf{F}, \mu)$ be a monotone measure space and μ be finite, continuous, and autocontinuous.

Prove that

$$\mu(A \cup A_n \cup B_n) \to \mu(A)$$

whenever $\{A_n\} \subset \mathbf{F}$, $\{B_n\} \subset \mathbf{F}$, $A \in \mathbf{F}$, $\mu(A_n) \to 0$, $\mu(B_n) \to 0$.

6.9. Let μ be a set function defined on $(X, \mathbf{F})$. A class $\mathbf{C}$ of sets in $\mathbf{F}$ is called a *chain* iff it is fully ordered by set inclusion. A chain $\mathbf{C}$ is called *μ-bounded* iff there exists $M > 0$ such that $|\mu(C)| \leq M$ for any $C \in \mathbf{C}$. A set function μ is called *local-uniformly autocontinuous from above* (or *from below*) iff it is autocontinuous from above (or from below), and for every μ-bounded chain $\mathbf{C} \subset \mathbf{F}$ and every $\varepsilon > 0$ there exists $\delta = \delta\,(\mathbf{C}, \varepsilon) > 0$ such that

$$\mu(E) - \varepsilon \leq \mu(E \cup F) \leq \mu(E) + \varepsilon$$
$$(\text{or } \mu(E) - \varepsilon \leq \mu(E - F) \leq \mu(E) + \varepsilon)$$

whenever $E \in \mathbf{C}$, $F \in \mathbf{F}$, $E \cap F = \emptyset$ (or $E \subset F$), $|\mu(F)| \leq \delta$; μ is called *local-uniformly autocontinuous* iff it is both local-uniformly autocontinuous from above and from below. Prove that, if μ is a monotone measure, then local-uniform autocontinuity is equivalent to autocontinuity for μ.

6.10. Let $X = \{1, 2, ...\}$ and $\mathbf{F} = \mathbf{P}(X)$. Construct an autocontinuous monotone measure on $(X, \mathbf{F})$ that is not uniformly autocontinuous. (Hint: $\mu(E) = (\sum_{i \in E} 1/i)^2$ for any $E \in \mathbf{F}$.)

6.11. By using the concept of quasi-measure, construct an example of a monotone measure on $([0, 1], \mathbf{B} \cap [0, 1])$ which is autocontinuous but not uniformly autocontinuous.

6.12. Can you construct a finite monotone measure on a countable measurable space that is autocontinuous but not uniformly autocontinuous? Justify your answer.

Chapter 7
Measurable Functions on Monotone Measure Spaces

7.1 Measurable Functions

In this chapter, let $(X, \mathbf{F})$ be a measurable space, μ: $\mathbf{F} \to [0, \infty]$ be a monotone measure (or continuous, semicontinuous monotone measure), and **B** be the Borel field on $(-\infty, \infty)$.

Definition 7.1. A real-valued function f: $X \to (-\infty, \infty)$ on X is **F**-*measurable* (or simply "*measurable*" when there is no confusion) iff

$$f^{-1}(B) = \{x|f(x) \in B\} \in \mathbf{F}$$

for any Borel set $B \in$ B. The set of all **F**-measurable functions is denoted by **G**.

Theorem 7.1. *If* $f : X \to (-\infty, \infty)$ *is a real-valued function, then the following statements are equivalent:*

(1) f is measurable;
(2) $\{x|f(x) \geq \alpha\} \in \mathbf{F}$ for any $\alpha \in (-\infty, \infty)$;
(3) $\{x|f(x) > \alpha\} \in \mathbf{F}$ for any $\alpha \in (-\infty, \infty)$;
(4) $\{x|f(x) \leq \alpha\} \in \mathbf{F}$ for any $\alpha \in (-\infty, \infty)$;
(5) $\{x|f(x) < \alpha\} \in \mathbf{F}$ for any $\alpha \in (-\infty, \infty)$.

Proof. (1) $\Rightarrow$ (2): $\{x|f(x) \geq \alpha\} = f^{-1}([\alpha, \infty))$, and $[\alpha, \infty)$ is a Borel set.
(2) $\Rightarrow$ (1): If $\{x|f(x) \geq \alpha\} \in \mathbf{F}$ for any $\alpha \in (-\infty, \infty)$, then $f^{-1}(B) \in \mathbf{F}$ for any $B \in \{[\alpha, \infty)|\alpha \in (-\infty, \infty)\}$. Denoting $\mathbf{A} = \{B|f^{-1}(B) \in \mathbf{F}\}$, $\mathbf{C} = \{[\alpha, \infty)|\alpha \in (-\infty, \infty)\}$, we have $\mathbf{A} \supset \mathbf{C}$. Given any $B \in \mathbf{A}$, it follows that $\overline{B} \in \mathbf{A}$, since $f^{-1}(\overline{B}) = \overline{f^{-1}(B)} \in \mathbf{F}$; that is, **A** is closed under the formation of complements. Similarly, given any $\{B_n\} \subset \mathbf{A}$, it follows that $\bigcup_{n=1}^{\infty} B \in \mathbf{A}$ since $f^{-1}(\bigcup_{n=1}^{\infty} B_n) = \bigcup_{n=1}^{\infty} f^{-1}(B_n) \in \mathbf{F}$, that is, **A** is closed under the formation of countable unions. Hence, **A** is a σ-algebra, and consequently, $\mathbf{A} \supset \mathbf{F}(\mathbf{C}) = \mathbf{B}$. This shows that f is a measurable function.

The proof of the rest is similar. □

Corollary 7.1. *If f is a measurable function, then* $\{x|f(x) = \alpha\} \in \mathbf{F}$ *for any* $\alpha \in (-\infty, \infty)$.

Z. Wang, G.J. Klir, *Generalized Measure Theory*,
DOI: 10.1007/978-0-387-76852-6_7, © Springer Science+Business Media, LLC 2009

Definition 7.2. Let R^n be the n-dimensional Euclidean space and let

$$\mathbf{S}^{(n)} = \left\{ \prod_{i-1}^{n} [a_i, b_i) | -\infty < a_i \leq b_i < \infty, i = 1, 2, \cdots, n \right\}$$

The σ-algebra $\mathbf{B}^{(n)} = \mathbf{F}(\mathbf{S}^{(n)})$ is called the *Borel field* on R^n and the sets in $\mathbf{B}^{(n)}$ are called *(n-dimensional) Borel sets. A function* f: $R^n \to R$ is called an *(n-ary) Borel function* iff it is a measurable function on the measurable space $(R^n, \mathbf{B}^{(n)})$.

Theorem 7.2. *Let $f_1, \ldots, f_n$ be measurable functions. If g: $R^n \to R$ is a Borel function, then $g(f_1, \ldots, f_n)$ is a measurable function.*

Proof. For any Borel set $B \subset (-\infty, \infty)$,

$$\begin{aligned}[g(f_1, \ldots, f_n)]^{-1}(B) &= \{x | g(f_1(x), \ldots, f_n(x)) \in B\} \\ &= \{x | (f_1(x), \ldots, f_n(x)) \in g^{-1}(B)\}.\end{aligned}$$

Since, for any $E = \prod_{i=1}^{n} [a_i, b_i) \in \mathbf{S}^{(n)}$,

$$\{x | (f_1(x), \cdots, f_n(x)) \in E\} = \bigcap_{i=1}^{n} \{x | f_i(x) \in [a_i, b_i)\} \in \mathbf{F}$$

by applying the method similarly used in the proof of Theorem 7.1, we have

$$\{x | (f_1(x), \cdots, f_n(x)) \in F\} \in \mathbf{F}$$

for any $F \in \mathbf{B}^{(n)}$. As g is a Borel function, $g^{-1}(B) \in \mathbf{B}^{(n)}$ for any Borel set $B \subset (-\infty, \infty)$. Thus, we have

$$\{x | (f_1(x), \cdots, f_n(x)) \in g^{-1}(B)\} \in \mathbf{F}$$

for any Borel set B $\subset (-\infty, \infty)$. Hence, $g(f_1, \ldots, f_n)$ is measurable. □

As a special case of Theorem 7.2, if f_1 and f_2 are measurable, and $\alpha \in (-\infty, \infty)$ is a constant, then $\alpha f_1, f_1 + f_2, f_1 - f_2, |f_1|, f_1 \cdot f_2, |f_1|^\alpha, f_1 \vee f_2, f_1 \wedge f_2$, and the constant α, all of these are measurable (this can also be proven directly). Furthermore, we have

$$\{x | f_1(x) = f_2(x)\} = \{x | f_1(x) - f_2(x) = 0\} \in \mathbf{F}.$$

Theorem 7.3. *If $\{f_n\}$ is a sequence of measurable functions, and*

$$h(x) = \sup_n \{f_n(x)\},$$

$$g(x) = \inf_n \{f_n(x)\},$$

for any $x \in X$, then h and g are measurable.

Proof. By using Theorem 7.1, for any $\alpha \in (-\infty, \infty)$,

$$\{x|h(x) > \alpha\} = \{x|\sup_n\{f_n(x)\} > \alpha\} = \bigcup_{n=1}^{\infty}\{x|f_n(x) > \alpha\} \in \mathbf{F}$$

and

$$\{x|g(x) \geq \alpha\} = \{x|\inf_n\{f_n(x)\} \geq \alpha\} = \bigcap_{n=1}^{\infty}\{x|f_n(x) \geq \alpha\} \in \mathbf{F}.$$

Thus, h and g are measurable. □

Corollary 7.2. *If $\{f_n\}$ is a sequence of measurable functions, and*

$$\bar{f}(x) = \overline{\lim_n} f_n(x),$$

$$\underline{f}(x) = \underline{\lim_n} f_n(x),$$

then $\bar{f}$ and $\underline{f}$ are measurable. Furthermore, if $\lim_n f_n$ exists, then, it is measurable as well.

Proof. Since

$$\bar{f}(x) = \inf_m \sup_{n\geq m}\{f_n(x)\} \quad \text{and} \quad \underline{f}(x) = \sup_m \inf_{n\geq m}\{f_n(x)\},$$

the conclusions issue from the above theorem. □

In this chapter, we consider only measurable functions that are nonnegative, and symbols $f, f_1, f_2, \ldots, f_n, \ldots$ are used to indicate nonnegative measurable functions. The class of all nonnegative measurable functions is denoted by **G**. Most results hereafter can be generalized, without any essential difficulty, to the case in which the measurable functions are extended real-valued.

7.2 "Almost" and "Pseudo-Almost"

The definition of a measurable function on a continuous monotone measure (or semicontinuous monotone measure) space $(X, \mathbf{F}, \mu)$ is identical with classical measure theory, and, consequently, it does not relate to the set function μ; however, aspects of the set function must be considered when properties of measurable functions are discussed. For example, if f is a measurable function on a finite monotone measure space, what is the meaning of the statement "f is equal to zero almost everywhere"?

In probability theory, the statement "a random variable ξ is equal to 0 with probability 1" is equivalent to the statement "a random variable ξ is not equal to 0 with probability 0," because the probability measures possess additivity; that is, if p is a probability measure, then

$$p(E) + p(\overline{E}) = p(E \cup \overline{E}) = 1$$

for any event E. Since the monotone measures generally lose the additivity, the concept "almost everywhere" splits naturally into two different concepts, "almost everywhere" and "pseudo-almost everywhere" on monotone measure space, as indicated in the following definition.

Definition 7.3. Let $A \in \mathbf{F}$, and let P be a proposition with respect to points in A. If there exists $E \in \mathbf{F}$ with $\mu(E) = 0$ such that P is true on $A - E$, then we say "P is *almost everywhere* true on A." If there exists $F \in \mathbf{F}$ with $\mu(A - F) = \mu(A)$ such that P is true on $A - F$, then we say "P is *pseudo-almost everywhere* true on A."

We denote "almost everywhere" and "pseudo-almost everywhere" by "a.e." and "p.a.e.," respectively, and denote "$\{f_n\}$ converges to f a.e." (or "$\{f_n\}$ converges to f p.a.e.") by "$f_n \xrightarrow{\text{a.e.}} f$" (or "$f_n \xrightarrow{\text{p.a.e.}} f$", respectively).

Example 7.1. Let $X = \{0,1\}$, $\mathbf{F} = \mathbf{P}(X)$, and

$$\mu(E) = \begin{cases} 1 & \text{if } E \neq \emptyset \\ 0 & \text{if } E = \emptyset \end{cases}$$

for any $E \in \mathbf{F}$. If we define a measurable function sequence on $(X, \mathbf{F}, \mu)$ as follows:

$$f_n(x) = \begin{cases} 1 - 1/n & \text{if } x = 1 \\ 1/n & \text{if } x = 0, \end{cases} \qquad n = 1, 2, \ldots,$$

then both $f_n \xrightarrow{\text{p.a.e.}} 0$ and $f_n \xrightarrow{\text{p.a.e.}} 1$, but neither $f_n \xrightarrow{\text{a.e.}} 0$ nor $f_n \xrightarrow{\text{a.e.}} 1$.

Example 7.2. Let $X = \{0, 1\}$, $\mathbf{F} = \mathbf{P}(X)$, and

$$\mu(E) = \begin{cases} 1 & \text{if } E = X \\ 0 & \text{if } E \neq X \end{cases}$$

for any $E \in \mathbf{F}$. For the measurable function sequence given in Example 7.1, we have both $f_n \xrightarrow{\text{a.e.}} 0$ and $f_n \xrightarrow{\text{a.e.}} 1$, but neither $f_n \xrightarrow{\text{p.a.e.}} 0$ nor $f_n \xrightarrow{\text{p.a.e.}} 1$.

From these examples, we observe that this case is vastly different from that in the classical measure theory: If both $f_n \xrightarrow{\text{a.e.}} f$ and $f_n \xrightarrow{\text{a.e.}} f'$, then $f = f'$ a.e. But now, in Example 7.1 and Example 7.2, the limit functions 1 and 0 are not equal everywhere, of course—they are neither equal a.e. nor equal p.a.e.

We should note that the situations of the concepts "a.e." and "p.a.e." are not quite symmetric. In fact, if a proposition P is true a.e. on $A \in \mathbf{F}$, then it is also true a.e. on any subset of A that belongs to **F**; but such a statement is not always valid when we replace "a.e." with "p.a.e.," as the following example shows.

Example 7.3. Let $X = \{a,b,c\}$, $\mathbf{F} = \mathbf{P}(X)$, μ be given by

$$\mu(E) = \begin{cases} |E| & \text{if } E \neq \{a,b\} \\ 3 & \text{if } E = \{a,b\} \end{cases}$$

for any $E \in \mathbf{F}$, and

$$f(x) = \begin{cases} 0 & \text{if } x \in \{a,b\} \\ 1 & \text{if } x = c. \end{cases}$$

It is easy to verify that μ is a monotone measure, and

$$\mu(\{x|\, f(x) = 0, x \in X\}) = \mu(\{a,b\}) = 3 = \mu(X).$$

So, $f = 0$ on X p.a.e. But

$$\mu(\{x|\, f(x) = 0, x \in \{a,\ c\}\}) = \mu(\{a\}) = 1 \neq \mu(\{a,\ c\}) = 2.$$

So, the statement "$f = 0$ on $\{a, c\}$ p.a.e." is not true.

The other related concepts, such as "almost uniform convergence" and "convergence in measure" for measurable function sequences, split on monotone measure spaces as well.

Definition 7.4. Let $A \in \mathbf{F}, f \in \mathbf{G}$, $\{f_n\} \subset \mathbf{G}$. If there exists $\{E_k\} \subset \mathbf{F}$ with $\lim_k \mu(E_k) = 0$ such that $\{f_n\}$ converges to f on $A - E_k$ uniformly for any fixed $k = 1, 2, \ldots$, then we say that $\{f_n\}$ converges to f on A *almost uniformly* and denote it by $f_n \xrightarrow{\text{a.u.}} f$. If there exists $\{F_k\} \subset \mathbf{F}$ with $\lim_k \mu(A - F_k) = \mu(A)$ such that $\{f_n\}$ converges to f on $A - F_k$ uniformly for any fixed $k = 1, 2, \ldots$, then we say that $\{f_n\}$ converges to f on A *pseudo-almost uniformly* and denote it by $f_n \xrightarrow{\text{p.a.u.}} f$.

Definition 7.5. Let $A \in \mathbf{F}, f \in \mathbf{G}$, and $\{f_n\} \subset \mathbf{G}$. If

$$\lim_n \mu(\{x||f_n(x) - f(x)| \geq \varepsilon\} \cap A) = 0$$

for any given $\varepsilon > 0$, then we say that $\{f_n\}$ converges *in* μ (or, converges *in measure* if there is no confusion) to f on A, and denote it by $f_n \xrightarrow{\mu} f$ on A. If

$$\lim_n \mu(\{x||\, f_n(x) - f(x)| < \varepsilon\} \cap A) = \mu(A)$$

for any given $\varepsilon > 0$, then we say $\{f_n\}$ converges *pseudo-in* μ (or, converges *pseudo-in measure*) to f on A, and denote it by $f_n \xrightarrow{\text{p}.\mu} f$ on A.

In the above three definitions, when $A = X$, we can omit "on A" from the statements.

Example 7.4. Let $X = [0, \infty)$, $\mathbf{F} = \mathbf{B}_+$, and μ be the Lebesgue measure, where $\mathbf{B}_+$ is the class of all Borel sets in $[0, \infty)$. If we take $f_n(x) = x/n, n = 1, 2, \ldots$, and $f(x) = 0$ for any $x \in X$, then we have

$$f_n \xrightarrow{\text{p.a.u.}} f,$$

but $\{f_n\}$ does not converge to f on X almost uniformly. Also, we have

$$f_n \xrightarrow{\text{p.}\mu} f,$$

but $\{f_n\}$ does not converge in μ to f on X.

7.3 Relation Among Convergences of Measurable Function Sequence

The new concepts introduced in Section 7.2 complicate the relation among the several convergences of the measurable function sequence on a continuous monotone measure space or a semicontinuous monotone measure space. If only three concepts (a.e. convergence, a.u. convergence, and convergence in measure) are considered in classical measure theory, we should discuss six implication relations among them. Three of these relations are described by Egoroff's theorem, Lebesgue's theorem, and Riesz's theorem. But now, in monotone measure theory, since each convergence concept splits into two, there are 30 implication relations we should discuss. Using the structural characteristics of set function, which are introduced in Chapter 6, we examine the most important relations in this section.

Theorem 7.4. *For any $A \in \mathbf{F}$ and any proposition P with respect to the points in A, P is true on A p.a.e. whenever P is true on A a.e. if and only if μ is null-additive.*

Proof. Sufficiency: Let μ be null-additive. If P is true on A a.e. then there exists $E \in \mathbf{F}$ with $\mu(E) = 0$ such that $P(x)$ is true for any $x \in A - E$. By null-additivity and Theorem 6.2 we have $\mu(A - E) = \mu(A)$. So P is true on A p.a.e.

Necessity: For any $A \in \mathbf{F}$, $E \in \mathbf{F}$ with $\mu(E) = 0$, take "$x \in A - E$" as a proposition $P(x)$. Obviously, P is true on A a.e. If it implies that P is true on A p.a.e., then there exists $F \in \mathbf{F}$ with $\mu(A - F) = \mu(A)$ such that $P(x)$ is true for any $x \in A - F$. That is, $x \in A - F$ implies $x \in A - E$ and therefore

$$A - E \supset A - F.$$

By the monotonicity of μ we have

$$\mu(A - E) = \mu(A)$$

and from Theorem 6.2 we know that μ is null-additive. □

Corollary 7.3. *Let $A \in \mathbf{F}, f \in \mathbf{G}, \{f_n\} \subset \mathbf{G}$, and μ be null-additive. If $f_n \xrightarrow{\text{a.e.}} f$ on A, then $f_n \xrightarrow{\text{p.a.e.}} f$ on A.*

Theorem 7.5. *Let $A \in \mathbf{F}, f \in \mathbf{G}, \{f_n\} \subset \mathbf{G}$, and μ be autocontinuous from below. If $f_n \xrightarrow{\text{a.}\mu\text{.}} f$ on A, then $f_n \xrightarrow{\text{p.a.e.}} f$ on A.*

Proof. If $f_n \xrightarrow{\text{a.u.}} f$ on A, then there exists $\{E_k\} \subset \mathbf{F}$ with $\lim_k \mu(E_k) = 0$ such that $\{f_n\}$ converges to f *on* $A - E_k$ uniformly for any $k = 1, 2, \ldots$. Since μ is autocontinuous from below, we have $\lim_k \mu(A - E_k) = \mu(A)$ and consequently, $f_n \xrightarrow{\text{p.a.u.}} f$ on A. □

Theorem 7.6. *For any $A \in \mathbf{F}$, and for any $f \in \mathbf{G}$* and $\{f_n\} \subset \mathbf{G}$, $f_n \xrightarrow{\text{p.}\mu} f$ *on A whenever $f_n \xrightarrow{\mu} f$ on A if and only if μ is autocontinuous from below.*

Proof. Sufficiency: Let μ be autocontinuous from below. If $f_n \xrightarrow{\mu} f$ on A, then for any given $\varepsilon > 0$, we have

$$\lim_n \mu(\{x||f_n(x) - f(x)| \geq \varepsilon\} \cap A) = 0.$$

Since μ is autocontinuous from below, we have

$$\begin{aligned}\lim_n \mu(\{x||f_n(x) - f(x)| < \varepsilon\} \cap A) &= \lim_n \mu(A - \{x||f_n(x) - f(x)| \geq \varepsilon\} \cap A)) \\ &= \mu(A).\end{aligned}$$

So, $f_n \xrightarrow{\text{p.}\mu} f$ on A.

Necessity: For any $A \in \mathbf{F}$ and any $\{B_n\} \subset \mathbf{F}$ with $\lim_n \mu(B_n) = 0$, we define a measurable function sequence $\{f_n\}$ by

$$f_n(x) = \begin{cases} 0 & \text{if } x \in \overline{B}_n \\ 1 & \text{if } x \in B_n \end{cases}$$

for any $n = 1, 2, \ldots$. It is easy to see that $f_n \xrightarrow{\mu} 0$ on A. If it implies $f_n \xrightarrow{\text{p.}\mu} 0$ on A, then for $\varepsilon = 1 > 0$, we have

$$\lim_n \mu(\{x|| f_n(x)| < 1\} \cap A) = \mu(A).$$

As

$$\{x||f_n(x)| < 1\} \cap A = \overline{B}_n \cap A = A - B_n,$$

so

$$\lim_n \mu(A - B_n) = \mu(A).$$

This shows that μ is autocontinuous from below. □

The validity of Theorems 7.4–7.6 is independent of the continuity of μ.

The following is a generalization of Egoroff's theorem from classical measure space to monotone measure space.

Theorem 7.7. *Let μ be a continuous monotone measure, $A \in \mathbf{F}$, and $\mu(A) < \infty$. If $f_n \to f$ on A everywhere, then both $f_n \xrightarrow{\text{a.u.}} f$ and $f_n \xrightarrow{\text{p.a.u.}} f$ on A.*

Proof. There is no loss of generality in assuming that $A = X$ and μ is finite.

If we denote

$$E_n^m = \bigcap_{i=n}^{\infty} \{x \,|\, |f_i(x) - f(x)| < 1/m\},$$

for any $m = 1, 2, \ldots$, then $E_1{}^m \subset E_2{}^m \subset, \ldots$. The set of all those points that are such that $\{f_n(x)\}$ converges to $f(x)$ is

$$\bigcap_{m=1}^{\infty} \bigcup_{n=1}^{\infty} E_n^m.$$

If $f_n \to f$ everywhere, then $\bigcup_{n=1}^{\infty} E_n^m = X$ for any $m = 1, 2, \ldots$. That is, $E_n^m \nearrow X$ as $n \to \infty$, and, therefore, $\overline{E_n^m} \searrow \emptyset$ as $n \to \infty$, for any fixed $m = 1, 2, \ldots$. Given $\varepsilon > 0$ arbitrarily, by using the continuity from above and the finiteness of μ, there exists n_1 such that $\mu(\overline{E_{n_1}^1}) < \varepsilon/2$; for this n_1, there exists n_2 such that

$$\mu(\overline{E_{n_1}^1} \cup \overline{E_{n_2}^2}) < \varepsilon/2 + \varepsilon/2^2 = \tfrac{3}{4}\varepsilon;$$

and so on. Generally, there exists $n_1, n_2, \ldots n_k$, such that $\mu(\bigcup_{i=1}^{k} \overline{E_{n_i}^i}) < \sum_{i=1}^{k} \varepsilon/2^i = (1 - 1/2^k)\varepsilon < \varepsilon$. Hence, we obtain a number sequence $\{n_i\}$ and a set sequence $\{\overline{E_{n_i}^i}\}$. By using the continuity from below of μ, we know that

$$\mu\left(\bigcup_{i=1}^{\infty} \overline{E_{n_i}^i}\right) \leq \varepsilon.$$

Now, we just need to prove that $\{f_n\}$ converges to f on $\bigcap_{i=1}^{\infty} E_{n_i}^i$ uniformly. For any given $\delta > 0$, we take $i_0 > 1/\delta$. If $x \in \bigcap_{i=1}^{\infty} E_{n_i}^i$, then, since $x \in E_{n_{i_0}}^{i_0}$, we have

$$|f_i(x) - f(x)| < 1/i_0 < \delta$$

whenever $i \geq n_{i_0}$. Thus, we have proved that $f_n \xrightarrow{\text{a.u.}} f$.

In a similar way, we can prove that $f_n \xrightarrow{\text{p.a.u.}} f$ on A. □

The following example shows that the result in Theorem 7.7 may not be true when $\mu(A) = \infty$.

Example 7.5. Let monotone measure space $(X, \mathbf{F}, \mu)$ and functions, $f, f_1, f_2, \ldots$ be the same as in Example 7.4. We have $\mu(X) = \infty$ and $f_n \to f$ on X everywhere. However, as pointed out in Example 7.4, $\{f_n\}$ does not converge to f on X almost uniformly.

Corollary 7.4. *Let μ be a continuous monotone measure, $A \in \mathbf{F}$, $\mu(A) < \infty$, and μ be null-additive. If $f_n \xrightarrow{\text{a.e.}} f$ on A, then both $f_n \xrightarrow{\text{a.u.}} f$ and $f_n \xrightarrow{\text{p.a.u.}} f$ on A.*

The following theorem gives an inverse conclusion of Corollary 7.4.

Theorem 7.8. *Let $A \in \mathbf{F}$. If $f_n \xrightarrow{\text{a.u.}} f$ (or $f_n \xrightarrow{\text{p.a.u.}} f$) on A; then $f_n \xrightarrow{\text{a.e.}} f$ (or $f_n \xrightarrow{\text{p.a.e.}} f$, respectively) on A.*

Proof. If $f_n \xrightarrow{\text{a.u.}} f$ on A, then there exists $\{E_k\} \subset \mathbf{F}$ with $\lim_k \mu(E_k) = 0$ such that $\{f_n\}$ converges to f on $A - E_k$ (even uniformly) for any $k = 1, 2, \ldots$. Take $E = \cap_{k=1}^{\infty} E_k$. Since $E \subset E_k$ for every k, by the monotonicity of μ, we have $\mu(E) = 0$. Thus, for any $x \in A - E$, there exists some E_k such that $x \in A - E_k$, and therefore, $\{f_n(x)\}$ converges to $f(x)$. This shows $f_n \xrightarrow{\text{a.e.}} f$ on A.

The proof that $f_n \xrightarrow{\text{p.a.e.}} f$ on A is similar. □

The validity of Theorem 7.8 is also independent of the continuity of μ.

Now, we give two forms of generalization on semicontinuous monotone measure spaces for Lebesgue's theorem in classical measure theory.

Theorem 7.9. *Let $A \in \mathbf{F}$. If $f_n \xrightarrow{\text{a.e.}} f$ on A, μ is continuous from above, and $\mu(A) < \infty$, then $f_n \xrightarrow{\mu} f$ on A; if $f_n \xrightarrow{\text{p.a.e.}} f$ on A and μ is continuous from below, then $f_n \xrightarrow{\text{p.}\mu} f$ on A.*

Proof. We only prove the second conclusion; the proof of the first one is similar.

If $f_n \xrightarrow{\text{p.a.e.}} f$ on A, then there exists $B \in \mathbf{F}$ with $B \subset A$ and $\mu(B) = \mu(A)$ such that for any $x \in B$, $\lim_n f_n(x) = f(x)$. Thus, for any given $\varepsilon > 0$ and $x \in B$, there exists $N(x)$ such that

$$|f_n(x) - f(x)| < \varepsilon$$

whenever $n \geq N(x)$. If we write

$$A_k = \{x | N(x) \leq k\} \cap B,$$

then

$$A_k \nearrow \bigcup_{k=1}^{\infty} A_k = B$$

Since

$$\{x | |f_n(x) - f(x)| < \varepsilon\} \cap A \supset A_n,$$

we have

$$B \supset \{x | |f_n(x) - f(x)| < \varepsilon\} \cap A \cap B \supset A_n \cap B = A_n \nearrow B$$

and, therefore,

$$\lim_n \mu(\{x||f_n(x) - f(x)| < \varepsilon\} \cap A \cap B) = \mu(B).$$

Consequently,

$$\begin{aligned}\mu(A) &\geq \lim_n \mu(\{x||f_n(x) - f(x)| < \varepsilon\} \cap A) \\ &\geq \lim_n \mu(\{x||f_n(x) - f(x)| < \varepsilon\} \cap A \cap B) \\ &= \mu(B) \\ &= \mu(A).\end{aligned}$$

This shows that $f_n \xrightarrow{\text{p.}\mu} f$ on A. □

The next theorem gives inverse conclusions to the above theorem. These conclusions generalize Riesz's theorem.

Theorem 7.10. *Let $A \in \mathbf{F}$, μ be a lower semicontinuous monotone measure that is autocontinuous from above. If $f_n \xrightarrow{\mu} f$ on A, then there exists some subsequence $\{f_{n_i}\}$ of $\{f_n\}$ such that both $\{f_{n_i}\} \xrightarrow{\text{a.e.}} f$ and $\{f_{n_i}\} \xrightarrow{\text{p.a.e.}} f$ on A.*

Proof. We may assume $A = X$ without any loss of generality. If $f_n \xrightarrow{\mu} f$, then we have

$$\lim_n \mu(\{x||f_n(x) - f(x)| \geq 1/k\}) = 0$$

for any $k = 1, 2,\ldots$. So, there exists n_k such that

$$\mu(\{x||f_{n_k}(x) - f(x)| \geq 1/k\}) < 1/k.$$

We may assume $n_{k+1} > n_k$ for any $k = 1, 2,\ldots$. If we write

$$E_k = \{x||f_{n_k}(x) - f(x)| \geq 1/k\},$$

then

$$\lim_k \mu(E_k) = 0.$$

Since μ is autocontinuous from above, by Theorem 6.10 there exists some subsequence $\{E_{k_i}\}$ of $\{E_k\}$ such that

$$\mu(\overline{\lim_i}\,\mu(E_{k_i}) = 0.$$

Now we shall prove that $f_{n_{k_i}}$ converges to f on $X - \overline{\lim_i}\ E_{k_i}$. In fact, for any $x \in X - \overline{\lim_i}\ E_{k_i}$, since $x \in \bigcup_{j=1}^{\infty} \bigcap_{i=j}^{\infty} \overline{E_{k_i}}$, there exists $j(x)$ such that $x \in \bigcap_{i=j(x)}^{\infty} \overline{E_{k_i}}$, namely,

$$|f_{n_{k_i}}(x) - f(x)| < 1/k_i,$$

for every $i \geq j(x)$. Thus, for any given $\varepsilon > 0$, taking i_0 such that $1/k_{n_0} < \varepsilon$, we have

$$|f_{n_{k_i}}(x) - f(x)| < \tfrac{1}{k_i} \leq \tfrac{1}{k_{i_0}} < \varepsilon$$

whenever $i \geq j(x) \vee i_0$. This shows that

$$f_{n_{k_i}} \xrightarrow{\text{a.e.}} f.$$

As μ is null-additive, by Theorem 7.4 we have

$$f_{n_{k_i}} \xrightarrow{\text{p.a.e.}} f$$

as well. □

Corollary 7.5. *Let $A \in \mathbf{F}$, μ be a continuous monotone measure that is autocontinuous from below, and $\mu(A) < \infty$. If $f_n \xrightarrow{\mu} f$ on A then there exists some subsequence $\{f_{n_i}\}$ of $\{f_n\}$ such that both $f_{n_i} \xrightarrow{\text{a.e.}} f$ and $f_{n_i} \xrightarrow{\text{p.a.e.}} f$ on A.*

Proof. Since the autocontinuity from below is equivalent to the autocontinuity from above when μ is a finite continuous monotone measure, if we regard A as X, the conclusion follows from Theorem 7.10. □

Theorem 7.11. *Let $A \in \mathbf{F}$. If $f_n \xrightarrow{\text{a.u.}} f$ (or $f_n \xrightarrow{\text{p.a.u.}} f$) on A, then $f_n \xrightarrow{\mu} f$ (or $f_n \xrightarrow{\text{p.}\mu} f$, respectively) on A.*

Proof. If $f_n \xrightarrow{\text{a.u.}} f$ on A, then for any $\varepsilon > 0$ and $\delta > 0$ there exist $E \in \mathbf{F}$ with $\mu(E) < \delta$ and n_0 such that

$$|f_n(x) - f(x)| < \varepsilon$$

whenever $x \in A - E$ and $n \geq n_0$. So we have

$$\mu(\{x \,|\, |f_n(x) - f(x)| \geq \varepsilon\} \cap A) \leq \mu(E \cap A) \leq \mu(E) < \delta$$

for any $n \geq n_0$. This shows that $f_n \xrightarrow{\mu} f$ on A.

In a similar way, we can prove that $f_n \xrightarrow{\text{p.a.u.}} f$ on A implies $f_n \xrightarrow{p.\mu} f$ on A. □

7.4 Convergences of Measurable Function Sequence on Possibility Measure Spaces

Let π be a possibility measure on a measurable space $(X, \mathbf{F})$, where $\mathbf{F} = \mathbf{P}(X)$. We call $(X, \mathbf{P}(X), \pi)$ a *possibility measure space*. Since π is a finite upper semicontinuous monotone measure that is uniformly autocontinuous, the previous discussion in Chapters 6 and 7 works for the possibility measure space, assuming that we replace $f_n \xrightarrow{\mu} f$ with $f_n \xrightarrow{\pi} f$. Furthermore, taking advantage of the maxitivity of possibility measures, we can obtain rather interesting results.

Theorem 7.12. *Let $A \subset X$. Then, $f_n \xrightarrow{\pi} f$ on A is equivalent to $f_n \xrightarrow{\text{a.u.}} f$ on A.*

Proof. There is no loss of generality in assuming that $A = X$. The fact that $f_n \xrightarrow{\text{a.u.}} f$ implies $f_n \xrightarrow{\pi} f$ is guaranteed by Theorem 7.10 since possibility measure π is continuous from below as well as autocontinuous. Hence, we only need to prove that $f_n \xrightarrow{\pi} f$ implies $f_n \xrightarrow{\text{a.u.}} f$.

If $f_n \xrightarrow{\pi} f$, then for any positive integer i we have

$$\pi(\{x \| f_n(x) - f(x)| \geq 1/i\}) \to 0$$

as $n \to \infty$. That is, for any positive integer k, there exists n_{ik} such that

$$\pi(\{x \| f_n(x) - f(x)| \geq 1/i\}) < 1/k$$

as $n \geq n_{ik}$. Taking

$$E_k = \bigcup_{i=1}^{\infty} \bigcup_{n \geq n_{ik}} \{x \| f_n(x) - f(x)| \geq 1/i\},$$

we have

$$\pi(E_k) = \sup_{i \geq 1, n \geq n_{ik}} \pi(\{x \| f_n(x) - f(x)| \geq 1/i\}) \leq 1/k.$$

Now, we show that $\{f_n\}$ converges to f uniformly on $\overline{E}_k$. For any $\varepsilon > 0$, take i such that $1/i < \varepsilon$. If $x \notin E_k$, then $x \in \{x \| f_n(x) - f(x)| < 1/i\}$ for any $n \geq n_{ik}$; that is, there exists $n_0 = n_{ik}$ such that

$$|f_n(x) - f(x)| < 1/i < \varepsilon,$$

where $n \geq n_0$. The proof is now complete. □

By using Theorem 7.8 we immediately obtain the following corollary.

Corollary 7.6. *Let $A \subset X$. Then, $f_n \xrightarrow{\pi} f$ on A implies $f_n \xrightarrow{\text{a.e.}} f$ on A.*

Since, in general, a possibility measure is not continuous from above, we cannot get the inverse proposition of Corollary 7.6 by using Theorem 7.9. This is shown by the following counterexample.

Example 7.6. Let $X = (0, 1]$ and let a possibility measure π be defined as

$$\pi(E) = \begin{cases} 1 & \text{if } E \neq \emptyset \\ 0 & \text{if } E = \emptyset. \end{cases}$$

We take

$$f_n(x) = \begin{cases} 0 & \text{if } x \in (1/n, \ 1] \\ 1 & \text{otherwise.} \end{cases}$$

Then, $f_n \to 0$ everywhere on X, but $\{f_n\}$ does not converge to zero in measure π. In fact, taking $\varepsilon = 1/2$, we have

$$\pi(\{x|f_n(x) > 1/2\}) = \pi((0, 1/n]) = 1$$

for any n.

Theorem 7.13. *Let $A \subset X$. Then, $f_n \xrightarrow{\text{a.u.}} f$ on A is equivalent to $|f_n - f| \wedge p \to 0$ on A uniformly, where p is the possibility profile of π.*

Proof. As in the proof of Theorem 7.12, we can assume that $A = X$.

Suppose $f_n \xrightarrow{\text{a.u.}} f$. Then, for any given $\varepsilon > 0$, there exists a set $E \subset X$ with $\pi(E) < \varepsilon$ such that $|f_n - f| \to 0$ on $\overline{E}$ uniformly; that is, there exists $n(\varepsilon)$ such that $|f_n(x) - f(x)| < \varepsilon$ for any $x \notin E$ whenever $n \geq n(\varepsilon)$. Since $\pi(E) < \varepsilon$ implies $p(x) < \varepsilon$ for any $x \in E$, we have

$$|f_n(x) - f(x)| \wedge p(x) < \varepsilon$$

whenever $n \geq n(\varepsilon)$. This shows that

$$|f_n - f| \wedge p \to 0$$

uniformly.

Conversely, suppose $|f_n - f| \wedge p \to 0$ uniformly. Then, for any given positive integer k there exists n_k such that

$$|f_n(x) - f(x)| \wedge p(x) < 1/k$$

for any x whenever $n > n_k$. Denoting $E_k = \cup_{n \geq n_k}\{x||f_n(x) - f(x)| \geq 1/k\}$, we have $p(x) < 1/k$ for any E_k. If we take $F_i = \cup_{k \geq i} E_k$, then

$$\pi(F_i) = \sup_{x \in F_i} p(x) \leq 1/i.$$

Now, we show that $f_n \to f$ uniformly on $\overline{F}_i$ for each $i = 1, 2, \ldots$. Given an arbitrary $\varepsilon > 0$, take $k \geq i$ such that $1/k < \varepsilon$. For any $x \notin F_i$, we have $x \notin E_k$ and, therefore, $x \notin \{x \mid |f_n(x) - f(x)| \geq 1/k\}$ whenever $n \geq n_k$. That is, $|f_n(x) - f(x)| < 1/k < \varepsilon$ whenever $n \geq n_k$. The proof is now complete. □

Summing up the results presented in this section, we can characterize the relations among several convergences of a measurable function sequence on possibility measure spaces as follows:

$$f_n \xrightarrow{\pi} f \quad \Leftrightarrow \quad f_n \xrightarrow{\text{a.u.}} f \quad \Leftrightarrow \quad |f_n - f| \wedge p \xrightarrow{\text{u.}} 0 \quad \Rightarrow f_n \xrightarrow{\text{a.e.}} f,$$

where the symbol "$\xrightarrow{\text{u.}}$" means "converge uniformly."

The concepts of pseudo-convergences of a function sequence are unimportant on the possibility measure space.

Notes

7.1. The paper by Wang [1984] contains early discussions on convergences of measurable function sequences on monotone measure (fuzzy measure) spaces. After introducing the concept of "pseudo-almost," Wang [1985a] derived more results regarding the relationship among several types of convergences of measurable function sequences on the basis of the concepts of pseudo-autocontinuity and converse autocontinuity.

7.2. Some results presented in this chapter were generalized to fuzzy σ-algebra by Qiao [1990].

7.3. The convergences of measurable function sequences on possibility measure spaces were studied by Wang [1987].

Exercises

7.1. Let $(X, \mathbf{F})$ be a measurable space and let f_1 and f_2 be measurable functions. Without using Theorem 7.2, prove that the following functions are measurable:

$$cf_1 (c \text{ is a constant}); f_1 - f_2; f_1 + f_2; f_1 \vee f_2; f_1 \wedge f_2; |f_1|; f_1^2; f_1 \cdot f_2.$$

7.2. Let f be a measurable function on $(X, \mathbf{F})$. Prove that

$$\{x \mid f(x) = \alpha\} \in \mathbf{F}$$

for any $\alpha \in (-\infty, \infty)$.

7.3. Let f be a function defined on $(X, \mathbf{F})$. If $\{x|f(x) = \alpha\} \in \mathbf{F}$ for any $\alpha \in (-\infty, \infty)$, can you correctly assert that f is measurable? If you can, give a proof; if you cannot, give an example to justify your conclusion.

7.4. Let $\{f_n\}$ be a sequence of measurable functions on $(X, \mathbf{F})$. Prove that

$$\{x|\overline{\lim_n} f_n(x) = \underline{\lim_n} f_n(x)\} \in \mathbf{F}.$$

7.5. Let $\mathbf{G}$ be the class of all nonnegative finite measurable functions on a monotone measure space $(X, \mathbf{F}, \mu)$. Both $\overset{\text{a.e.}}{=}$ (almost everywhere equality) and $\overset{\text{p.a.e.}}{=}$ (pseudo-almost everywhere equality) are binary relations on $\mathbf{G}$. Prove that these relations are reflexive and symmetric, but not transitive in general.

7.6. Prove that the relation $\overset{\text{a.e.}}{=}$ is an equivalence relation on $\mathbf{G}$ (see Exercise 7.5) if and only if μ is weakly null-additive (see Exercise 6.1).

7.7. Can you find a condition such that the statement "P is true on A p.a.e." implies the statement "P is true on A a.e."?

7.8. Construct an example of a measurable function f defined on a monotone measure space $(X, \mathbf{F}\ \mu)$ in which "$f \overset{\text{a.e.}}{=} 0$" is true, but "$f \overset{\text{p.a.e}}{=} 0$" is not true.

7.9. Construct an example of a semicontinuous fuzzy measure space $(X, \mathbf{F}, \mu)$ and a sequence of measurable functions $\{f_n\}$ such that $\{f_n\}$ converges to some measurable function f almost everywhere, but does not converge to f pseudo-almost uniformly.

7.10. Let $X = (0, 1]$, $\mathbf{F}$ be the class of all Borel sets in X, and $\mu = m^2$, where m is the Lebesgue measure. Assume we order all rational numbers in X as follows:

$$x_1 = 1, x_2 = 1/2, x_3 = 1/3, x_4 = 2/3, x_5 = 1/4, x_6 = 3/4, x_7 = 1/5, x_8 = 2/5,\ x_9 = 3/5, x_{10} = 4/5, x_{11} = 1/6, x_{12} = 5/6, x_{13} = 1/7, x_{14} = 2/7, \ldots.$$

Furthermore, we define a sequence of measurable functions $\{f_n\}$ on $(X, \mathbf{F}, \mu)$ by

$$f_n(x) = \begin{cases} 1 & \text{if } |x - x_n| < 1/(2n)^{1/2} \\ 0 & \text{otherwise} \end{cases}$$

for $n = 1, 2, \ldots$. Prove that:

(a) μ is autocontinuous;

(b) $f_n \xrightarrow{\mu} 0$;

(c) f_n does not converge to 0 at any point in X.

Can you find a subsequence $\{f_{n_i}\}$ of $\{f_n\}$ such that $\{f_{n_i}\}$ converges to zero both almost everywhere and pseudo-almost everywhere?

7.11. Prove that if μ is a finite continuous monotone measure and $f_n \to f$ everywhere, then $f_n \xrightarrow{\text{p.a.u.}} f$.

7.12. Prove that if μ is a finite and null-additive continuous monotone measure, then $f_n \xrightarrow{\text{a.e.}} f$ implies $f_n \xrightarrow{\text{p.a.u.}} f$.

Chapter 8
Integration

8.1 The Lebesgue Integral

Let $(X, \mathbf{F}, \mu)$ be a measure space. That is, X is a nonempty set, $\mathbf{F}$ is a σ-algebra of subsets of X, and $\mu : \mathbf{F} \to [0, \infty]$ is a classical measure, which is nonnegative and σ-additive with $\mu(A) < \infty$ for at least one $A \in \mathbf{F}$. X is called the universal set and may not be finite. In this chapter we suppose that μ is σ-finite, that is, there exists $\{A_i\} \subset \mathbf{F}$, such that $\mu(A_i) < \infty$ for every $i = 1, 2, \ldots$ and $\cup_{i=1}^{\infty} A_i = X$.

Definition 8.1. Function $s : X \to (-\infty, \infty)$ having a form $\sum_{i=1}^{m} a_i \chi_{A_i}$ is called a *simple function*, where each a_i is a real constant, $A_i \in \mathbf{F}$, and χ_{A_i} is the characteristic function of A_i, $i = 1, 2, \ldots, m$.

Any simple function is measurable since each characteristic function of a measurable set is measurable and the linear combination of measurable functions is also measurable. For any given nonnegative measurable function $f : X \to [0, \infty)$, there exist some nondecreasing sequences of simple functions whose limit is f. For example, we can take

$$s_j = \sum_{i=1}^{j \cdot 2^j} \frac{i-1}{2^j} \chi_{A_{ji}},$$

where $A_{ji} = \{x | \frac{i-1}{2^j} \le f(x) < \frac{i}{2^j}\}$, $i = 1, 2, \ldots, j \cdot 2^j$, for $j = 1, 2, \ldots$. It is not difficult to verify that $\{s_j\}$ is nondecreasing and $\lim_{j \to \infty} s_j(x) = f(x)$ for every $x \in X$.

Such sequences are not unique. However, we may use any one of these sequences to define the Lebesgue integral of f on X with respect to μ as follows.

Definition 8.2. Let f be a nonnegative measurable function on X. The *Lebesgue integral* of f with respect to measure μ is defined as

$$\int f \, d\mu = \lim_{j \to \infty} \int s_j \, d\mu = \lim_{j \to \infty} \sum_{i}^{m_j} a_{ji} \, \mu(A_{ji}),$$

Z. Wang, G.J. Klir, *Generalized Measure Theory*,
DOI: 10.1007/978-0-387-76852-6_8, © Springer Science+Business Media, LLC 2009

where each $s_j = \sum_i^{m_j} a_{ji}\chi_{A_{ji}}$ is a simple function and $\{s_j\}$ is nondecreasing sequence with $\lim_{j\to\infty} s_j = f$.

Such a definition is unambiguous due to the σ-additivity of μ. That is, for any two sequences of nondecreasing simple functions, $\{s_j\}$ and $\{t_j\}$, with $\lim_{j\to\infty} s_j = \lim_{j\to\infty} t_j = f$, we have $\lim_{j\to\infty} \int s_j\, d\mu = \lim_{j\to\infty} \int t_j\, d\mu$. So, the Lebesgue integral is well defined for any nonnegative measurable function.

From Definition 8.2, if function f itself is a simple function expressed as $\sum_{i=1}^{m} a_i\chi_{A_i}$, then $\int f\, d\mu = \sum_{i=1}^{m} a_i\mu(A_i)$.

For any given measurable set A, replacing A_{ji} by $A_{ji} \cap A$ in Definition 8.2, we may define the Lebesgue integral of f with respect to μ on A, denoted as $\int_A f\, d\mu$. When $A = X$, we return to $\int f\, d\mu$ as given in Definition 8.2. As a special case where A is a Borel set on the real line, μ is the Lebesgue measure, and f is a Borel measurable function on A, the integral $\int_A f\, d\mu$ is called simply the Lebesgue integral of f on A.

Example 8.1. Let X be the closed unit interval [0, 1], $\mathbf{F}$ be the set of all Borel sets in [0,1], and μ be the Lebesgue measure. We have known that the set of all rational number in [0, 1], denoted by Q_0, is a countable set, and, therefore, $\mu(Q_0) = 0$. Function f is defined on [0, 1] as

$$f(x) = \begin{cases} 0 & \text{if } x \in Q_0 \\ 1 & \text{otherwise.} \end{cases}$$

Then, f is a nonnegative measurable function. Function f is not continuous at any point in [0, 1] and is not integrable on [0, 1] in the Riemann sense, that is, $\int_0^1 f(x)dx$ does not exist. However, its Lebesgue integral exists and it is calculated as follows:

$$\int_{[0,1]} f\, d\mu = 0 \cdot \mu([0,1] - Q_0) + 1 \cdot \mu(Q_0) = 0 \cdot 1 + 1 \cdot 0 = 0.$$

The Lebesgue integral has also an equivalent definition:

$$\int f\, d\mu = \sup\left\{ \sum_{i=1}^{m} a_i\mu(A_i) \,\middle|\, \sum_{i=1}^{m} a_i\chi_{A_i} \le f; a_i \in [0,\infty), A_i \in \mathbf{F},\ i = 1,2,\cdots,m; m \ge 1 \right\}.$$

For any measurable function that may be not nonnegative, let

$$f^+(x) = \begin{cases} f(x) & \text{if } f(x) \geq 0 \\ 0 & \text{if } f(x) < 0 \end{cases}$$

and

$$f^-(x) = \begin{cases} -f(x) & \text{if } f(x) \leq 0 \\ 0 & \text{if } f(x) > 0. \end{cases}$$

Both f^+ and f^- are nonnegative measurable functions. Then, the Lebesgue integral of f on X with respect to μ is defined by

$$\int f d\mu = \int f^+ d\mu - \int f^- d\mu$$

if not both terms on the right-hand side are infinite. In case $\mu(X) < \infty$ and f is lower bounded, integral $\int f d\mu$ can be also defined by

$$\int f d\mu = \int (f - m) d\mu + m \cdot \mu(X),$$

where m is a lower bound of f, i.e., $f(x) - m \geq 0, \forall x \in X$.

Consider any real-valued function f that is measurable on an interval $I = [a, b]$ with respect to the Borel field (the class of all Borel sets). If the Riemann integral $\int_a^b f(x)dx$ exists, then the corresponding Lebesgue integral exists as well, and

$$\int_I f d\mu = \int_a^b f(x)dx,$$

where μ is the Lebesgue measure. That is, the Lebesgue integral is a generalization of the Riemann integral.

If set function $\mu : \mathbf{F} \to (-\infty, \infty)$ is a finite classical signed measure, then μ can be decomposed as $\mu = \mu^+ - \mu^-$, where both μ^+ and μ^- are finite classical measures. Thus, the Lebesgue integral of measurable function f with respect to finite classical signed measure μ can be defined as

$$\int f d\mu = \int f d\mu^+ - \int f d\mu^-$$

if not both terms on the right-hand side are infinite.

When μ is finite, the Lebesgue integral has another equivalent definition:

$$\int f d\mu = \int_{-\infty}^0 [\mu(F_\alpha) - \mu(X)] d\alpha + \int_0^\infty \mu(F_\alpha) d\alpha \quad (8.1)$$

if not both terms on the right-hand side are infinite, where $F_\alpha = \{x | f(x) \geq \alpha\}$ for $\alpha \in (-\infty, \infty)$. Set F_α is measurable since function f is measurable. So, $\mu(F_\alpha)$ is well defined for all $\alpha \in (-\infty, \infty)$. The integrals on the right-hand side of the equality are Riemann integrals. By the finiteness of μ we know that $\mu(F_\alpha)$ is a function of α with bounded variation. Hence, these two Riemann integrals are well defined. When $f > 0$, Eq. (8.1) is reduced to

$$\int f d\mu = \int_0^\infty \mu(F_\alpha) d\alpha.$$

This equation is called the *transformation theorem* of the Lebesgue integral. It can be proved by showing that each summation $\sum_i^{m_j} a_{ji}\mu(A_{ji})$ in Definition 8.2 is just a Darboux's sum for the Riemann integral $\int_0^\infty \mu(F_\alpha)d\alpha$. In the special case that X is the real line, this equality has an intuitive geometrical meaning: Both integrals represent the area between the curve of f and the x-axis.

8.2 Properties of the Lebesgue Integral

Let $(X, \mathbf{F}, \mu)$ be a σ-finite measure space. Since μ is additive, the concepts of "almost everywhere" and "pseudo-almost everywhere" coincide with each other. In this case we have the following properties for the Lebesgue integral. The proof of the theorem is omitted here.

Theorem 8.1. *Let f and g be measurable functions on $(X, \mathbf{F}, \mu)$, A and B be measurable sets, and a be a real constant.*

(1) $\int_A 1\, d\mu = \mu(A)$;

(2) $\int_A f d\mu = \int f \cdot \chi_A d\mu$;

(3) *if* $f \leq g$ *on* A, *then* $\int_A f d\mu \leq \int_A g\, d\mu$;

(4) $\int_A |f| d\mu = 0$ *if and only if* $\mu(\{x | f(x) \neq 0\} \cap A) = 0$, *i.e.*, $f = 0$ *on* A *almost everywhere;*

(5) *if* $A \subset B$, *then* $\int_A f d\mu \leq \int_B f d\mu$;

(6) $\int_A af d\mu = a \int_A f d\mu$;

(7) $\int_A (f + g) d\mu = \int_A f d\mu + \int_A g\, d\mu$;

(8) *if* $A \cap B = \emptyset$, *then* $\int_A f d\mu + \int_B f d\mu = \int_{A \cup B} f d\mu$.

The properties showing in (6) and (7) of Theorems 8.1 are called the linearity of the Lebesgue integral.

Now, we turn to sequences of Lebesgue integrals. Let $\{f_n\}$ be a sequence of measurable functions and f be a measurable function on measure space $(X, \mathbf{F}, \mu)$. As is mentioned above, the statement " $\{f_n\}$ converges to f almost everywhere" coincides with the statement " $\{f_n\}$ converges to f pseudo-almost everywhere" due to the additivity of μ. Similarly, statement " $\{f_n\}$ converges to f in measure" coincides with statement " $\{f_n\}$ converges to f pseudo-in measure." So, in this section we only use the statements " $\{f_n\}$ converges to f almost everywhere" and " $\{f_n\}$ converges to f in measure," denoted by $f_n \xrightarrow{a.e.} f$ and $f_n \xrightarrow{\mu} f$, respectively, as in classical measure theory.

The principal convergence theorem for the sequence of Lebesgue integrals is the following bounded convergence theorem.

Theorem 8.2. *Let A be a measurable set. If $f_n \xrightarrow{\mu} f$ on A and there exists a nonnegative measurable function g satisfying $\int_A g\, d\mu < \infty$ such that $|f_n| \le g$ on A almost everywhere for all $n = 1, 2, \ldots$, then*

$$\lim_{n\to\infty} \int_A f_n\, d\mu = \int_A f\, d\mu.$$

Proof. Without any loss of generality, we may assume that $\mu(A) = M < \infty$ and $A = X$. For any given $\varepsilon > 0$, since $\int g d\mu = \int_0^\infty \mu(G_\alpha) d\alpha < \infty$ where $G_\alpha = \{x | g(x) \ge \alpha\}$ for $\alpha \in [0, \infty)$, there exists $N > 0$ such that $\int_{\frac{N}{2}}^\infty \mu(G_\alpha) d\alpha < \frac{\varepsilon}{4}$. From $f_n \xrightarrow{\mu} f$, by the Riesz theorem we know that there exists a subsequence of $\{f_n\}$, denoted by $\{f_{n_i}\}$, such that $f_{n_i} \xrightarrow{\text{a.e.}} f$. Furthermore, from the condition that $|f_n| \le g$ almost everywhere for all $n = 1, 2, \ldots$, we have $|f| \le g$ almost everywhere. Thus, $|f_n - f| \le 2g$ for all $n = 1, 2, \ldots$ almost everywhere. Since $f_n \xrightarrow{\mu} f$, for $\frac{\varepsilon}{4M}$, we can find n_0 such that $\mu(\{x | |f_n(x) - f(x)| \ge \frac{\varepsilon}{4M}\}) \le \frac{\varepsilon}{4N}$ for all $n \ge n_0$. Hence

$$\begin{aligned}
\left|\int f_n\, d\mu - \int f d\mu\right| &= \left|\int (f_n - f) d\mu\right| \le \int |f_n - f| d\mu = \int_0^\infty \mu(\{x | |f_n(x) - f(x)| \ge \alpha\}) d\alpha \\
&\le \int_0^{\frac{\varepsilon}{4M}} M\, d\alpha + \int_{\frac{\varepsilon}{4M}}^N \mu(\{x | |f_n(x) - f(x)| \ge \alpha\})\, d\alpha \\
&\quad + \int_N^\infty \mu(\{x | |f_n(x) - f(x)| \ge \alpha\})\, d\alpha \\
&\le \frac{\varepsilon}{4} + \int_0^N \mu(\{x | |f_n(x) - f(x)| \ge \frac{\varepsilon}{4M}\}) d\alpha + \int_N^\infty \mu(\{x | 2g(x) \ge \alpha\}) d\alpha \\
&\le \frac{\varepsilon}{4} + \int_0^N \frac{\varepsilon}{4N}\, d\alpha + \int_N^\infty \mu(G_{\alpha/2}) d\alpha \\
&\le \frac{\varepsilon}{4} + \frac{\varepsilon}{4} + 2\int_N^\infty \mu(G_{\alpha/2}) d(\alpha/2) \\
&\le \frac{\varepsilon}{4} + \frac{\varepsilon}{4} + 2\int_{\frac{N}{2}}^\infty \mu(G_\alpha) d\alpha \\
&\le \frac{\varepsilon}{4} + \frac{\varepsilon}{4} + \frac{\varepsilon}{2} \\
&= \varepsilon
\end{aligned}$$

for every $n \ge n_0$. This means that $\lim_{n\to\infty} \int f_n d\mu = \int f d\mu$. □

In case μ is finite, $f_n \xrightarrow{\text{a.e.}} f$ implies $f_n \xrightarrow{\mu} f$. So, we can obtain the following corollary where the finiteness of μ has been replaced with the σ-finiteness by the traditional stratagem.

Corollary 8.1. *Under the same condition assumed in Theorem 8.2, if* $f_n \xrightarrow{a.e.} f$ *and* $|f_n| \leq g$ *on A almost everywhere for all* $n = 0, 1, \ldots$, *then*

$$\lim_{n\to\infty} \int_A f_n \, d\mu = \int_A g \, d\mu.$$

8.3 Lebesgue Integrals on Finite Sets

Let us consider the case when X is finite, that is, $X = \{x_1, x_2, ..., x_n\}$, and take $\mathbf{P}(X)$ as $\mathbf{F}$. Denoting $\mu(\{x_i\})$ by w_i for $i = 1, 2, \ldots$, we have

$$\int f \, d\mu = \sum_{i=1}^{n} w_i f(x_i)$$

since

$$f = \sum_{i=1}^{n} f(x_i) \cdot \chi_{\{x_i\}}$$

is a simple function. This is exactly the weighted sum of $\{f(x_1), f(x_2), ..., f(x_n)\}$, where w_i, $i = 1, 2, ..., n$, are weights (w_i may not be in the unit closed interval). When $\sum_{i=1}^{n} w_i = 1$ and $0 \leq w_i \leq 1$ for each $i = 1, 2, ..., n$, the right-hand side of this equality is just the weighted average. Conversely, any weighted sum (including the weighted average) can be expressed as a Lebesgue integral. In fact, for any given weighted sum $\sum_{i=1}^{n} w_i f(x_i)$, the class of all singletons and the empty set $\emptyset$, $\mathbf{S} = \{\{x_i\} | x_i \in X\} \cup \{\emptyset\}$, is a semiring. If we define $\mu : \mathbf{S} \to (-\infty, \infty)$ by $\mu(\{x_i\}) = w_i$ and $\mu(\emptyset) = 0$, then set function μ is a signed measure on semiring $\mathbf{S}$. Such a signed measure can be uniquely extended to a signed measure on the ring generated by $\mathbf{S}$, that is, on $\mathbf{P}(X)$. By the additivity of the signed measure, the extension can be presented as

$$\mu(A) = \sum_{x_i \in A} \mu(\{x_i\}) \quad \forall A \subset X.$$

Thus, we have

$$\int f \, d\mu = \sum_{i=1}^{n} \mu(\{x_i\}) \cdot f(x_i) = \sum_{i=1}^{n} w_i f(x_i).$$

Example 8.2. Three workers, x_1, x_2, and x_3, are hired for manufacturing a certain kind of wooden toys separately. Their individual efficiencies are 5, 6, and 7 toys per day respectively. This week, they work for 6, 3, and 4 days respectively. Let $X = \{x_1, x_2, x_3\}$ and

$$f(x) = \begin{cases} 6 & \text{if } x = x_1 \\ 3 & \text{if } x = x_2 \\ 4 & \text{if } x = x_3. \end{cases}$$

Since these three workers work separately, taking measure μ with $\mu(\{x_1\}) = 5$, $\mu(\{x_2\}) = 6$, and $\mu(\{x_3\}) = 7$, the total amount of the manufactured toys by them in this week is

$$\int f d\mu = \sum_{i=1}^{3} \mu(\{x_i\}) \cdot f(x_i) = 5 \cdot 6 + 6 \cdot 3 + 7 \cdot 4 = 76.$$

The weighted sum is the simplest and most common aggregation tool in information fusion. Example 8.2 shows a typical information fusion problem, where $X = \{x_1, x_2, x_n\}$ is the set of all information sources, $f(x_i)$ is the numerical amount of information received from source x_i, and $\mu(\{x_i\})$ is the importance of source x_i, $i = 1, 2, 3$, respectively. Such a model is linear and can be used only when there is no interaction among the contribution rates from information sources towards the fusion attribute or the interaction can be ignored, like the case of Example 8.2, in which the three workers work separately.

The traditional linear models are also widely used in data mining, such as in multiregressions and classifications. They can be expressed in terms of the Lebesgue integral as well.

Example 8.3. The traditional linear multiregression has the form

$$y = a_0 + a_1 f(x_1) + a_2 f(x_2) + \cdots + a_n f(x_n) + N(0, \sigma^2),$$

where y is the value of the objective attribute Y, $f(x_i)$ is the observation value of predictive attribute x_i for each $i = 1, 2, ..., n$, a_i (where $i = 0, 2, ..., n$) are regression coefficients, and $N(0, \sigma^2)$ is a normally distributed random perturbation with mean zero and standard deviation σ. This linear multiregression form can be rewritten as

$$y = c + \int f d\mu + N(0, \sigma^2),$$

where f is a function on $X = \{x_1, x_2, ..., x_n\}$, μ is a signed measure on $\mathbf{P}(X)$ with $\mu(\{x_i\}) = a_i$, $i = 1, 2, ..., n$, and constant $c = a_0$.

Example 8.4. We consider a traditional classification problem now. Suppose that there are two classes of n-dimensional data in classifying space (n-dimensional Euclidean sample space):

$$C_1 = \{f_{1j} = (t_{1j_1}, t_{1j_2}, ..., t_{1j_n}) | j = 1, 2, ..., l_1\}$$

and

$$C_2 = \{f_{2j} = (t_{2j_1}, t_{2j_2}, ..., t_{2j_n}) | j = 1, 2, ..., l_2\}.$$

We want to find a classifying boundary $a_1t_1 + a_2t_2 + \cdots + a_nt_n = c$ in the n-dimensional Euclidean sample space such that the misclassification rate is minimized, that is,

$$\min \left[\left| \{f_{1j} | a_1 t_{1j_1} + a_2 t_{1j_2} + \cdots + a_n t_{1j_n} > c\} \right| + \left| \{f_{2j} | a_1 t_{2j_1} + a_2 t_{2j_2} + \cdots + a_n t_{2j_n} \le c\} \right| \right],$$

where $a_1, a_2, ..., a_n$, and c are unknown parameters, while symbol $|A|$ denotes the cardinality of a set A. This minimization problem may have infinitely many solutions, though they may be very close to each other. Usually, some additional criteria are employed to obtain the "best" solution among them. Equation $a_1t_1 + a_2t_2 + \cdots + a_nt_n = c$ presents an $(n - 1)$-dimensional hyper-plane in the n-dimensional sample space. It divides the sample space into two subsets that form a partition of the sample space. Linear function $p : R^n \to R$ with

$$p(t_1, t_2..., t_n) = a_1t_1 + a_2t_2 + \cdots + a_nt_n$$

can be understood as a projection that projects each point in the sample space onto a straight line. Real number c then is the classifying boundary on the straight line. If we use $f(x_i)$ to denote t_i, $i = 1, 2, ... n$, then the projection is just the Lebesgue integral $\int f d\mu$ where μ is the signed measure determined by $\mu(\{x_i\}) = a_i$.

8.4 A General View of Integration on Finite Sets

Let $(X, \mathbf{P}(X))$ be a measurable space, where $X = \{x_1, x_2, ..., x_n\}$, and f be a function on X. When μ is a signed measure on $\mathbf{P}(X)$, the Lebesgue integral of f with respect to μ can be expressed by a weighted sum of the values of μ as

$$\int f d\mu = \sum_{i=1}^{n} f(x_i) \cdot \mu(\{x_i\}).$$

This expression is not unique. More generally, the Lebesgue integral can be expressed as

$$\int f d\mu = \sum_{A \subset X} a_A \cdot \mu(A),$$

where the values a_A satisfy for all $x \in X$ the constraint

$$\sum_{x \in A \subset X} a_A = f(x)$$

due to the additivity of μ. Here, $\{a_A | A \subset X\}$ can be regarded as a decomposition of function f.

However, when μ is not additive, but is a signed general measure, the value of $\sum_{A \subset X} a_A \cdot \mu(A)$ depends on the chosen decomposition $\{a_A | A \subset X\}$. Each decomposition of f represents a specified manner of integration. Thus, we may introduce a general concept of integrals with respect to signed general measures.

Let μ be a signed general measure on $(X, \mathbf{P}(X))$ and f be a nonnegative function on X.

Definition 8.3. A set function $\pi : \mathbf{P}(X) - \{\emptyset\} \to [0, \infty)$ is called a *partition* of f if

$$f(x) = \sum_{x \in A \subset X} \pi(A) \quad \forall x \in X.$$

Taking the characteristic function of a crisp set or the membership function of a fuzzy set as f, it is easy to see that the concept of partition in Definition 8.3 is a generalization of the classical partition for crisp sets and the fuzzy partition for fuzzy sets.

Definition 8.4. A type of integral with respect to μ is a rule r by which, for any given f, a partition π of f can be obtained. Regarding both π and μ on $\mathbf{P}(X) - \{\emptyset\}$ as $(2^n - 1)$-dimensional vectors, the value of the integral of f under rule r, denoted by (r) $\int f d\mu$, is the inner product of π and μ, that is, (r) $\int f d\mu = \pi \cdot \mu = \sum_{A \subset X} \pi(A) \cdot \mu(A)$, where (r) is used to indicate the type of integral.

The above definition provides a flexible aggregation tool in information fusion and data mining. It is generally called an "*r-integral*," or simply, "an integral" when the partitioning rule r has been chosen and there is no confusion.

Given a signed general measure μ, we may induce a signed measure μ' by assigning $\mu'(\{x_i\}) = \mu(\{x_i\})$ for all $i = 1, 2, ..., n$. Thus, the Lebesgue integral with respect to signed measure μ' can be regarded as one of the various types of integral with respect to a signed general measure μ. Its corresponding partitioning rule can be described as follows: decomposing function f in such a way that set function π vanishes at all sets $A \epsilon \mathbf{P}(X)$ except singletons. Geometrically, under this rule, any function is divided vertically, no matter what kind of signed general measure is given. This is the simplest partition of f.

Example 8.5. We use the data given in Example 8.2. The Lebesgue integral of f corresponds to a partition illustrated in Fig. 8.1. Geometrically, nonnegative function f is divided vertically. In this manner the value of the integral of function f with respect to signed general measure μ only depends on the values of μ at singletons. So, if we introduce a classical measure μ' determined by $\mu'(\{x_i\}) = \mu(\{x_i\})$ for $i = 1, 2, 3$, the value of the integral for f under this vertically partitioning rule, denoted by (v) $\int f d\mu$, is just the value of the Lebesgue integral of f with respect to μ', that is,

$$(\mathrm{v})\int f d\mu = \sum_{i=1}^{3} \mu(\{x_i\}) \cdot f(x_i) = \int f d\mu' = \sum_{i=1}^{3} \mu'(\{x_i\}) \cdot f(x_i) = 76.$$

We have several different types of r-integrals. They are discussed in the next two chapters. In general, r-integrals are not linear with respect to their integrands. However, r-integrals possess some properties similar to Lebesgue integrals, as shown in the next theorem.

Theorem 8.3. *Let μ be a general measure on $(X, \mathbf{P}(X))$ and f be a nonnegative function on X. Then,*

(1) (r) $\int f d\mu \geq 0$;
(2) (r) $\int f d\mu = 0$ *if for any set A with $\mu(A) > 0$, there exists $x \in A$ such that* $f(x) = 0$.

Proof. (1) follows directly from Definition 8.4. To prove (2), we use a proof by contradiction as follows. Assume that (r) $\int f d\mu > 0$. Then, from Definition 8.3 there exists at least one term in the summation that is greater than zero, say,

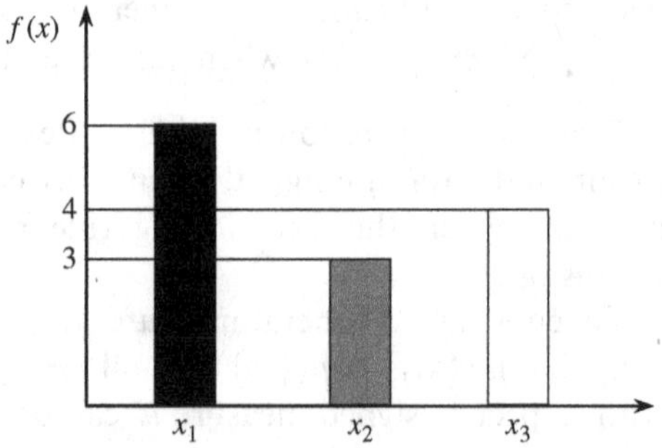

Fig. 8.1 The partition of f corresponding to the Lebesgue integral

$\pi(A) \cdot \mu(A) > 0$; that is, both $\pi(A) > 0$ and $\mu(A) > 0$. From $f(x) = \sum_{x \in A \subset X} \pi(A)$ for every $x \in X$, we know that $f(x) > 0$ for every $x \in A$. This contradicts the fact that there exists $x \in A$ such that $f(x) = 0$. So, the assumption is wrong and we have (r) $\int f d\mu = 0$. □

Notes

8.1. We recommend the excellent books by Chae [1995] and Burk [1998] for more information on Lebesgue integration, including the history of relevant ideas that led to its development. One of those ideas is the Darboux sum, which is mentioned in Section 8.1.

8.2. The general view of integration on finite sets, which is discussed in Section 8.4, was first introduced in [Wang et al., 2006a]. For other general views of integrals with respect to monotone measures, see [Benvenuti et al., 2002] and [Struk, 2006].

Exercises

8.1. Let $(X, \mathbf{F}, \mu)$ be a measure space and $f: X \to [0, \infty)$ be a measurable function on X. Verify that $\{s_j\}$ is nondecreasing and $\lim_{j \to \infty} s_j(x) = f(x)$ for every $x \in X$, where

$$s_j = \sum_{i=1}^{j \cdot 2^j} \frac{i-1}{2^j} \chi_{A_{ji}}$$

in which

$$A_{ji} = \{x | \frac{i-1}{2^j} \leq f(x) < \frac{i}{2^j}\}, i = 1, 2, ..., j \cdot 2^j$$

for $j = 1, 2, \ldots$.

8.2. Let $\{s_j\}$ and $\{t_j\}$ be two sequences of nondecreasing simple functions on measure space $(X, \mathbf{F}, \mu)$ with $\lim_{j \to \infty} s_j = \lim_{j \to \infty} t_j = f$. Prove that $\lim_{j \to \infty} \int s_j d\mu = \lim_{j \to \infty} \int t_j d\mu$.

8.3. Let X be the unit closed interval [0, 1], $\mathbf{F}$ be the set of all Borel sets in [0, 1], and μ be the Lebesgue measure. Calculate the Lebesgue integral $\int f d\mu$ where $f(x) = x$ for $x \in [0, 1]$. Compare the result with the value of the Riemann integral $\int_0^1 f(x) dx$.

8.4. Let X be the set of all positive integers, **F** be the power set of X, and measure μ be determined by $\mu(\{x_i\}) = \frac{1}{2^i}$, $i = 1, 2, \ldots$. Calculate the Lebesgue integral $\int g \, d\mu$ where $g(x_i) = \frac{1}{2^i}$, $i = 1, 2, \ldots$.

8.5. Let μ be the Lebesgue measure. Prove that, for any real-valued measurable function f on an interval $I = [a, b]$, if the Riemann integral $\int_a^b f(x)dx$ exists, then the corresponding Lebesgue integral $\int_I f \, d\mu$ exists, too, and $\int_I f \, d\mu = \int_a^b f(x)dx$.

8.6. Prove the properties of the Lebesgue integral listed in Theorem 8.1.

8.7. Use a counterexample to show that the conclusion in Theorem 8.2 may not be true if the condition "there exists a nonnegative measurable function g satisfying $\int_A g \, d\mu < \infty$ such that $|f_n| \le g$ on A almost everywhere for all $n = 1, 2, \ldots$" fails.

8.8. Let $(X, \mathbf{F}, \mu)$ be a measure space and $f : X \to [0, \infty)$ be a measurable function on X. Prove that $\int f \, d\mu = \int_0^\infty \mu(F_\alpha) \, d\alpha$ where $F_\alpha = \{x | f(x) \ge \alpha\}$ for $\alpha \in [0, \infty)$. (Hint: Use the Darboux sum for the Riemann integral.)

8.9. Generalize the conclusion in Exercise 8.8 to

$$\int f d\mu = \int_{-\infty}^{0} [\mu(F_\alpha) - \mu(X)] d\alpha + \int_0^\infty \mu(F_\alpha) d\alpha,$$

where f is lower bounded, μ is finite, and $F_\alpha = \{x | f(x) \ge \alpha\}$ for $\alpha \in (-\infty, \infty)$. (Hint: Use properties (1) and (7) in Theorem 8.1.)

8.10. Give the measure μ determined by the data shown in Example 8.2.

8.11. When X is countable, under the condition given in Exercise 8.8, show that $\int f \, d\mu = \sum_{A \subset X} a_A \cdot \mu(A)$, where constants a_A satisfy the constraint

$$\sum_{x \in A \subset X} a_A = f(x) \quad \forall x \in X.$$

Chapter 9
Sugeno Integrals

9.1 Definition

In this chapter, we assume that $(X, \mathbf{F})$ is a measurable space, where $X \in \mathbf{F}$, $\mu : \mathbf{F} \to [0, \infty]$ is a continuous monotone measure, and $\mathbf{G}$ is the class of all finite nonnegative measurable functions defined on $(X, \mathbf{F})$. For any given $f \in \mathbf{G}$, we write $F_\alpha = \{x | f(x) \geq \alpha\}$, $F_{\alpha+} = \{x | f(x) > \alpha\}$, where $\alpha \in [0, \infty]$. Let the sets F_α and $F_{\alpha+}$ be called an *α-level set* and a *strict α-level set* of f, respectively.

Since the range of functions that we consider in this chapter is $[0, \infty)$, we use the following convention:

$$\inf_{x \in \emptyset} f(x) = \infty.$$

Definition 9.1. Let $A \in \mathbf{F}, f \in \mathbf{G}$. The *Sugeno integral* of f on A with respect to μ, which is denoted by $⨍_A f d\mu$, is defined by

$$⨍_A f d\mu = \sup_{\alpha \in [0,\infty]} [\alpha \wedge \mu(A \cap F_\alpha)].$$

When $A = X$, the Sugeno integral may also be denoted by $⨍ f d\mu$.

Sometimes, the Sugeno integral is also referred to in the literature as the *fuzzy integral.*

From now on, we use the convention that the appearance of a symbol $⨍_A f d\mu$ implies that $A \in \mathbf{F}$ and $f \in \mathbf{G}$. If $X = (-\infty, \infty)$, $\mathbf{F}$ is the Borel field $\mathbf{B}$, μ is the Lebesgue measure, and f: $X \to [0, \infty)$ is a unimodal continuous function, then the geometric significance of $⨍ f d\mu$ is the edge's length of the largest square between the curve of $f(x)$ and the x-axis (see Fig. 9.1).

Lemma 9.1. (1) *Both F_α and $F_{\alpha+}$ are nonincreasing with respect to α, and $F_{\alpha+} \supset F_\beta$ when $\alpha < \beta$.*

(2) $\lim_{\beta \to \alpha-} F_\beta = \lim_{\beta \to \alpha-} F_{\beta+} = F_\alpha \supset F_{\alpha+} = \lim_{\beta \to \alpha+} F_\beta = \lim_{\beta \to \alpha+} F_{\beta+}.$

Z. Wang, G.J. Klir, *Generalized Measure Theory*,
DOI: 10.1007/978-0-387-76852-6_9,

Fig 9.1 Geometric interpretation of the Sugeno integral under special conditions

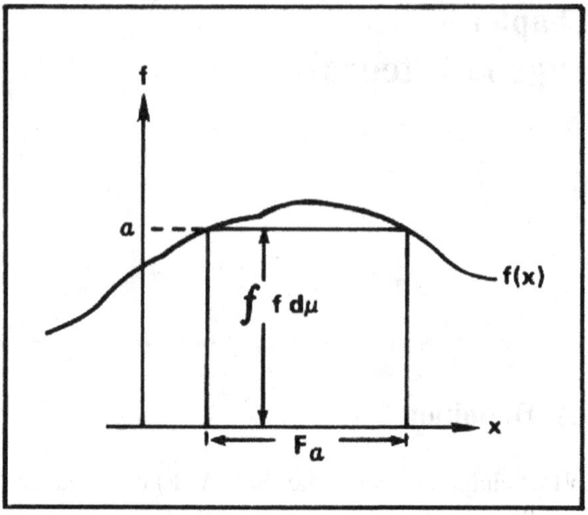

Proof. (1) is evident. (2) follows from the following facts:

$$\begin{aligned}\bigcap_{\beta<\alpha}\{x|f(x)\geq\beta\} &= \bigcap_{\beta<\alpha}\{x|f(x)>\beta\}\\ &= \{x|f(x)\geq\alpha\}\supset\{x|f(x)>\alpha\}\\ &= \bigcup_{\beta>\alpha}\{x|f(x)\geq\beta\} = \bigcup_{\beta>\alpha}\{x|f(x)>\beta\}.\end{aligned}$$ □

Theorem 9.1.

$$\begin{aligned}⨍_A f\,d\mu &= \sup_{\alpha\in[0,\infty)}[\alpha\wedge\mu(A\cap F_\alpha)] = \sup_{\alpha\in[0,\infty]}[\alpha\wedge\mu(A\cap F_{\alpha+})]\\ &= \sup_{\alpha\in[0,\infty)}[\alpha\wedge\mu(A\cap F_{\alpha+})] = \sup_{E\in\mathbf{F}(f)}[(\inf_{x\in E}f(x))\wedge\mu(A\cap E)]\\ &= \sup_{E\in\mathbf{F}}[(\inf_{x\in E}f(x))\wedge\mu(A\cap E)],\end{aligned}$$

where **F**(*f*) *is the σ-algebra generated by f, the smallest σ-algebra such that f is measurable.*

Proof. (1) Since $F_\alpha = F_{\alpha+} = \emptyset$ when $\alpha = \infty$, equations

$$⨍_A f\,d\mu = \sup_{\alpha\in[0,\infty)}[\alpha\wedge\mu(A\cap F_\alpha)]$$

and

$$\sup_{\alpha\in[0,\infty]} [\alpha \wedge \mu(A \cap F_{\alpha+})] = \sup_{\alpha\in[0,\infty)} [\alpha \wedge \mu(A \cap F_{\alpha+})]$$

are evident.

(2) We prove now that

$$\sup_{\alpha\in[0,\infty)} [\alpha \wedge \mu(A \cap F_{\alpha})] = \sup_{\alpha\in[0,\infty)} [\alpha \wedge \mu(A \cap F_{\alpha+})].$$

On the one hand, by Lemma 9.1 and the monotonicity of μ, we have

$$\mu(A \cap F_{\alpha}) \geq \mu(A \cap F_{\alpha+})$$

for any $\alpha \in [0, \infty)$. Hence,

$$\sup_{\alpha\in[0,\infty)} [\alpha \wedge \mu(A \cap F_{\alpha})] \geq \sup_{\alpha\in[0,\infty)} [\alpha \wedge \mu(A \cap F_{\alpha+})].$$

On the other hand, for any $\varepsilon > 0$ and $\alpha \in (0, \infty)$, taking $\alpha' \in ((\alpha - \varepsilon) \vee 0, \alpha)$, we have

$$\alpha \wedge \mu(A \cap F_{\alpha}) \leq (\alpha' + \varepsilon) \wedge \mu(A \cup F_{\alpha'+});$$

hence, we have

$$\begin{aligned}\sup_{\alpha\in[0,\infty)} [\alpha \wedge \mu(A \cap F_{\alpha})] &= \sup_{\alpha\in[0,\infty)} [\alpha \wedge \mu(A \cap F_{\alpha+})] \\ &\leq \sup_{\alpha'\in(0,\infty)} [(\alpha' + \varepsilon) \wedge \mu(A \cap F_{\alpha'+})] \\ &\leq \sup_{\alpha'\in(0,\infty)} [\alpha' \wedge \mu(A \cap F_{\alpha'+})] + \varepsilon \\ &= \sup_{\alpha\in[0,\infty)} [\alpha \wedge \mu(A \cap F_{\alpha+})] + \varepsilon.\end{aligned}$$

Since ε may be close to zero arbitrarily, we obtain

$$\sup_{\alpha\in[0,\infty)} [\alpha \wedge \mu(A \cap F_{\alpha})] \leq \sup_{\alpha\in[0,\infty)} [\alpha \wedge \mu(A \cap F_{\alpha+})].$$

Consequently, we have

$$\sup_{\alpha\in[0,\infty)} [\alpha \wedge \mu(A \cap F_\alpha)] = \sup_{\alpha\in[0,\infty)} [\alpha \wedge \mu(A \cap F_{\alpha+})].$$

(3) It remains to prove that

$$⨍_A f\,d\mu = \sup_{E\in\mathbf{F}(f)} [(\inf_{x\in E} f(x)) \wedge \mu(A \cap E)] = \sup_{E\in\mathbf{F}} [(\inf_{x\in E} f(x)) \wedge \mu(A \cap E)].$$

First, for any $\alpha \in [0, \infty]$, since $\inf_{x\in F_\alpha} f(x) \geq \alpha$, noting $F_\alpha \in \mathbf{F}(f)$, we have

$$[\alpha \wedge \mu(A \cap F_\alpha)] \leq \sup_{E\in\mathbf{F}(f)} [(\inf_{x\in E} f(x)) \wedge \mu(A \cap E)]$$

and, therefore, we have

$$⨍_A f\,d\mu = \sup_{\alpha\in[0,\infty]} [\alpha \wedge \mu(A \cap F_\alpha)] \leq \sup_{E\in\mathbf{F}(f)} [(\inf_{x\in E} f(x)) \wedge \mu(A \cap E)].$$

Next, since f is $\mathbf{F}$-measurable, we have $\mathbf{F}(f) \subset \mathbf{F}$ and, therefore, we have

$$\sup_{E\in\mathbf{F}(f)} [(\inf_{x\in E} f(x)) \wedge \mu(A \cap E)] \leq \sup_{E\in\mathbf{F}} [(\inf_{x\in E} f(x)) \wedge \mu(A \cap E)].$$

Finally, for any given $E \in \mathbf{F}$, if we take $\alpha' = \inf_{x\in E} f(x)$, then $E \subset F_{\alpha}'$. It follows that

$$\mu(A \cap E) \leq \mu(A \cap F_{\alpha'})$$

by the monotonicity of μ and, therefore,

$$[\inf_{x\in E} f(x)] \wedge \mu(A \cap E) \leq \alpha' \wedge \mu(A \cap F_{\alpha'}) \leq \sup_{\alpha\in[0,\infty]} [\alpha \wedge \mu(A \cap F_\alpha)] = ⨍_A f\,d\mu$$

for any $E \in \mathbf{F}$. Consequently, we have

$$\sup_{E\in\mathbf{F}} [(\inf_{x\in E} f(x)) \wedge \mu(A \cap E)] \leq ⨍_A f\,d\mu$$ □

To simplify the calculation of the Sugeno integral, for a given $(X, \mathbf{F}, \mu)$, $f \in \mathbf{G}$ and $A \in \mathbf{F}$, we write

$$\Gamma = \{\alpha | \alpha \in [0, \infty], \mu(A \cap F_\alpha) > \mu(A \cap F_\beta) \text{ for any } \beta > \alpha\}.$$

It is easy to see that

$$⨍_A f\,d\mu = \sup_{\alpha\in\Gamma}\,[\alpha \wedge \mu(A\cap F_\alpha)].$$

Example 9.1. Consider the monotone measure space given in Example 7.3. Let

$$f(x) = \begin{cases} 3 & \text{if } x = a \\ 2.5 & \text{if } x = b \\ 2 & \text{if } x = c. \end{cases}$$

Then

$$⨍ f\,d\mu = [3\wedge\mu(\{a\})]\vee[2.5\wedge\mu(\{a,b\})]\vee[2\wedge\mu(X)] = 1\vee 2.5\vee 2 = 2.5.$$

Example 9.2. Let $X = [0,1]$, **F** be the class of all Borel sets in X, $\mu = m^2$, where m is the Lebesgue measure, $f(x) = x/2$. We have

$$F_\alpha = \{x|f(x)\geq\alpha\} = [2\alpha, 1].$$

Since $\Gamma = [0, 1/2)$, we only need to consider $\alpha \in [0, 1/2)$. So, we have

$$⨍ f\,d\mu = \sup_{\alpha\in[0,1/2)}\,[\alpha\wedge\mu(F_\alpha)] = \sup_{\alpha\in[0,1/2)}\,[\alpha\wedge(1-2\alpha)^2].$$

In this expression, $(1-2\alpha)^2$ is a decreasing continuous function of α when $\alpha \in [0, 1/2)$. Hence, the supremum will be attained at the point which is one of the solutions of the equation

$$\alpha = (1-2\alpha)^2,$$

that is, at $\alpha = 1/4$. Consequently, we have

$$⨍ f\,d\mu = 1/4.$$

9.2 Properties of the Sugeno Integral

The following theorem gives the most elementary properties of the Sugeno integral.

Theorem 9.2.
(1) *If* $\mu(A) = 0$, *then* $⨍_A f\,d\mu = 0$ *for any* $f \in \mathbf{G}$;
(2) *if* μ *is continuous from below and* $⨍_A f\,d\mu = 0$, *then* $\mu(A \cap \{x | f(x) > 0\}) = 0$;
(3) *if* $f_1 \le f_2$, *then* $⨍_A f_1 d\mu \le ⨍_A f_2 d\mu$;
(4) $⨍_A f d\mu = ⨍ f \cdot \chi_A d\mu$, *where* χ_A *is the characteristic function of* A;
(5) $⨍_A a\,d\mu = a \wedge \mu(A)$ *for any constant* $a \in [0, \infty)$;
(6) $⨍_A (f + a)d\mu \le ⨍_A f\,d\mu + ⨍_A a\,d\mu$ *for any constant* $a \in [0, \infty)$.

Proof. We only need to prove (2) and (6); the remaining properties can be obtained directly from the definition of the Sugeno integral.

For (2), we use a proof by contradiction. Assume

$$\mu(A \cap \{x | f(x) > 0\}) = c > 0.$$

Since

$$A \cap \{x | f(x) \ge 1/n\} \nearrow A \cap \{x | f(x) > 0\}.$$

by using the continuity from below of μ, we have

$$\lim_n \mu(A \cap \{x | f(x) \ge 1/n\}) = c.$$

So, there exists n_0 such that

$$\mu(A \cap F_{1/n_0}) = \mu(A \cap \{x | f(x) \ge 1/n_0\}) \ge c/2.$$

Consequently, we have

$$⨍_A f\,d\mu = \sup_{\alpha \in [0,\infty]} [\alpha \wedge \mu(A \cap F_\alpha)] \ge 1/n_0 \wedge c/2 > 0.$$

This contradicts $⨍_A f d\mu = 0$.

For (6), from Theorem 9.1, we have

$$\begin{aligned}
⨍_A (f + a)d\mu &= \sup_{E \in \mathbf{F}} \left\{ [\inf_{x \in E} (f(x) + a)] \wedge \mu(A \cap E) \right\} \\
&\le \sup_{E \in \mathbf{F}} \left\{ [((\inf_{x \in E} f(x)) \wedge \mu(A \cap E)) + (a \wedge \mu(A \cap E))] \right\} \\
&\le \sup_{E \in \mathbf{F}} \left\{ [((\inf_{x \in E} f(x)) \wedge \mu(A \cap E)) + (a \wedge \mu(A))] \right\} \\
&= \sup_{E \in \mathbf{F}} [(\inf_{x \in E} f(x)) \wedge \mu(A \cap E)] + (a \wedge \mu(A)) \\
&= ⨍_A f\,d\mu + ⨍_A a\,d\mu.
\end{aligned}$$

The proof is now complete. □

Corollary 9.1.
(7) *If* $A \supset B$, *then* $⨍_A f\,d\mu \geq ⨍_B f\,d\mu$;
(8) $⨍_A (f_1 \vee f_2) d\mu \geq ⨍_A f_1 d\mu \vee ⨍_A f_2 d\mu$;
(9) $⨍_A (f_1 \wedge f_2) d\mu \leq ⨍_A f_1 d\mu \wedge ⨍_A f_2 d\mu$;
(10) $⨍_{A \cup B} f d\mu \geq ⨍_A f\,d\mu \vee ⨍_B f d\mu$;
(11) $⨍_{A \cap B} f\,d\mu \leq ⨍_A f\,d\mu \wedge ⨍_B f\,d\mu$.

Proof. Property (7) can be obtained from properties (3) and (4) of Theorem 9.2; properties (8) and (9) come from (3); properties (10) and (11) follow directly from (7). □

Properties (1)–(4) [and, therefore, (7)–(11)] are similar to those of the classical Lebesgue integral, but (5) and (6) are somewhat different from the classical ones. We should note that, in general, the Sugeno integral lacks some important properties that the Lebesgue integral possesses. For instance, the Lebesgue integral has linearity, that is,

$$\int_A (f_1 + f_2)\, d\mu = \int_A f_1\, d\mu + \int_A f_2\, d\mu$$

and

$$\int_A af d\mu = a \int_A f d\mu,$$

but the Sugeno integral does not. We can see this in the following example.

Example 9.3. Let $X = [0, 1]$, **F** be the class of all Borel sets in X (namely, $\mathbf{B} \cap [0, 1]$), and μ be the Lebesgue measure. We take $f(x) = x$ for any $x \in X$, and $a = 1/2$. Then we have

$$⨍ af\, d\mu = ⨍ \tfrac{x}{2} d\mu = 1/3$$

and

$$a ⨍ f d\mu = \tfrac{1}{2} ⨍ x\, d\mu = \tfrac{1}{2} \times \tfrac{1}{2} = \tfrac{1}{4}.$$

Consequently, we have

$$⨍ af\, d\mu \neq a ⨍ f\,d\mu$$

(see Fig. 9.2).

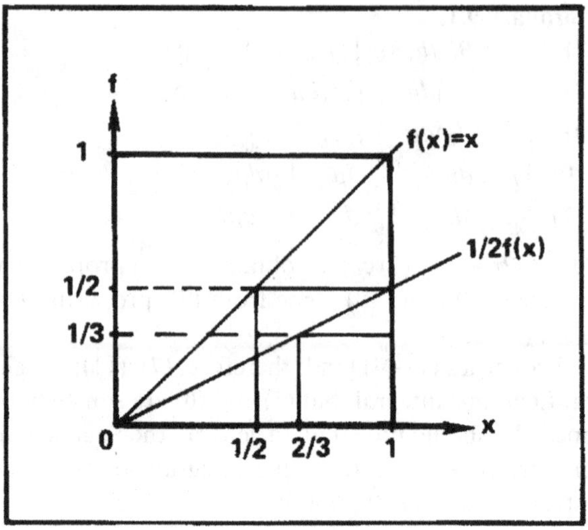

Fig 9.2 Illustration to Example 9.3

Lemma 9.2. *Let* $A \in \mathbf{F}$, $a \in [0, \infty)$, $f_1 \in \mathbf{G}$, *and* $f_2 \in \mathbf{G}$,. *If* $|f_1 - f_2| \leq a$ *on* A, *then we have*

$$\left| \fint_A f_1 \, d\mu - \fint_A f_2 \, d\mu \right| \leq a.$$

Proof. Since $f_1 \leq f_2 + a$ on A, using the properties (3), (5), and (6) of the Sugeno integral (Theorem 9.2), we have

$$\fint_A f_1 \, d\mu \leq \fint_A (f_2 + a) \, d\mu \leq \fint_A f_2 \, d\mu + \fint_A a \, d\mu = \fint_A f_2 \, d\mu + [a \wedge \mu(A)]$$
$$\leq \fint_A f_2 \, d\mu + a.$$

Similarly, from $f_2 \leq f_1 + a$ on A, we have

$$\fint_A f_2 \, d\mu \leq \fint_A f_1 \, d\mu + a.$$

Consequently, we have

$$\left| \fint_A f_1 \, d\mu - \fint_A f_2 \, d\mu \right| \leq a.$$

□

Lemma 9.3. $\fint_A f d\mu \leq \alpha \vee \mu(A \cap F_{\alpha+}) \leq \alpha \vee \mu(A \cap F_\alpha)$ *for any* $\alpha \in [0, \infty]$.

Proof. For any $\alpha \in [0, \infty]$, using Theorem 9.1 and Lemma 9.1, we have

$$\begin{aligned}\fint_A f\, d\mu &= \sup_{\alpha' \in [0,\alpha]} [\alpha' \wedge \mu(A \cap F_{\alpha'+})] \vee \sup_{\alpha' \in (\alpha,\infty]} [\alpha' \wedge \mu(A \cap F_{\alpha'+})] \\ &\leq \sup_{\alpha' \in [0,\alpha]} \alpha' \vee \sup_{\alpha' \in (\alpha,\infty]} \mu(A \cap F_{\alpha'+}) \leq \alpha \vee \mu(A \cap F_{\alpha+}) \leq \alpha \vee \mu(A \cap F_\alpha).\end{aligned}$$

This completes the proof. □

Lemma 9.4. $\fint_A f d\mu = \infty$ *if and only if* $\mu(A \cap F_\alpha) = \infty$ *for any* $\alpha \in [0, \infty)$.
Proof. Necessity: If $\fint_A f d\mu = \infty$, then it follows from Lemma 9.3 that

$$\alpha \vee \mu(A \cap F_\alpha) = \infty$$

So, if $\alpha \in [0, \infty)$, then

$$\mu(A \cap F_\alpha) = \infty$$

Sufficiency: It follows directly from Definition 9.1. □

Lemma 9.5. *For any* $\alpha \in [0, \infty)$ *we have*

(1) $\fint_A f d\mu \geq \alpha \Leftrightarrow \mu(A \cap F_\beta) \geq \alpha$ *for any* $\beta < \alpha \Leftarrow \mu(A \cap F_\alpha) \geq \alpha$;
$\fint_A f d\mu < \alpha \Leftrightarrow$ *there exists* $\beta < \alpha$ *such that* $\mu(A \cap F_\beta) < \alpha \Rightarrow \mu(A \cap F_\alpha) < \alpha \Rightarrow \mu(A \cap F_{\alpha+}) < \alpha$.

(2) $\fint_A f d\mu \leq \alpha \Leftrightarrow \mu(A \cap F_{\alpha+}) \leq \alpha \Leftarrow \mu(A \cap F_\alpha) \leq \alpha$;
$\fint_A f du > \alpha \Leftrightarrow \mu(A \cap F_{\alpha_+}) > \alpha \Rightarrow \mu(A \cap F_\alpha) > \alpha$.

(3) $\fint_A f d\mu = \alpha \Leftrightarrow$ *for any* $\beta < \alpha$, $\mu(A \cap F_\beta) \geq \alpha \geq \mu(A \cap F_{\alpha+}) \Leftarrow \mu(A \cap F_\alpha) = \alpha$.
When $\mu(A) < \infty$, *we have*

(4) $\fint_A f d\mu \geq \alpha \Leftrightarrow \mu(A \cap F_\alpha) \geq \alpha$.

(5) $\fint_A f d\mu = \alpha \Leftrightarrow \mu(A \cap F_\alpha) \geq \alpha \geq \mu(A \cap F_{\alpha+})$.

Proof. (1) We only need to consider the case when $\alpha \in (0, \infty)$. If $\mu(A \cap F_\beta) \geq \alpha$ for any $\beta < \alpha$, then

$$\begin{aligned}\fint_A f d\mu &= \sup_{\beta \in [0,\infty)} [\beta \wedge \mu(A \cap F_\beta)] \geq \sup_{\beta \in [0,\alpha)} [\beta \wedge \mu(A \cap F_\beta)] \geq \sup_{\beta \in [0,\alpha)} [\beta \wedge \alpha] \\ &= \sup_{\beta \in [0,\alpha)} \beta = \alpha.\end{aligned}$$

Conversely, if there exists $\beta < \alpha$ such that $\mu(A \cap F_\beta) < \alpha$, then, by Lemma 9.3,

$$\fint_A f d\mu \leq \beta \vee \mu(A \cap F_\beta) < \alpha.$$

Thus, we have proved the equivalence relation in (1). The other implication relations issue from Lemma 9.1 and the monotonicity of μ.
(2) If $\mu\,(A \cap F_{\alpha+}) \le \alpha$, by Lemma 9.3,

$$\fint_A f d\mu \le \alpha \vee \mu(A \cap F_{\alpha+}) = \alpha.$$

Conversely, by using Lemma 9.1 and the continuity from below of μ, we have

$$\lim_{\beta \to \alpha+} \mu(A \cap F_\beta) = \mu(A \cap F_{\alpha+}).$$

If $\mu(A \cap F_{\alpha+}) > \alpha$, then there exists $\alpha_0 > \alpha$ such that $\mu(A \cap F_{\alpha_0}) > \alpha$. So, from Definition 9.1 we have

$$\fint_A f d\mu \ge \alpha_0 \wedge \mu(A \cap F_{\alpha_0}) > \alpha.$$

Thus, the equivalence relations in (2) have been proved; the remaining properties can be obtained in the same way as in (1).
(3) This property is directly obtained by combining (1) and (2).
(4) When $\mu(A) < \infty$, we have

$$\lim_{\beta \to \alpha-} \mu(A \cap F_\beta) = \mu(A \cap F_\alpha).$$

So $\mu(A \cap F_\alpha) \ge \alpha$ if and only if $\mu(A \cap F_\beta) \ge \alpha$ for any $\beta < \alpha$; therefore, (4) follows directly from (1).
(5) Similarly, this property follows directly from (3). □

In classical measure theory if two measurable functions f_1 and f_2 are equal a.e., then their integrals are equal. What about the Sugeno integral on monotone measure space? The answer is negative, as is illustrated by the following example.

Example 9.4. Let $X = \{0, 1\}$, $\mathbf{F} = \mathbf{P}(X)$,

$$\mu(E) = \begin{cases} 1 & \text{if } E = X \\ 0 & \text{if } E \ne X. \end{cases}$$

If

$$f_1(x) = \begin{cases} 1 & \text{if } x = 1 \\ 0 & \text{if } x = 0, \end{cases}$$

$$f_2(x) = 1,$$

then $f_1 = f_2$ a.e., but $\fint f_1\, d\mu = 0$ and $\fint f_2\, d\mu = 1$.

However, we have an important theorem for Sugeno integrals.

Theorem 9.3. *$⨍ f_1 d\mu = ⨍ f_2 d\mu$ whenever $f_1 = f_2$ a.e. if and only if μ is null-additive.*

Proof. Sufficiency: If μ is null-additive, then from

$$\mu(\{x|f_1(x) \neq f_2(x)\}) = 0,$$

we know that

$$\mu(\{x|f_2(x) \geq \alpha\}) \leq \mu(\{x|f_1(x) \geq \alpha\} \cup \{x|f_1(x) \neq f_2(x)\}) = \mu(\{x|f_1(x) \geq \alpha\})$$

for any $\alpha \in [0, \infty]$. The converse inequality holds as well. So, we have

$$\mu(\{x|f_1(x) \geq \alpha\}) = \mu(\{x|f_2(x) \geq \alpha\})$$

for any $\alpha \in [0, \infty]$ and, therefore, from Definition 9.1, we have

$$⨍ f_1\, d\mu = ⨍ f_2\, d\mu.$$

Necessity: For any $E \in \mathbf{F}$, $F \in \mathbf{F}$ with $\mu(F) = 0$, if $\mu(E) = \infty$, then, by the monotonicity of μ, $\mu(E \cup F) = \infty = \mu(E)$. Now, we assume $\mu(E) < \infty$, and use a proof by contradiction to show that $\mu(E \cup F) = \mu(E)$. If this equality is not true (that is, if $\mu(E \cup F) > \mu(E)$), we take $a \in (\mu(E), \mu(E \cup F))$, and

$$f_1(x) = \begin{cases} a & \text{if } x \in E \\ 0 & \text{if } x \notin E \end{cases} \quad \text{and} \quad f_2(x) = \begin{cases} a & \text{if } x \in E \cup F \\ 0 & \text{if } x \notin E \cup F; \end{cases}$$

then $\mu(\{x|f_1(x) \neq f_2(x)\}) = \mu(F - E) \leq \mu(F) = 0$, that is,

$$f_1 = f_2 \quad \text{a.e.}$$

So, it should hold that

$$⨍ f_1\, d\mu = ⨍ f_2\, d\mu.$$

But now we have

$$⨍ f_1\, d\mu = a \wedge \mu(E) = \mu(E)$$

and

$$⨍ f_2\, d\mu = a \wedge \mu(E \cup F) = a \neq \mu(E).$$

Thus, we get a contradiction. □

Corollary 9.2. *If μ is null-additive, then $⨍_A f_1\, d\mu = ⨍_A f_2\, d\mu$ whenever $f_1 = f_2$ a.e. on A.*

Proof. If $f_1 = f_2$ a.e. on A, then $f_1\chi_A = f_2\chi_A$ a.e. From Theorem 9.3 and Theorem 9.2(4), we get the conclusion. □

Corollary 9.3. *If μ is null-additive, then for any $f \in \mathbf{G}$,*

$$⨍_{A\cup B} f\, d\mu = ⨍_A f\, d\mu$$

whenever $A \in \mathbf{F}$, $B \in \mathbf{F}$ with $\mu(B) = 0$.

Proof. The conclusion follows directly from

$$f \cdot \chi_{A\cup B} = f \cdot \chi_A \text{ a.e.}$$ □

Analogously, we also can obtain a condition under which $⨍ f_1 d\mu = ⨍ f_2\, d\mu$ whenever $f_1 = f_2$ p.a.e. (see Wang [1985a]).

In Chapter 7, we discuss several convergences of measurable function sequences on monotone measure spaces. In classical measure theory there are some concepts of convergence of measurable function sequences that concern the integral. One of them is the mean convergence. Since the Sugeno integral has been defined for measurable functions, we can introduce a concept of s-mean convergence on monotone measure spaces as follows.

Definition 9.2. Let $\{f_n\} \subset \mathbf{G}, f \in \mathbf{G}$. We say that $\{f_n\}$ *s-mean converges* to f iff

$$\lim_n ⨍ |f_n - f| d\mu = 0.$$

However, the following theorem shows that such a convergence concept is not necessary.

Theorem 9.4. *The s-mean convergence is equivalent to the convergence in measure on monotone measure spaces.*

Proof. if $f_n \xrightarrow{\mu} f$, then for any given $\varepsilon > 0$, there exists n_0 such that

$$\mu(\{x \mid |f_n(x) - f(x)| \geq \varepsilon/2\}) < \varepsilon$$

whenever $n \geq n_0$. So, by Lemma 9.5(1) we know

$$⨍ |f_n - f| d\mu < \varepsilon.$$

This shows that $\{f_n\}$ s-mean converges to f. Conversely, if $f_n \xrightarrow{\mu} f$ does not hold, then there exist $\varepsilon > 0$, $\delta > 0$, and a sequence $\{n_i\}$ such that

$$\mu(\{x||f_{n_i}(x) - f(x)| \geq \varepsilon\}) > \delta$$

for any n_i, $i = 1, 2, \ldots$. From Definition 9.1, we may directly conclude that

$$\int | f_{n_i} - f | d\mu \geq \varepsilon \wedge \mu(\{x|| f_{n_i}(x) - f(x)| \geq \varepsilon\}) \geq \varepsilon \wedge \delta > 0$$

for any n_i, $i = 1, 2, \ldots$. This shows that $\{f_n\}$ does not s-mean converge to f. □

9.3 Convergence Theorems of the Sugeno Integral Sequence

Under a given condition, if a measurable function sequence converges to some measurable function in a certain sense, then the corresponding integral sequence converges to the integral of the limit function. That is, the symbols of the limit and the integral can be exchanged. This is the convergence theorem. In classical measure theory there are the monotone convergence theorem, the uniform convergence theorem, and Lebesgue's dominated convergence theorem, all of which are well known. For the Sugeno integral sequence there are a lot of convergence theorems as well. In this section we will give several convergence theorems of Sugeno integral sequence under some conditions as weak as possible. In these theorems we assume that the monotone measure μ is continuous, $\{f_n\} \subset \mathbf{G}$, and $f \in \mathbf{G}$. In addition, we use for both function sequences and number sequences the symbols $\searrow$, $\nearrow$, and $\to$, respectively, to denote the concepts of decreasingly converges to, increasingly converges to, and we write $F_\alpha^n = \{x|f_n(x) \geq \alpha\}$, $F_{\alpha+}^n = \{x|f_n(x) > \alpha\}$.

Lemma 9.6. *If $f_n \searrow f$, then $F_\alpha^n \searrow \bigcap_{n=1}^\infty F_\alpha^n = F_\alpha$ and $F_{\alpha+} \subset F_{\alpha+}^n \searrow \bigcap_{n=1}^\infty F_{\alpha+}^n \subset F_\alpha$. If $f_n \nearrow f$, then $F_{\alpha+}^n \nearrow \bigcup_{n=1}^\infty F_{\alpha+}^n = F_{\alpha+}$ and $F_\alpha \supset F_\alpha^n \nearrow \bigcup_{n=1}^\infty F_\alpha^n \supset F_{\alpha+}$.*

Proof. We only prove that $F_{\alpha+} \subset F_{\alpha+}^n \searrow \bigcap_{n=1}^\infty F_{\alpha+}^n \subset F_\alpha$ when $f_n \searrow f$. The rest is similar. Let $f_n \searrow f$. Since $f_n \geq f$ for any $x \in X$,

$$x \in F_{\alpha+} \Rightarrow f(x) > \alpha \Rightarrow f_n(x) > \alpha \Rightarrow x \in F_{\alpha+}^n.$$

So $F_{\alpha+} \subset F_{\alpha+}^n$. For the same reason, $\{F_{\alpha+}^n\}$ is nonincreasing with respect to n, so $F_{\alpha+}^n \searrow \bigcap_{n=1}^\infty F_{\alpha+}^n$ Finally, we have

$$x \in \bigcap_{n=1}^{\infty} F_{\alpha+}^{n} \Rightarrow x \in F_{\alpha+}^{n} \text{ for any } n$$

$$\Rightarrow f_n(x) > \alpha \text{ for any } n$$

$$\Rightarrow f(x) \geq \alpha$$

$$\Rightarrow f(x) \in F_{\alpha}.$$

So $\cap_{n=1}^{\infty} F_{\alpha+}^{n} \subset F_{\alpha}$. □

Theorem 9.5. *Let* $A \in \mathbf{F}$. *If* $f_n \searrow f$ *on* A, *and there exists* n_0 *such that*

$$\mu\left(\left\{x \middle| f_{n_0}(x) > \int_A f\, d\mu\right\} \cap A\right) < \infty,$$

or if $f_n \nearrow f$, *then*

$$\lim_n \int_A f_n d\mu = \int_A f d\mu.$$

Proof. We can assume that $A = X$ without any loss of generality. Write $\int f d\mu = c$, and let $f_n \searrow f$ with n_0 such that

$$\mu(\{x | f_{n_0}(x) > c\}) < \infty.$$

If $c = \infty$, by the monotonicity of Sugeno integral (see Theorem 9.2(3)), we have

$$\int f_n\, d\mu \geq \int f d\mu = \infty;$$

that is, the conclusion of this theorem holds. If $c < \infty$, then

$$\int f_n\, d\mu \geq c$$

for any $n = 1, 2, \ldots$, and, therefore,

$$\lim_n \int f_n\, d\mu \geq c.$$

Now we use reduction to absurdity to prove that the equality holds. If we assume

$$\lim_n \int f_n\, d\mu > c$$

then there exists $c' > c$ such that

$$\lim_n \oint f_n \, d\mu > c'$$

and, therefore,

$$\oint f_n \, d\mu > c'$$

for any n. From Lemma 9.5(2) we know that

$$\mu(F_c^n) > c'$$

for any n. Since there exists n_0 such that

$$\mu(F_{c'}^{n_0}) = \mu(\{x | f_{n_0}(x) \geq c'\}) \leq \mu(\{x | f_{n_0}(x) > c\}) < \infty,$$

by applying the continuity from above of μ, from Lemma 9.6, we get

$$\mu(F_{c'}) = \lim_n \mu(F_{c'}^n) \geq c'.$$

By Lemma 9.5(1), we know that

$$\oint f \, d\mu \geq c' > c.$$

This contradicts $\oint f \, d\mu = c$. Consequently, we have

$$\lim_n \oint f_n \, d\mu = c = \oint f \, d\mu.$$

When $f_n \nearrow f$, the proof is similar to the above. □

Corollary 9.4. *Let $A \in \mathbf{F}$. If $f_n \searrow f$ on A, and there exist n_0 and a constant $c \leq \oint_A f \, d\mu$ such that*

$$\mu(\{x | f_{n_0}(x) > c\} \cap A) < \infty,$$

then

$$\oint_A f_n \, d\mu \searrow \oint_A f d\mu.$$

Corollary 9.5. *If* $f_n \searrow f$ *and* μ *is finite, then*

$$\fint_A f_n \, d\mu \searrow \fint_A f d\mu.$$

Corollary 9.6. *Let* μ *be null-additive.*
(1) *If* $f_n \searrow f$ *a.e., and there exists* n_0 *and a constant* $c \leq \fint f \, d\mu$ *such that*

$$\mu(\{x | f_{n_0}(x) > c\}) < \infty,$$

then

$$\fint f_n \, d\mu \searrow \fint f d\mu.$$

(2) *If* $f_n \nearrow f$ a.e., *then*

$$\fint f_n d\mu \nearrow \fint f d\mu.$$

The above theorem may be called the "convergence theorem of Sugeno integral sequence for the sequence of monotone measurable functions," or the *monotone convergence theorem*, for short.

A result similar to Fatou's lemma in classical measure theory may be obtained from Theorem 9.5 as follows.

Theorem 9.6. *Let* $A \in \mathbf{F}$. *If* $f(x) = \underline{\lim}_n f_n(x)$ *for any* $x \in A$, *then*

$$\fint_A f d\mu \leq \underline{\lim}_n \fint_A f_n \, d\mu.$$

Proof. If we write $g_n(x) = \inf_{i \geq n} f_i(x)$ for all $x \in A$, then $g_n \nearrow f$ on A. By Theorem 9.5, we obtain

$$\lim_n \fint_A g_n d\mu = \fint_A f d\mu.$$

Since $g_n \leq f_n$ on A, we have

$$\fint_A g_n d\mu \leq \fint_A f_n d\mu,$$

and, therefore,

$$\lim_n \fint_A g_n \, d\mu \le \underline{\lim}_n \fint_A f_n \, d\mu.$$

Consequently, we have

$$\fint_A f d\mu \le \underline{\lim}_n \fint_A f_n d\mu.$$ □

In Theorem 9.5, when $\{f_n\}$ is a nonincreasing sequence, the condition that there exists n_0 such that

$$\mu\left(\left\{x \middle| f_{n_0}(x) > \fint_A f d\mu\right\} \cap A\right) < \infty$$

cannot be abandoned casually; without this condition, the conclusion of this theorem might not hold. We can see this from the following example.

Example 9.5. Let $X = [0, \infty)$, **F** be the class of all Borel sets that are in X (namely, $\mathbf{F} = \mathbf{B} \cap X$), and μ be the Lebesgue measure. Take $f_n(x) = x/n$ for any $x \in X$ and any $n = 1, 2, \ldots$; then $f_n \searrow f \equiv 0$. Such a measurable function sequence $\{f_n\}$ does not satisfy the condition given in Theorem 9.5. In fact, we have

$$\mu\left(\left\{x \middle| f_n(x) > \fint f d\mu\right\}\right) = \mu(\{x | f_n(x) > 0\}) = \mu(X) = \infty$$

for any $n = 1, 2, \ldots$. Consequently, $\fint f_n d\mu = \infty$ for any $n = 1, 2, \ldots$, but $\fint f d\mu = 0$; that is,

$$\lim_n \fint f_n d\mu \ne \fint f d\mu.$$

Making use of the monotone convergence theorem we can give a convergence theorem of the Sugeno integral sequence for the measurable function sequence, which is convergent everywhere.

Theorem 9.7. *Let $A \in \mathbf{F}$. If $f_n \to f$ on A, and there exists n_0 and a constant $c \le \fint_A f d\mu$ such that*

$$\mu(\{x | \sup_{n \ge n_0} f_n > c\} \cap A) < \infty,$$

then

$$\fint_A f_n d\mu \to \fint_A f d\mu.$$

Proof. Without any loss of generality, we can assume that $A = X$. Let $h_n = \sup_{i \geq n} f_i, g_n = \inf_{i \geq n} f_i$; then h_n and g_n, $n = 1, 2, \ldots$, are measurable, and $h_n \searrow f$, $g_n \nearrow f$. Since $g_n \leq f_n \leq h_n$, we have

$$⨍ g_n d\mu \leq ⨍ f_n d\mu \leq ⨍ h_n \, d\mu,$$

and, therefore,

$$\lim_n ⨍ g_n \, d\mu \leq \varliminf_n ⨍ f_n d\mu \leq \varlimsup_n ⨍ f_n d\mu \leq \lim_n ⨍ h_n d\mu.$$

Noting that

$$\mu(\{x | h_{n_0}(x) > c\}) < \infty,$$

where $c \leq ⨍ f \, d\mu$, from Theorem 9.5 and Corollary 9.4, we get

$$\lim_n ⨍ g_n d\mu = \lim_n ⨍ h_n d\mu = ⨍ f \, d\mu.$$

So

$$\varliminf_n ⨍ f_n \, d\mu = \varlimsup_n ⨍ f_n \, d\mu = ⨍ f \, d\mu.$$ □

This theorem can be called the *everywhere convergence theorem.*

For a measurable function sequence which is convergent a.e., we have the following theorem.

Theorem 9.8. *$⨍_A f_n d\mu \to ⨍_A f \, d\mu$ whenever $A \in \mathbf{F}$, $f_n \xrightarrow{\text{a.e.}} f$ on A and there exists n_0 and a constant $c \leq ⨍_A f \, d\mu$ such that*

$$\mu(\{x | \sup_{n \geq n_0} f_n(x) > c\} \cap A) < \infty,$$

if and only if μ is null-additive.

Proof. It follows directly from Theorem 9.3 and Theorem 9.7. □

Corollary 9.7. *Let μ be finite and subadditive. If $f_n \xrightarrow{\text{a.e.}} f$, then we have*

$$⨍ f_n \, d\mu \to ⨍ f \, d\mu.$$

Theorem 9.8 can be called the *a.e. convergence theorem.*

A proposition analogous to Lebesgue's dominated convergence theorem in classical measure theory does not always hold for the Sugeno integral.

Example 9.6. Let $X = [0, \infty)$, $\mathbf{F} = \mathbf{B} \cap X$, and μ be the Lebesgue measure. Take

$$f_n(x) = \begin{cases} 1 & \text{if } x > n \\ 0 & \text{if } x \in [0, n]; \end{cases}$$

then $f_n \searrow f \equiv 0$. Observe that $0 \le f_n(x) \le 1$ for any $x \in X$ and any $n = 1, 2, \ldots$, and

$$⨍ 1\, d\mu = 1 < \infty.$$

In our case, however,

$$⨍ f_n\, d\mu = 1, \quad n = 1, 2, \ldots$$

and

$$⨍ f\, d\mu = 0.$$

Consequently, we have

$$\lim_n ⨍ f_n\, d\mu \neq ⨍ f\, d\mu.$$

In this example, the function sequence $\{f_n\}$ does not satisfy the finiteness condition on μ given in Theorem 9.5.

The most interesting convergence theorem of Sugeno integral sequence is for the measurable function sequence that converges in measure. Such a theorem is called the *convergence in measure theorem* and is given as follows.

Theorem 9.9. *$⨍_A f_n d\mu \to ⨍_A f\, d\mu$ whenever $A \in \mathbf{F}$, $\{f_n\} \subset \mathbf{G}$, $f \in \mathbf{G}$ and $f_n \xrightarrow{\mu} f$ on A, if and only if μ is autocontinuous.*

Proof. Sufficiency: Without any loss of generality, we can assume $A = X$. Let *μ be autocontinuous, $f_n \xrightarrow{\mu} f$ and let $c = ⨍ f\, d\mu$.*

(1) In the case when $c < \infty$, by Lemma 9.5(3), for any given $\varepsilon > 0$ we have $\mu(F_{c-\varepsilon}) \ge c$ and $\mu(F_{c+\varepsilon}) \le c$. On the one hand, it is easy to see that

$$F^n_{c+2\varepsilon} \subset F_{c+\varepsilon} \cup \{x \,|\, |f_n(x) - f(x)| \ge \varepsilon\}.$$

Since $f_n \xrightarrow{\mu} f$, we have

$$\mu(\{x||f_n(x) - f(x)| \geq \varepsilon\}) \to 0.$$

An application of autocontinuity from above yields that

$$\mu(F_{c+\varepsilon} \cup \{x||f_n(x) - f(x)| \geq \varepsilon\}) \to \mu(F_{c+\varepsilon}).$$

So there exists n_0 such that

$$\mu(F^n_{c+2\varepsilon}) \leq \mu(F_{c+\varepsilon} \cup \{x||f_n(x) - f(x)| \geq \varepsilon\}) \leq \mu(F_{c+\varepsilon}) + \varepsilon \leq c + \varepsilon \leq c + 2\varepsilon$$

whenever $n \geq n_0$. It follows, by Lemma 9.5(2), that

$$⨍ f_n \, d\mu \leq c + 2\varepsilon$$

for any $n \geq n_0$. On the other hand, to prove a converse inequality we only need to consider the case when $c > 0$. For any given $\varepsilon \in (0, c/2)$, we have

$$F^n_{c-2\varepsilon} \supset F_{c-\varepsilon} - \{x||f_n(x) - f(x)| \geq \varepsilon\}.$$

Since $f_n \xrightarrow{\mu} f$ and μ is autocontinuous from below, there exists n'_0 such that

$$\mu(F^n_{c-2\varepsilon}) \geq \mu(F_{c-\varepsilon}) - \varepsilon \geq c - 2\varepsilon$$

whenever $n \geq n'_0$. It follows from Lemma 9.5(1) that

$$⨍ f_n \, d\mu \geq c - 2\varepsilon$$

for any $n \geq n'_0$. Hence, $\lim_n ⨍ f_n \, d\mu$ exists, and

$$⨍ f_n \, d\mu \to c.$$

(2) Otherwise, $c = \infty$. In this case, from Lemma 9.4, $\mu(F_\alpha) = \infty$ for any $\alpha \in [0, \infty)$. For any given $N > 0$, we have

$$F^n_N \supset F_{N+1} - \{x||f_n(x) - f(x)| \geq 1\}.$$

Since $f_n \xrightarrow{\mu} f$ and μ is autocontinuous from below, there exists n_0 such that

$$\mu(F^n_N) \geq \mu(F_{N+1} - \{x||f_n(x) - f(x)| \geq 1\}) \geq N$$

whenever $n \geq n_0$. It follows from Lemma 9.5(1) that

$$⨍ f_n \, d\mu \geq N$$

for any $n \geq n_0$. This shows that

$$⨍ f_n \, d\mu \to \infty = c.$$

Necessity: For any $B \in \mathbf{F}$ and $\{B_n\} \subset \mathbf{F}$ with $\mu(B_n) \to 0$ we are going to prove that $\mu(B \cup B_n) \to \mu(B)$. Benefiting from the monotonicity of μ, we only need to consider the case when $\mu(B) < \infty$. Take $a > \mu(B)$ and

$$f(x) = \begin{cases} a & \text{if } x \in B \\ 0 & \text{if } x \notin B, \end{cases}$$

$$f_n(x) = \begin{cases} a & \text{if } x \in B \cup B_n \\ 0 & \text{if } x \notin B \cup B_n, \end{cases}$$

for any $n = 1, 2, \ldots$. Then, for any given $\varepsilon > 0$ we have

$$\{x \mid |f_n(x) - f(x)| \geq \varepsilon\} \subset B_n$$

for any $n = 1, 2, \ldots$. So $f_n \xrightarrow{\mu} f$. By the hypothesis of this proposition it should hold that

$$⨍ f_n \, d\mu \to ⨍ f \, d\mu.$$

Since

$$⨍ f_n \, d\mu = a \wedge \mu(B \cup B_n),$$

and

$$⨍ f \, d\mu = a \wedge \mu(B) = \mu(B),$$

we get

$$\mu(B \cup B_n) \to \mu(B).$$

That is, by Theorem 6.12 μ is autocontinuous. □

Making use of Theorems 9.4 and 9.9, we can immediately get the following convergence theorem of the Sugeno integral sequence for the measurable function sequence which *s*-mean converges (this theorem can be called the *s-mean convergence theorem*).

Theorem 9.10. *$⨍ f_n\, d_\mu \to ⨍ f\, d\mu$ whenever $\{f_n\} \subset \mathbf{G}$, $f \in \mathbf{G}$, and $\{f_n\}$ s-mean converges to f if and only if μ is autocontinuous.*

Example 9.7. (X, **F**, μ) is given in Example 3.5 and Example 6.3. μ is not autocontinuous from above. Take $f(x) = \chi_{\{1\}}(x)$, $f_n(x) = \chi_{\{1,\, n\}}(x)$ for $x \in X$ and $n = 1, 2, \ldots$. Then, for any given $\varepsilon \in (0, 1)$ we have

$$\mu(\{x \mid |f_n(x) - f(x)| \geq \varepsilon\}) = \mu(\{n\}) = 2^{-n} \to 0,$$

namely, $f_n \xrightarrow{\mu} f$. But $⨍ f\, d\mu = 1/2$ and $⨍ f_n\, d\mu = 1$ for any $n = 1, 2, \ldots$. $⨍ f_n\, d\mu$ does not tend to $⨍ f\, d\mu$.

Definition 9.3. Let (X, **F**, μ) be a monotone measure space, $f \in \mathbf{G}$. f is called *Sugeno integrable* (with respect to μ) iff $⨍ f\, d\mu < \infty$.

If we write

$\mathbf{L}^1(\mu) = \{f \mid f \in \mathbf{G}, f$ is Sugeno integrable with respect to $\mu\}$,

then we have the following theorem.

Theorem 9.11. *Let $A \in \mathbf{F}$, μ be uniformly autocontinuous. If $f_n \xrightarrow{\mu} f$ on A, then*

(1) $⨍_A f\, d\mu = \infty \Leftrightarrow$ there exists n_0 such that $⨍_A f_n\, d\mu = \infty$ for any $n \geq n_0$;

(2) $⨍_A f\, d\mu < \infty \Leftrightarrow$ there exists n_0 such that $⨍_A f_n\, d\mu < \infty$ for any $n \geq n_0$.

When $A = X$, we can rewrite the above propositions as

(1′) $f \notin \mathbf{L}^1(\mu) \Leftrightarrow$ there exists n_0 such that $f_n \notin \mathbf{L}^1(\mu)$ for any $n \geq n_0$;

(2′) $f \in \mathbf{L}^1(\mu) \Leftrightarrow$ there exists n_0 such that $f_n \in \mathbf{L}^1(\mu)$ for any $n \geq n_0$.

Proof. Without any loss of generality, we can assume $A = X$.

(1) Since the uniform autocontinuity implies the autocontinuity, from $f_n \xrightarrow{\mu} f$, by Theorem 9.9 we have

$$⨍ f_n\, d\mu \to ⨍ f d\mu.$$

So, if there exists n_0 such that $⨍ f_n\, d\mu = \infty$ for any $n \geq n_0$, we get $⨍ f\, d\mu = \infty$. Conversely, if $⨍ f\, d\mu = \infty$, by Lemma 9.4, $\mu(F_{\alpha+1}) = \infty$ for any $\alpha \in [0, \infty)$. Since $f_n \xrightarrow{\mu} f$ and μ is uniformly autocontinuous, there exists n_0 such that

$$\mu(F_{\alpha+1} - \{x \mid |f_n(x) - f(x)| \geq 1\}) = \infty$$

for any $\alpha \in [0, \infty)$ whenever $n \geq n_0$. From

$$F_\alpha^n \supset F_{\alpha+1} - \{x||f_n(x) - f(x)| \geq 1\}$$

for any $\alpha \in [0, \infty)$, we have

$$\mu(F_\alpha^n) \geq \mu(F_{\alpha+1} - \{x||f_n(x) - f(x)| \geq 1\}) = \infty$$

for any $\alpha \in [0, \infty)$ whenever $n \geq n_0$. Consequently, we have

$$⨍ f_n \, d\mu = \infty$$

for any $n \geq n_0$.

(2) An application of reduction to absurdity can show the implication $\Leftarrow$. As to the implication $\Rightarrow$, we can get it from

$$⨍ f_n \, d\mu \to ⨍ f d\mu < \infty. \qquad \square$$

At last, we give a convergence theorem of Sugeno integral sequence for the measurable function sequence that converges uniformly (it can be called the *uniform convergence theorem*). The symbol $f_n \xrightarrow{\text{u.}} f$ on A will denote that $\{f_n\}$ converges to f on A uniformly.

Theorem 9.12. *Let $A \in \mathbf{F}$. If $f_n \xrightarrow{\text{u.}} f$ on A, then*

$$⨍_A f_n \, d\mu \to ⨍_A f d\mu.$$

Proof. For any given $\varepsilon > 0$, since $f_n \xrightarrow{\text{u.}} f$ on A, there exists n_0 such that

$$|f_n - f| \leq \varepsilon$$

on A whenever $n \geq n_0$. By Lemma 9.2 we have

$$\left| ⨍_A f_n \, d\mu - ⨍_A f d\mu \right| \leq \varepsilon$$

for any $n \geq n_0$. This shows

$$⨍_A f_n \, d\mu \to ⨍_A f d\mu. \qquad \square$$

9.4 Transformation Theorem for Sugeno Integrals

In this section, we discuss how to transform a Sugeno integral $⨍_A f d\mu$, which is defined on a monotone measure space $(X, \mathbf{F}, \mu)$, into another Sugeno integral $⨍ g\, dm$ defined on the Lebesgue measure space $([0, \infty], \bar{\mathbf{B}}_+, m)$, where $\bar{\mathbf{B}}_+$ is the class of all Borel sets in $[0, \infty]$ and m is the Lebesgue measure.

Theorem 9.13. *For any $A \in \mathbf{F}$,*

$$⨍_A f d\mu = ⨍ \mu(A \cap F_\alpha) dm,$$

where $F_\alpha = \{x | f(x) \geq \alpha\}$ and m is the Lebesgue measure.

Proof. Denote $g(\alpha) = \mu(A \cap F_\alpha)$. From Lemma 9.1, we know that $g(\alpha)$ is decreasing with respect to α. For any $\alpha \in [0, \infty]$, denote

$$\mathbf{B}_\alpha = \{E | \sup E = \alpha, E \in \bar{\mathbf{B}}_+\}.$$

Then, $\{\mathbf{B}_\alpha | \alpha \in [0, \infty]\}$ is a partition of $\bar{\mathbf{B}}_+$ and $\sup_{E \in \mathbf{B}_\alpha} m(E) = \alpha$. Thus, from Theorem 9.1,

$$⨍ \mu(A \cap F_\alpha)\, dm = ⨍ g(\alpha)\, dm = \sup_{E \in \bar{\mathbf{B}}_+} [\inf_{\beta \in E} g(\beta) \wedge m(E)]$$

$$= \sup_{\alpha \in [0,\infty]} \sup_{E \in \mathbf{B}_\alpha} [\inf_{\beta \in E} g(\beta) \wedge m(E)].$$

Since $g(\beta)$ is decreasing, we have

$$g(\alpha-) \geq \inf_{\beta \in E} g(\beta) \geq g(\alpha)$$

for any $E \in \mathbf{B}_\alpha$, where $g(\alpha -) = \lim_{\beta \to \alpha-} g(\beta)$. So, on the one hand, we have

$$⨍ \mu(A \cap F_\alpha) dm \geq \sup_{\alpha \in [0,\infty]} [g(\alpha) \wedge \sup_{E \in \mathbf{B}_\alpha} m(E)] = \sup_{\alpha \in [0,\infty]} [g(\alpha) \wedge \alpha]$$

$$= \sup_{\alpha \in [0,\infty]} [\alpha \wedge \mu(A \cap F_\alpha)] = ⨍_A f d\mu;$$

on the other hand, for any given $\varepsilon > 0$,

$$\begin{aligned}
⨍\mu(A \cap F_\alpha)dm &\le \sup_{\alpha\in[0,\infty]} [g(\alpha-) \wedge \sup_{E\in \mathbf{B}_\alpha} m(E)] = \sup_{\alpha\in[0,\infty]} [g(\alpha-) \wedge \alpha] \\
&\le \sup_{\alpha\in[\varepsilon,\infty]} [\alpha \wedge g(\alpha-)] \vee \varepsilon \le \sup_{\alpha\in[\varepsilon,\infty]} [\alpha \wedge g(\alpha-\varepsilon)] \vee \varepsilon \\
&\le \sup_{(\alpha-\varepsilon)\in[0,\infty]} [(\alpha-\varepsilon) \wedge g(\alpha-\varepsilon)] + \varepsilon \\
&= \sup_{(\alpha-\varepsilon)\in[0,\infty]} [(\alpha-\varepsilon) \wedge \mu(A \cap F_{\alpha-\varepsilon})] + \varepsilon = ⨍_A f d\mu + \varepsilon.
\end{aligned}$$

Since ε may be close to zero arbitrarily, we obtain

$$⨍\mu(A \cap F_\alpha)\, dm = ⨍_A f d\mu.$$

The proof is now complete. □

9.5 Monotone Measures Defined by Sugeno Integrals

In this section we discuss how to define a monotone measure by using the Sugeno integral of a given measurable function with respect to another given monotone measure.

Theorem 9.14. *Let $(X, \mathbf{F}, \mu)$ be a monotone measure space, $f \in \mathbf{G}$. Then the set function ν defined by*

$$\nu(A) = ⨍_A f d\mu$$

for any $A \in \mathbf{F}$ is a monotone measure on $(X, \mathbf{F})$. Furthermore, if μ is continuous then ν is continuous from below; in addition, if μ is finite, then ν is a finite continuous monotone measure on $(X, \mathbf{F})$.

Proof. From Theorem 9.2, we know that $\nu(\emptyset) = 0$, and ν is monotone. Furthermore, we need to prove that ν is continuous from below when μ is continuous. Let $\{E_n\}$ be an increasing set sequence in $\mathbf{F}$, $E_n \nearrow E \in \mathbf{F}$. Then, we have

$$f \cdot \chi_{E_n} \nearrow f \cdot \chi_E$$

From Theorem 9.5, we have

$$\lim_n \nu(E_n) = \lim_n ⨍_{E_n} f d\mu = \lim_n ⨍ f \cdot \chi_{E_n} d\mu = ⨍ f \cdot \chi_E d\mu = ⨍_E f d\mu = \nu(E).$$

If μ is finite, for any given decreasing set sequence $\{E_n\}$ in **F** with $E_n \searrow E \in \mathbf{F}$, from

$$f \cdot \chi_{E_n} \searrow f \cdot \chi_E$$

and Theorem 9.5, we have also

$$\lim_{n} \nu(E_n) = \nu(E).$$

That is, ν is continuous from above. Consequently, ν is a continuous monotone measure. The finiteness of ν follows from

$$\nu(X) = ⨍ f d\mu \leq \mu(X) < \infty. \qquad \square$$

The following example shows that the set function ν may be not continuous from above when μ is not finite even it is continuous.

Example 9.8. Let $X = [0, \infty)$, **F** be the class of all Borel sets in $[0, \infty)$, μ be the Lebesgue measure, $f(x) \equiv 1$. Taking $E_n = [n, \infty)$, $n = 1, 2, \ldots$, we have $E_n \searrow \emptyset$, and

$$\nu(E_n) = ⨍_{E_n} f(x)\, d\mu = ⨍_{[n,\infty)} 1 \cdot d\mu = 1$$

for $n = 1, 2, \ldots$, but

$$\nu(\emptyset) = ⨍_{\emptyset} f(x) d\mu = 0.$$

So, ν is not continuous from above.

It is natural to ask whether ν is absolutely continuous with respect to μ. Unfortunately, generally speaking, the answer is negative. We can see this in the following example.

Example 9.9. Let $X = \{a, b\}$, $\mathbf{F} = \mathbf{P}(X)$, and μ be a monotone measure on $(X, \mathbf{F})$ with

$$\mu(E) = \begin{cases} 0 & \text{if } E = \emptyset \\ 1 & \text{otherwise.} \end{cases}$$

Taking

$$f(x) = \begin{cases} 0 & \text{if } x = a \\ 1 & \text{if } x = b, \end{cases}$$

we get a monotone measure ν by the expression

$$\nu(E) = ⨍_E f\,d\mu$$

for any $E \in \mathbf{F}$. Now, for $\varepsilon = 1/2 > 0$, take $F = X$, $E = \{a\}$. Even though $F \supset E$, and $\mu(F) - \mu(E) = 0$, which is less than any positive number $\delta > 0$, we have

$$\nu(F) - \nu(E) = ⨍ f\,d\mu - ⨍_{\{a\}} f\,d\mu = 1 - 0 = 1 > \varepsilon.$$

ν is not absolutely continuous with respect to μ.

However, if we introduce a weaker concept than the concept of absolute continuity given in Section 5.3, a weak absolute continuity, we will have a positive answer about the above-mentioned question.

Definition 9.4. Let μ and ν be two monotone measures on **C**. We say that ν is *weakly absolutely continuous* with respect to μ, in symbol $\nu \tilde{\ll} \mu$ iff, for any $\varepsilon > 0$, there exists $\delta > 0$, such that $\nu(E) < \varepsilon$ whenever, $E \in \mathbf{C}$ and $\mu(E) < \delta$.

It is evident that if μ and ν are two monotone measures on $(X, \mathbf{F})$, then $\nu \ll \mu$ implies $\nu \tilde{\ll} \mu$.

As well as the absolute continuity given before, the weak absolute continuity is a generalization of the concept of absolute continuity given in classical measure theory.

Theorem 9.15. *Let $(X, \mathbf{F}, \mu)$ be a finite monotone measure space and $f \in \mathbf{G}$. If ν is defined by*

$$\nu(E) = ⨍_E f\,d\mu$$

for any $E \in \mathbf{F}$, then $\nu \tilde{\ll} \mu$.

Proof. For any given $\varepsilon > 0$, take $\delta = \varepsilon$. Thus, for any $E \in \mathbf{F}$ with $\mu(E) < \delta = \varepsilon$, we have

$$\nu(E) = ⨍_E f\,d\mu \le \mu(E) < \varepsilon.$$

That is, $\nu \tilde{\ll} \mu$. □

9.6 More Results on Sugeno Integrals with Respect to a Monotone Measure

In this section, we consider Sugeno integrals with respect to monotone measures that may be not continuous. Let μ: $\mathbf{F} \to [0, \infty]$ be a monotone measure. We have seen that when μ is continuous from below, $⨍_A f\,d\mu = 0$ implies

$\mu(A \cap \{x|f(x)>0\}) = 0$ in Theorem 9.2(2). The following is a counterexample where μ is not continuous from below.

Example 9.10. Let $X = (0,1]$, $\mathbf{F} = \mathbf{B}_{(0,1]}$, where $\mathbf{B}_{(0,1]}$ is the class of all Borel sets in the interval (0,1], and let μ be defined on $\mathbf{F}$ as follows:

$$\mu(E) = \begin{cases} 1 & \text{if } E = X \\ 0 & \text{otherwise} \end{cases}$$

for any $E \in \mathbf{F}$. μ is nonnegative, monotone, and continuous from above, but not continuous from below. Take a measurable function $f(x) = x$, $x \in X$. Since $\mu(F_\alpha) = 0$ for any $\alpha \in (0, 1]$, we have

$$⨍ f d\mu = 0.$$

But, $\mu(X) = 1 \neq 0$.

As for Lemma 9.5, when μ is a monotone measure, checking the proof carefully, we find: conclusion (1) is still valid for μ; conclusions (2) and (3) are valid when μ is continuous from below; conclusion (4) is valid when μ is continuous from above; the validity of conclusion (5) needs both the continuity from above and the continuity from below of μ.

For a monotone measure μ, instead of Lemma 9.5, we may have the following lemma.

Lemma 9.7. *For any* $\alpha \in [0, \infty)$, *we have*

(1) $⨍_A f d\mu \geq \alpha \Leftrightarrow \mu(A \cap F_\beta) \geq \alpha$ *for any* $\beta < \alpha \Leftarrow \mu(A \cap F_\alpha) \geq \alpha$;
$⨍_A f d\mu < \alpha \Leftrightarrow$ *there exists* $\beta < \alpha$ *such that* $\mu(A \cap F_\beta) < \alpha \Rightarrow \mu(A \cap F_\alpha) < \alpha \Rightarrow \mu(A \cap F_{\alpha+}) < \alpha$;

(2) $⨍_A f d\mu \leq \alpha \Leftrightarrow \mu(A \cap F_\gamma) \leq \alpha$ *for any* $\gamma > \alpha \Leftarrow \mu(A \cap F_\alpha) \leq \alpha$;
$⨍_A f d\mu > \alpha \Leftrightarrow$ *there exists* $\gamma > \alpha$ *such that* $\mu(A \cap F_\alpha) > \alpha$.

(3) $⨍_A f d\mu = \alpha \Leftrightarrow \mu(A \cap F_\beta) \geq \alpha \geq \mu(A \cap F_\gamma)$ *for any* $\beta < \alpha$ *and* $\gamma > \alpha$.

Proof. We only need to prove some parts of (2). If $\mu(A \cap F_\gamma) \leq \alpha$ for any $\gamma > \alpha$, then

$$⨍_A f d\mu = \sup_{\gamma \in [0,\infty]} [\gamma \wedge \mu(A \cap F_\gamma)] = \sup_{\gamma \in [0,\alpha]} [\gamma \wedge \mu(A \cap F_\gamma)] \vee \sup_{\gamma \in (\alpha,\infty]} [\gamma \wedge \mu(A \cap F_\gamma)]$$

$$\leq \sup_{\gamma \in [0,\alpha]} \gamma \vee \sup_{\gamma \in (\alpha,\infty]} \mu(A \cap F_\gamma) \leq \alpha \vee \alpha = \alpha.$$

Conversely, if there exists $\gamma > \alpha$ such that $\mu(A \cap F_\gamma) > \alpha$, then

$$⨍_A f d\mu \geq \gamma \wedge \mu(A \cap F_\gamma) > \alpha. \qquad \square$$

The results given in Section 9.3 are mostly dependent on the continuity of μ. However, if μ is a possibility measure defined on $(X, \mathbf{P}(X))$, which we denote by π, then we have the following convergence theorem.

Theorem 9.16. *Let* $(X, \mathbf{P}(X), \pi)$ *be a possibility measure space. If* $f_n \xrightarrow{\pi} f$ *on* A, *then*

$$\fint_A f_n \, d\pi \to \fint_A f d\pi.$$

Proof. Since any possibility measure is maxitive (and, therefore, autocontinuous) and continuous from below, part (1) of sufficiency in the proof of Theorem 9.9 works (now, $c = \fint f \, d\pi \leq 1 < \infty$). □

From Theorem 7.12 and Theorem 9.16, we obtain the following corollary.

Corollary 9.8. *Let* $(X, \mathbf{P}(X), \pi)$ *be a possibility measure space. If* $f_n \xrightarrow{\text{a.u.}} f$ on A, then

$$\fint_A f_n \, d\pi \to \fint_A f \, d\pi.$$

Unfortunately, since possibility measures do not possess continuity from above in general, it is impossible to establish an everywhere (or, a.e.) convergence theorem of Sugeno integral sequence on a possibility measure space; that is, while $f_n \to f$ everywhere, $\fint f_n \, d\pi \to \fint f \, d\pi$ may not be true.

Example 9.11. Consider the possibility measure space $(X, \mathbf{P}(X), \pi)$ and the measurable function sequence $\{f_n\}$ given in Example 7.4. We have that $f_n \to f = 0$ everywhere on X, and $\fint f_n \, d\pi = 1$ for any $n = 1, 2, \ldots$, but $\fint f \, d\pi = 0$.

Notes

9.1 The concept of a Sugeno integral (fuzzy integral) for a measurable function $f : \mathbf{F} \to [0,1]$ on a normalized monotone measure space $(X, \mathbf{F}, \mu)$ was introduced by Sugeno [1974], who also discussed some elementary properties of this integral. Further investigations of the integral were also pursued by Batle and Trillas [1979], Wierzchon [1982], Dubois and Prade [1980], Grabisch et al. [1992], and other researchers. A generalization of the integral, as presented in this chapter, was introduced by Ralescu and Adams [1980] and Wang [1984].

9.2 As shown by Wang [1984], there is no essential difficulty to define the Sugeno integral (fuzzy integral) of a nonnegative extended real-valued measurable function (i.e., measurable function $f: \mathbf{F} \to [0, \infty]$) on monotone measure space. In fact, most properties and results presented in this chapter hold also for such an integral.

9.3 Ralescu and Adams [1980] introduced an equivalent definition of a Sugeno integral (fuzzy integral) by using simple measurable nonnegative functions—an idea similar to the definition of the Lebesgue integral.

9.4 The earliest monotone convergence theorem of Sugeno integral sequences was conjectured by Sugeno [1974]. It was proven by Ralescu and Adams [1980] by using an equivalent definition of a Sugeno integral (Note 9.3). They also introduced a theorem on convergence in monotone measure. Its proof is based on a rather strong condition of subadditivity. Wang [1984] improved this result by using autocontinuity, a far weaker condition, and proved that this condition is necessary and sufficient for the convergence by using a new concept of local uniform autocontinuity. In this chapter we use a brief proof, which is due to Wang [1984].

9.5 Most of the results given in Sections 9.1–9.3 were previously published by Wang [1984, 1985a].

9.6 Using the concepts of "pseudo a.e.," pseudo-null-additivity, and pseudo-autocontinuity, we can also obtain some convergence theorems similar to those presented in Section 9.3 [Wang, 1985a].

9.7 The transformation theorem of Sugeno integral (fuzzy integral) was proposed by Ralescu and Adams [1980]. The brief and effective proof given in Section 9.4 was obtained by Wang and Qiao [1990].

9.8 The Sugeno integral can also be defined on a monotone measure space with a fuzzy σ-algebra consisting of fuzzy sets. Many important results have already been obtained for this generalization. [Qiao,1990, 1991; Wang and Qiao, 1990; Zhang, 1992a, b].

9.9 Suárez and Gill [1986] introduce two families of nonlinear integrals based on the concepts of t-seminorms and t-semiconorms and show that Sugeno integral is a special case. Weber [1986] discusses the Sugeno integral in the context of integrals based on decomposable measures.

Exercises

9.1. Calculate the value of the Sugeno integral $⨍ f\,d\mu$,where μ and f are given as follows:

(a) $X = \{a, b\}, \mathbf{F} = \mathbf{P}(X)$.

$$\mu(E) = \begin{cases} 0 & \text{if } E = \emptyset \\ 0.5 & \text{if } E = \{a\} \\ 0.7 & \text{if } E = \{b\} \\ 1 & \text{if } E = X, \end{cases} \qquad f(x) = \begin{cases} 0.8 & \text{if } x = a \\ 0.4 & \text{if } x = b; \end{cases}$$

(b) $(X, \mathbf{F}, \mu)$ is as given in (a), but

$$f(x) = \begin{cases} 0.8 & \text{if } x = a \\ 0.9 & \text{if } x = b; \end{cases}$$

(c) $X = \{a, b, c, d\}$, $\mathbf{F} = \mathbf{P}(X)$, μ is a λ-measure with $\mu(\{a\}) = 1/15$, $\mu(\{b\}) = 1/4$, $\mu(\{c\}) = 1/5$, $\lambda = 1$, and

$$f(x) = \begin{cases} 2/3 & \text{if } x = a \\ 1/5 & \text{if } x = b \\ 1/2 & \text{if } x = c \\ 1 & \text{if } x = d; \end{cases}$$

(d) $(X, \mathbf{F}, \mu)$ is the same as in (c), but

$$f(x) = \begin{cases} 1/2 & \text{if } x = a \\ 1/3 & \text{if } x = b \\ 1/4 & \text{if } x = c \\ 1/5 & \text{if } x = d; \end{cases}$$

(e) $X = [0, 1]$, $\mathbf{F}$ is the class of all Borel sets in X, μ is the Lebesgue measure, and $f(x) = x^2$;

(f) $X = (-\infty, \infty)$, $\mathbf{F}$ is the Borel field, $\mu = m^2$, where m is the Lebesgue measure, and $f(x) = 1/(1 + x^2)$.

9.2. Let $(X, \mathbf{F}, \mu)$ be a monotone measure space and let $f\colon \mathbf{F} \to [0, \infty]$ be a nonnegative extended real-valued function. We can define the Sugeno integral of f with respect to μ, just as we do in Definition 9.1, and use the same symbol $⨍ f\, d\mu$. Prove that

$$⨍ f\, d\mu = \sup_{\alpha \in [0,\infty)} [\alpha \wedge \mu(F_\alpha)].$$

9.3. Prove the following:

(a) $⨍_A f\, d\mu \le \mu(A)$;

(b) $⨍_A (a \wedge f)\, d\mu = \alpha \wedge ⨍_A f\, d\mu$, where a is a nonnegative constant.

9.4. Let μ be maxitive. Prove that

$$⨍_A (f_1 \vee f_2)\, d\mu = ⨍_A f_1\, d\mu \vee ⨍_A f_2\, d\mu$$

for any $A \in \mathbf{F}$ and $f_1 \in \mathbf{G}$, $f_2 \in \mathbf{G}$.

9.5. Give an example to show that the equality in Exercise 9.4 may not be true when μ is not maxitive.

9.6. Give examples to show that the equalities

$$\fint (f_1 + f_2) d\mu = \fint f_1 d\mu + \fint f_2 d\mu$$

and

$$\fint cf\, d\mu = c \fint f\, d\mu \quad (c \text{ is a constant})$$

may be true or may not be true. This means that, in general, the Sugeno integral is not linear in the classical sense. However, if we use the supremum and the infimum instead of addition and multiplication in the expression of linearity, respectively, and if we call it a *maxitive linearity*, then the Sugeno integral is maxitively linear when μ is maxitive (see Exercise 9.3(b) and 9.4).

9.7. Give an example to show that

$$\fint_A f\, d\mu \geq \alpha \Rightarrow \mu(A \cap F_\alpha) \geq \alpha$$

may not be true when $\mu(A) = \infty$.

9.8. Prove that

$$F_\alpha^n \searrow \cap_{n=1}^{\infty} F_\alpha^n = F_\alpha$$

when $f_n \searrow f$, and prove that

$$F_{\alpha+}^n \nearrow \bigcup_{n=1}^{\infty} F_{\alpha+}^n = F_{\alpha+}$$

and

$$F_\alpha \supset F_\alpha^n \nearrow \bigcup_{n=1}^{\infty} F_\alpha^n \supset F_{\alpha+}$$

when $f_n \nearrow f$, where

$$F_\alpha^n = \{x | f_n(x) \geq \alpha\}, \quad F_{\alpha+}^n = \{x | f_n(x) > \alpha\}, \quad F_\alpha = \{x | f(x) \geq \alpha\},$$
$$F_{\alpha+} = \{x | f(x) > \alpha\}.$$

9.9. Prove the following:

(a) $\int_A \sup_n f_n d\mu \geq \sup_n \int_A f_n d\mu$;

(b) $\int_A \inf_n f_n d\mu \leq \inf_n \int_A f_n d\mu$;

(c) $\int_{\bigcup_n A_n} f_n d\mu \geq \sup_n \int_{A_n} f d\mu$;

(d) $\int_{\bigcap_n A_n} f d\mu \leq \inf_n \int_{A_n} f d\mu$.

9.10. Give an example to show that the equality

$$\{x|\lim_n f_n \geq \alpha\} = \lim_n (x|f_n \geq \alpha\}$$

may not be true, where α is a nonnegative constant and $\{f_n\}$ is an increasing sequence of functions in **G**.

9.11. Let μ and ν be two continuous monotone measures on $(X, \mathbf{F})$. Prove that, if $\nu \lll \mu$, then $\nu(E) = 0$ whenever $E \in \mathbf{F}$ and $\mu(E) = 0$ (this statement can also be regarded as a generalized form of classical absolute continuity). Can you give an example to show that the converse proposition is not true?

9.12. Give an example to show that the weak absolute continuity is really weaker than the absolute continuity.

Chapter 10
Pan-Integrals

10.1 Pan-Additions and Pan-Multiplications

Lebesgue's integral involves two binary operations, common addition and common multiplication of real numbers, while the Sugeno integral involves different binary operations, logical addition (maximum) and logical multiplication (minimum) of real numbers.

A natural idea is to consider an appropriate class of two binary operations in terms of which a generalized theory of integration could be formulated and developed, a theory under which both theories based upon Lebesgue's integral and the sugeno integral would be subsumed.

Let $R_+ = [0, \infty)$, $\overline{R}_+ = [0, \infty]$, $\mathbf{B}_+ = \mathbf{B} \cap R_+$, and $a, b, c, d, a_i, b_i, a_t \in \overline{R}_+ (i = 1, 2, \ldots, t \in T$, where T is any given index set).

Definition 10.1. Let $\oplus$ be a binary operation on $\overline{R}_+$. The pair $(\overline{R}_+, \oplus)$ is called a *commutative isotonic semigroup* and $\oplus$ is called a *pan-addition* on $\overline{R}_+$ iff $\oplus$ satisfies the following requirements:
(PA1) $a \oplus b = b \oplus a$;
(PA2) $(a \oplus b) \oplus c = a \oplus (b \oplus c)$;
(PA3) $a \leq b \Rightarrow a \oplus c \leq b \oplus c$ for any c;
(PA4) $a \oplus 0 = a$;
(PA5) $\lim_n a_n$ and $\lim_n b_n$ exist $\Rightarrow \lim_n (a_n \oplus b_n)$ exists, and $\lim_n (a_n \oplus b_n) = \lim_n a_n \oplus \lim_n b_n$.

From (PA1) and (PA3), it follows that

$$\text{(PA3}'\text{)}\ a \leq b \text{ and } c \leq d \Rightarrow a \oplus c \leq b \oplus d.$$

Because of (PA2), we may write $\oplus_{i=1}^n a_i$ for a $a_1 \oplus a_2 \oplus \cdots \oplus a_n$. We also use a similar symbol $\oplus_{t \in T} a_t$, where T is a finite index set. Furthermore, if T is an infinite index set, we define $\oplus_{t \in T} a_t = \sup_{T' \subset T} \oplus_{t \in T'} a_t$, where T' is finite.

Definition 10.2. Let $\otimes$ be a binary operation on $\overline{R}_+$. The triple $(\overline{R}_+, \oplus, \otimes)$, where $\oplus$ is a pan-addition on $\overline{R}_+$, is called a *commutative isotonic semiring* with respect to $\oplus$ and $\otimes$ iff:
(PM1) $a \otimes b = b \otimes a$;

Z. Wang, G.J. Klir, *Generalized Measure Theory*,
DOI: 10.1007/978-0-387-76852-6_10,

(PM2) $(a \otimes b) \otimes c = a \otimes (b \otimes c)$;
(PM3) $(a \oplus b) \otimes c = (a \otimes c) \oplus (b \otimes c)$;
(PA4) $a \leq b \Rightarrow a \otimes c \leq b \otimes c$ for any c;
(PM5) $a \neq 0$ and $b \neq 0 \Leftrightarrow a \otimes b \neq 0$;
(PM6) there exists $I \in \overline{R}_+$, such that $I \otimes a = a$, for any $a \in \overline{R}_+$;
(PM7) $\lim_n a_n$ and $\lim_n b_n$ exist and are finite $\Rightarrow \lim_n (a_n \otimes b_n) = \lim_n a_n \otimes \lim_n b_n$.

The operation $\otimes$ is called a *pan-multiplication* on $\overline{R}_+$, and the number I is called the *unit element* of $(\overline{R}_+, \oplus, \otimes)$. From (PM1) and (PM4), we derive
(PM4′) $a \leq b$ and $c \leq d \Rightarrow a \otimes c \leq b \otimes d$.

It is easy to see that (PM5) implies that $a \otimes 0 = 0$ and $0 \otimes a = 0$ for any $a \in \overline{R}_+$.

Example 10.1. $\overline{R}_+$ with the common addition and the common multiplication of real numbers is a commutative isotonic semiring. It is denoted by $(\overline{R}_+, +, \cdot)$ and its unit element is 1.

Example 10.2. $\overline{R}_+$ with the logical addition $\vee$ and the logical multiplication $\wedge$ of real numbers is a commutative isotonic semiring. It is denoted by $(\overline{R}_+, \vee, \wedge)$ and its unit element is ∞.

Example 10.3. $\overline{R}_+$ with the logical addition $\vee$ and the common multiplication of real numbers is a commutative isotonic semiring. It is denoted by $(\overline{R}_+, \vee, \cdot)$ and its unit element is 1.

Definition 10.3. If $(X, \mathbf{F}, \mu)$ is a continuous monotone measure space, and $(\overline{R}_+, \oplus, \otimes)$ is a commutative isotonic semiring, then $(X, \mathbf{F}, \mu, \overline{R}_+, \oplus, \otimes)$ is called a *pan-space*.

10.2 Definition of Pan-Integrals

Definition 10.4. Let $(X, \mathbf{F}, \mu, \overline{R}_+, \oplus, \otimes)$ be a pan-space and $E \subset X$. The function defined on X given by

$$\chi_E(x) = \begin{cases} I & \text{if } x \in E \\ 0 & \text{otherwise} \end{cases}$$

is called the *pan-characteristic function* of E, where I is the unit element of $(\overline{R}_+, \oplus, \otimes)$

Definition 10.5. Let $(X, \mathbf{F})$ be a measurable space. A partition $\{E_i\}$ of X is called *measurable* iff $E_i \in \mathbf{F}$ for every i.

Definition 10.6. Let $(X, \mathbf{F}, \mu, \overline{R}_+, \oplus, \otimes)$ be a pan-space. A function on X given by

$$s(x) = \bigoplus_{i=1}^{n} [a_i \otimes \chi_{E_i}(x)]$$

is called a *pan-simple measurable function*, where $a_i \in R_+$, $i = 1, 2, \ldots, n$, and $\{E_i \mid i = 1, 2, \ldots, n\}$ is a measurable partition of X.

In the rest of this chapter we restrict the discussion to a given pan-space $(X, \mathbf{F}, \mu, \overline{R}_+, \oplus, \otimes)$.

The set of all pan-simple measurable functions is denoted by **Q**. Obviously, $\mathbf{Q} \subset \mathbf{G}$. For any

$$s(x) = \oplus_{i=1}^{n}[a_i \otimes \chi_{E_i}(x)] \in \mathbf{Q},$$

let $e_S = \{(a_i, E_i) | i = 1, \cdots, n\}$. We write

$$P(e_S|A) = \bigoplus_{i=1}^{n}[a_i \otimes \mu(A \cap E_i)],$$

where $A \in \mathbf{F}$. To simplify the notation, we use $P(s|A)$ to replace $P(e_S\,|A)$in the following discussion if there is no confusion. However, we should remember that the value of $P(s|A)$ depends on the expression of s (i.e., e_S).

Given $f_1, f_2 \in \mathbf{G}$, we write $f_1 \leq f_2$ if $f_1(x) \leq f_2(x)$ for every $x \in X$.

Definition 10.7. Let $f \in \mathbf{G}$ and $A \in \mathbf{F}$. The *pan-integral* of f on A with respect to μ, which is denoted by $(p) \int x f d\mu$, is given by

$$(p)\int_A f d\mu = \sup_{0 \leq s \leq f, s \in \mathbf{Q}} P(e_s|A).$$

When $A = X$, we simply write $(p)\int f d\mu$ instead of $(p) \int_X f\, d\mu$.

Theorem 10.1. *Let $f \in \mathbf{G}$, $A \in \mathbf{F}$, and let $\hat{\mathbf{P}}$ denote the set of all measurable partitions of X. Then,*

$$(p)\int_A f d\mu = \sup_{\mathbf{E} \in \widehat{\mathbf{P}}}\left\{\bigoplus_{E \in \mathbf{E}}[(\inf_{x \in E} f(x)) \otimes \mu(A \cap E)]\right\}.$$

Proof. On the one hand, for any given $\mathbf{E} \in \hat{\mathbf{P}}$ and any chosen finite part $\{E_i | i = 1, 2, \ldots, n\} \subset \mathbf{E}$ we take

$$s(x) = \bigoplus_{i=1}^{n}\left\{[\inf_{x \in E_i} f(x)] \otimes + \chi_{E_i}(x)\right\}.$$

Then, $s(x) \in \mathbf{Q}$ and $s \leq f$. So,

$$\bigoplus_{i=1}^{n}\left[(\inf_{x \in E_i} f(x)) \otimes \mu(A \cap E_i)\right] \leq (p)\int_A f\, d\mu,$$

and, therefore,

$$\sup_{\mathbf{E} \in \widehat{\mathbf{P}}}\left\{\bigoplus_{E \in \mathbf{E}}[(\inf_{x \in E} f(x)) \otimes \mu(A \cap E)]\right\} \leq (p)\int_A f\, d\mu.$$

If, on the other hand, $s(x) = \oplus_{i=1}^{n}[a_i \otimes \chi_{E_i}(x)] \in \mathbf{Q}$, then $\{E_i | i = 1, 2, \ldots, n\} \in \hat{\mathbf{P}}$. Moreover, if $s \le f$, then $a_i \le \inf_{x \in E_i} f(x)$. Thus, we have

$$P(s|\ A) \le \bigoplus_{i=1}^{n}[(\inf_{x\in E_i} f(x)) \otimes \mu(A \cap E_i)] \le \sup_{\mathbf{E}\in\hat{\mathbf{P}}}\left\{ \bigoplus_{E\in\mathbf{E}}[(\inf_{x\in E} f(x)) \otimes \mu(A\cap E)]\right\}.$$

From this, it follows that

$$(p)\int_A f\, d\mu \le \sup_{\mathbf{E}\in\hat{\mathbf{P}}}\left\{ \bigoplus_{E\in\mathbf{E}}[(\inf_{x\in E} f(x)) \otimes \mu(A\cap E)]\right\}.$$

Consequently, we have

$$(p)\int_A f\, d\mu = \sup_{\mathbf{E}\in\hat{\mathbf{P}}}\left\{ \bigoplus_{E\in\mathbf{E}}[(\inf_{x\in E} f(x)) \otimes \mu(A\cap E)]\right\}.$$

The proof is now complete. □

Theorem 10.2. *When $\oplus$ is the logical addition $\vee$ and $\otimes$ is the logical multiplication $\wedge$, we have*

$$(p)\int_A f\, d\mu = ⨍_A f\, d\mu$$

for any $f \in \mathbf{G}$ and $A \in \mathbf{F}$; that is, the pan-integral and the Sugeno integral coincide.

Proof. Since $\{E,\ \overline{E}\}$ is a measurable partition of X for any $E \in \mathbf{F}$; the inequality

$$(p)\int_A f d\mu \ge ⨍_A f\, d\mu$$

follows from Theorem 9.1 and Theorem 10.1 directly. Conversely, for any given $\varepsilon > 0$ and any $\mathbf{E} \in \hat{\mathbf{P}}$, there exists $E_0 \in \mathbf{E}$ such that

$$\begin{aligned}\bigoplus_{E\in\mathbf{E}}[(\inf_{x\in E} f(x)) \otimes \mu(A\cap E)] &= \sup_{E\in\mathbf{E}}[(\inf_{x\in E} f(x)) \wedge \mu(A\cap E)] \\ &\le [\inf_{x\in E_0} f(x)] \wedge \mu(A\cap E_0) + \varepsilon \le ⨍_A f d\mu + \varepsilon.\end{aligned}$$

Thus, we have

$$(p)\int_A f d\mu \le ⨍_A f d\mu + \varepsilon.$$

Since ε may be arbitrarily close to zero, we have

$$(p)\int_A f d\mu \le ⨍_A f\, d\mu.$$

Consequently, we have

$$(p)\int_A f\,d\mu = \int_A f\,d\mu. \qquad \square$$

Theorem 10.3. *Let μ be σ–additive. When $\oplus$ is the common addition and $\otimes$ is the common multiplication, we have*

$$(p)\int_A f\,d\mu = \int_A f\,d\mu$$

for any $f \in \mathbf{G}$ and $A \in \mathbf{F}$; that is, the pan-integral and Lebesgue's integral coincide.

Proof. When $\oplus$ is $+$ and $\otimes$ is $\cdot$, the concept of pan-simple function coincides with the concept of nonnegative simple function employed in classical measure theory. From the definition of Lebesgue's integral,

$$\int_A f\,d\mu = \lim_n P(s_n|A)$$

for any $f \in \mathbf{G}$ and $A \in \mathbf{F}$, where $\{s_n\}$ is a sequence of nondecreasing nonnegative simple functions whose limit is f. It is easy to see that

$$(p)\int_A f\,d\mu \geq \int_A f\,d\mu.$$

Conversely, we can choose $\{s_n\} \subset \mathbf{Q}$ such that $s_n \leq f$, $n = 1, 2, \ldots$, and

$$\lim_n P(s_n|A) = (p)\int_A f\,d\mu.$$

Taking $\bar{s}_n = \sup_{i \leq n} s_i$, we have $\bar{s}_n \in \mathbf{Q}$, $n = 1, 2, \ldots$, and $\bar{s}_n \nearrow f$. Since $P(s|A)$ is nondecreasing with respect to s under the conditions given in the theorem, we have

$$\lim_n P(\bar{s}_n|A) = (p)\int_A f\,d\mu.$$

Owing to the monotonicity of $\{\bar{s}_n\}$, we also have

$$\lim_n P(\bar{s}_n|A) = \int_A f\,d\mu.$$

Hence, we have

$$(p)\int_A f\,d\mu = \int_A f\,d\mu. \qquad \square$$

We can also see from Theorem 10.1 that, under the conditions given in Theorem 10.3, the pan-integral is just Riemann's integral, provided that the function f is continuous.

10.3 Properties of Pan-Integral

Some interesting properties, which are similar to those of the Sugeno integral and the Lebesgue integral, are derived in this section.

Theorem 10.4. *For any $f \in \mathbf{G}$ and $A \in \mathbf{F}$,*

$$(p)\int_A f\,d\mu = (p)\int f \otimes \chi_A\,d\mu.$$

Proof. Taking $s' = s \otimes \chi_a$ for any given $s \in \mathbf{Q}$ satisfying $s \leq f$, we have $s' \in \mathbf{Q}$ and $s' \leq f \otimes \chi_A$. By using the commutative law (PM1) and the distributive law (PM3) of pan-multiplication, we have

$$s' = \bigoplus_{i=1}^{n}(a_i \otimes \chi_A \otimes \chi_{E_i}) = \bigoplus_{i=1}^{n}(a_i \otimes \chi_{A \cap E_i}).$$

This means that

$$P(s'|X) = \bigoplus_{i=1}^{n}[a_i \otimes \mu(A \cap E_i)].$$

Hence, we have

$$P(s|A) = P(s'|X) \leq (p)\int f \otimes \chi_A\,d\mu$$

and, therefore,

$$(p)\int_A f\,d\mu \leq (p)\int f \otimes \chi_A\,d\mu.$$

Conversely, for any given $s(x) = \oplus_{i=1}^{n}[a_i \otimes \chi_{E_i}(x)] \in \mathbf{Q}$ satisfying $s \leq f \otimes \chi_A$, we omit, without any loss of generality, those terms in which $a_i = 0$ and, therefore, we may assume that $a_i > 0$, $i = 1, 2, \ldots, n$. From $s \leq f \otimes \chi_A$, we deduce $E_i \subset A$, $i = 1, 2, \ldots, n$. Thus,

$$P(s|X) = \bigoplus_{i=1}^{n}[a_i \otimes \mu(E_i)] = \bigoplus_{i=1}^{n}[a_i \otimes \mu(A \cap E_i)] = P(s|X) \leq (p)\int_A f\,d\mu.$$

So, we have

$$(p)\int_A f\,d\mu \geq (p)\int f \otimes \chi_A\,d\mu.$$

Finally, we have

$$(p)\int_A f\,d\mu = (p)\int f \otimes \chi_A\,d\mu. \qquad \square$$

The properties listed in the following theorem can be easily obtained from Definition 10.7, Theorem 10.1, and Theorem 10.4.

Theorem 10.5. *Let $f, g \in \mathbf{G}$, $A, B \in \mathbf{F}$, and $a \in R_+$. Then we have the following:*
(1) if $f = 0$ on A a.e., then $(p)\int_A f\,d\mu = 0$;
(2) if $\mu(A) = 0$, then $(p)\int_A f\,d\mu = 0$;
(3) if $f \leq g$ on A, then $(p)\int_A f\,d\mu \leq (p)\int_A g\,d\mu$;
(4) if $A \subset B$, then $(p)\int_A f\,d\mu \leq (p)\int_B f\,d\mu$;
(5) $(p)\int_A a\,d\mu \geq a \otimes \mu(A)$.

The following example shows that the equality in (5) of Theorem 10.5 may not hold.

Example 10.4. Let $(X, \mathbf{F}, \mu)$ be the same monotone measure space as given in Exercise 9.1 (a). Taking $+$ and $\times$ as $\oplus$ and $\otimes$, respectively, we have

$$(p)\int 1d\mu = 1 \cdot \mu(\{a\}) + 1 \cdot \mu(\{b\}) = 0.5 + 0.7 = 1.2,$$

but

$$1 \cdot \mu(X) = 1.$$

Theorem 10.6. *Let $f \in \mathbf{G}$ and $A \in \mathbf{F}$. If $(p)\int_A f\,d\mu = 0$, then*

$$\mu(A \cap \{x | f(x) > 0\}) = 0$$

Proof. Denoting $B_n = A \cap \{x \mid f(x) > 1/n\}$, we have

$$B_n \nearrow \bigcup_{n=1}^{\infty} B_n = A \cap \{x | f(x) > 0\}.$$

By using Theorem 10.5, we obtain

$$0 = (p)\int_A f\,d\mu \geq (p)\int_{B_n} f\,d\mu \geq (p)\int_{B_n} \frac{1}{n}\,d\mu \geq \frac{1}{n} \otimes \mu(B_n) \geq 0.$$

From (PM5), we have

$$\mu(B_n) = 0, n = 1, 2, \ldots;$$

therefore,

$$\mu(A \cap \{x \big| f(x) > 0\}) = \lim_n \mu(B_n) = 0. \qquad \square$$

10.4 A Transformation Theorem

A transformation theorem for the Sugeno integral is presented in Chapter 9. Now, we consider a similar theorem, wherein the pan-addition is the common addition and the pan-multiplication is the common multiplication.

Theorem 10.7. *On a pan-space* $(X, \mathbf{F}, \mu, \overline{R}_+, +, \times)$*, if* μ *is superadditive, then*

$$(p)\int \mu(A \cap F_\alpha)dm \geq (p)\int_A f d\mu,$$

where $f \in \mathbf{G}$*,* $A \in \mathbf{F}$*, m is the Lebesgue measure on* $\overline{R}_+$*, and*

$$F_\alpha = \{x | f(x) \geq \alpha\},$$

for any $\alpha \in R_+$.

Proof. There is no loss of generality in assuming that $A = X$. From Theorem 10.1, we infer

$$(p)\int f d\mu = \sup_{\mathbf{E} \in \widehat{\mathbf{P}}}\left\{ \underset{E \in \mathbf{E}}{\oplus}([\inf_{x \in E} f(x)] \otimes \mu(E))\right\} = \sup_{\mathbf{E} \in \hat{\mathbf{P}}} \sup_{\mathbf{D} \subset \mathbf{E}}\left\{ \sum_{E \in \mathbf{D}}([\inf_{x \in E} f(x)] \cdot \mu(E))\right\},$$

where $\mathbf{D}$ is any finite subclass of $\mathbf{E}$. For any chosen

$$\mathbf{D} = \{E_1, E_2, \ldots, E_n\},$$

taking $\alpha_i = \inf_{x \in E_i} f(x)$, $i = 1, 2, \ldots, n$, and $\alpha_{n+1} = 0$, we may assume that

$$\alpha_1 \geq \alpha_2 \geq \cdots \geq \alpha_n \geq \alpha_{n+1} = 0$$

(otherwise, we just need to rearrange the order of $E_1, E_2, \ldots, E_n$). Let $B_i = (\alpha_i + 1, \alpha_i]$, $i = 1, 2, \ldots, n$. Accordingly, we have

$$m(B_i) = \alpha_i - \alpha_{i+1}, i = 1, 2, \ldots, n.$$

For any $\alpha \in B_i$, since $\alpha \le \alpha_i$, we may infer

$$F_\alpha \supset \bigcup_{j \le i} E_j.$$

Using the superadditivity of μ, one may derive

$$\mu(F_\alpha) \ge \sum_{j \le i} \mu(E_j).$$

Therefore, we obtain

$$\inf_{\alpha \in B_i} \mu(F_\alpha) \ge \sum_{j \le i} \mu(E_j).$$

Thus, we have

$$(p)\int \mu(F_\alpha)dm \ge \sum_{i=1}^{n} [\inf_{\alpha \in B_i} \mu(F_\alpha) \cdot m(B_i)] \ge \sum_{i=1}^{n} \left[\sum_{j \le i} \mu(E_j) \cdot (\alpha_i - \alpha_{i+1})\right]$$

$$= \sum_{i=1}^{n} \alpha_i \cdot \mu(E_i) = \sum_{i=1}^{n} (\inf_{x \in E_i} f(x)) \cdot \mu(E_i).$$

Finally, we have

$$(p)\int \mu(F_\alpha)dm \ge (p)\int f d\mu.$$ □

When μ is subadditive, a similar result is expressed by the following theorem.

Theorem 10.8. *On a pan-space* $(X, \mathbf{F}, \mu, \overline{R}_+, +, \times)$, *if* μ *is subadditive, then we have*

$$(p)\int \mu(A \cap F_\alpha)dm \le (p)\int_A f d\mu,$$

where $f \in F$, $A \in \mathbf{F}$, m *is the Lebesgue measure on* $\overline{R}_+$, *and*

$$F_\alpha = \{x | f(x) \ge \alpha\},$$

for any $\alpha \in R_+$.

Proof. Assume that $A = X$. As in the proof of Theorem 10.7, for any finite subclass of any partition of $\overline{R}_+$, $\{A_i | i = 1, 2, \ldots, k\}$, we denote $\beta_i = \sup A_i$ and $\beta_0 = \infty$, and we may assume that $\beta_0 \ge \beta_1 \ge \beta_2 \ge \ldots \ge \beta_k$.

Let $E_i = F_{\beta_i} - F_{\beta_{i-1}}, i = 1, 2, \ldots, k$; then, $\{E_i | i = 1, 2, \ldots, k\} \cup \{\bigcap_{i=1}^{k} \overline{E}_i\}$ is a partition of X. If there exists some $\alpha \in R_+$ such that $\mu(F_\alpha) = \infty$, then $(p)\int d\mu = \infty$. In this case the conclusion of the theorem is obviously true. So, we can assume that $\mu(F_\alpha) < \infty$ for any $\alpha \in \overline{R}_+$. Since $F_{\beta_i} = \bigcup_{j \le i} E_j$ and μ is subadditive, it follows that

$$\inf_{\alpha \in A_i} \mu(F_\alpha) \le \mu(F_{\beta_i}) \le \sum_{j \le i} \mu(E_j)$$

and

$$\sum_{j \ge i} m(A_j) \le \beta_i \le \inf_{x \in F_{\beta_i}} f(x) \le \inf_{x \in E_i} f(x)$$

for $i = 1, 2, \ldots, k$. Thus, we have

$$\sum_{i=1}^{k} (\inf_{\alpha \in A_i} \mu(F_\alpha) \cdot m(A_i)) \le \sum_{i=1}^{k} \left\{ \left[\sum_{j \le i} \mu(E_j) \right] \cdot m(A_i) \right\}$$
$$= \sum_{i=1}^{k} \left[\mu(E_i) \cdot \sum_{j \ge i} m(A_j) \right] \le \sum_{i=1}^{k} [\inf_{x \in E_i} f(x) \cdot \mu(E_i)]$$
$$\le (p) \int f d\mu.$$

Consequently, we have

$$(p) \int \mu(F_\alpha) dm \le (p) \int f d\mu. \qquad \square$$

As a direct corollary of Theorem 10.7 and Theorem 10.8, we obtain the following result.

Theorem 10.9. *In the symbols of Theorem 10.7, if μ is additive, then we have*

$$(p) \int \mu(A \cap F_\alpha) dm = (p) \int_A f d\mu.$$

In the above three theorems, the pan-integral $(p) \int \mu(A \cap F_\alpha)\, dm$ is called the *Choquet integral* of f with respect to μ on A, denoted by $(c) \int_A f\, d\mu$, and discussed in Chapter 11. Both Theorem 10.7 and Theorem 10.8 show the relation between the pan-integral and the Choquet integral on a pan-space (X, **F**, μ, $\overline{R}_+$, $+$, $\times$).

Notes

10.1. The concept of a pan-integral was introduced and its properties discussed by Qingji Yang in his dissertation in 1983. A revised version was published in [Yang, 1985]. Further investigations of pan-integral on pan-additive monotone (fuzzy) measure spaces were pursued by Yang and Song [1985] and Wang et al. [1996c]. See also [Mesiar and Rybárik, 1995].

10.2. Similar to the pan-integral on pan-additive monotone measure spaces, Weber [1984] and Murofushi and Sugeno [1989, 1991b] developed the theory of $\perp$-decomposable measures and fuzzy t-conorm integrals.

Exercises

10.1. Calculate the value of the pan-integral for each pair of monotone measure spaces $(X, \mathbf{F}, \mu)$ and measurable functions f given in Exercise 9.1 provided that the concerned commutative isotonic semiring is $(\overline{R}_+, +, \times)$.

10.2. Repeat Exercise 10.1 for the commutative isotonic semiring $(\overline{R}_+, \vee, \times)$

10.3. Using Theorem 10.1 directly, show that the pan-integral is a generalization of Riemann's integral.

10.4. Prove Theorem 10.5.

10.5. Prove the following inequalities for any $A, B \in \mathbf{F}$ and $f, g \in \mathbf{G}$:

(a) $(p) \int_{A \cup B} f \, d\mu \geq (p) \int_A f \, d\mu \vee (p) \int_B f \, d\mu$;

(b) $(p) \int_{A \cap B} f \, d\mu \leq (p) \int_A f \, d\mu \wedge (p) \int_B f \, d\mu$;

(c) $(p) \int_A (f \vee g) \, d\mu \geq (p) \int_A f \, d\mu \vee (p) \int_A g \, d\mu$;

(d) $(p) \int_A (f \wedge g) \, d\mu \leq (p) \int_A f \, d\mu \wedge (p) \int_A g \, d\mu$.

10.6. Show that $(p) \int_A f \, d\mu = (p) \int f \, d\mu'$, where $\mu'(E) = \mu(A \cap E)$ for any $E \in \mathbf{F}$.

10.7. Calculate the value of the Choquet integral for each pair of monotone measure spaces $(X, \mathbf{F}, \mu)$ and measurable functions f given in Exercise 9.1.

Chapter 11
Choquet Integrals

11.1 Choquet Integrals for Nonnegative Functions

Let $(X, \mathbf{F}, \mu)$ be a monotone measure space. That is, X is a nonempty set, $\mathbf{F}$ is a σ-algebra of subsets of X, and μ: $\mathbf{F} \to [0, \infty]$ is a monotone measure. Also, let $A \in \mathbf{F}$ and f be a nonnegative measurable function on $(X, \mathbf{F})$. We have seen that the Lebesgue integral of f with respect to μ may not be well defined due to the nonadditivity of μ. Indeed, for two sequences of nondecreasing simple functions $\{s_j\}$ and $\{t_j\}$ with $\lim_{j\to\infty} s_j = \lim_{j\to\infty} t_j = f$, where $s_j = \sum_{i=1}^{m_s(j)} a_{ji}\chi_{A_{ji}}$ and $t_j = \sum_{i=1}^{m_t(j)} b_{ji}\chi_{B_{ji}}$ for each j, it is possible that

$$\lim_{j\to\infty} \sum_{i=1}^{m_s(j)} a_{ji}\mu(A_{ji}) \neq \lim_{j\to\infty} \sum_{i=1}^{m_t(j)} b_{ji}\mu(B_{ji}).$$

Fortunately, there are some equivalent definitions of the Lebesgue integral that may yet be valid with respect to monotone measures. One of them is the Riemann integral, as is shown in Section 8.1. When it is used in this way, the integral is usually referred to as a *Choquet integral.*

Definition 11.1. The *Choquet integral* of a nonnegative measurable function f with respect to monotone measure μ on measurable set A, denoted by $(\mathrm{C})\int_A f d\mu$, is defined by the formula

$$(\mathrm{C})\int_A f d\mu = \int_0^\infty \mu(F_\alpha \cap A) d\alpha,$$

where $F_\alpha = \{x | f(x) \geq \alpha\}$ for $\alpha \in [0, \infty)$. When $A = X$, $(\mathrm{C})\int_X f d\mu$ is usually written as $(\mathrm{C})\int f d\mu$.

Since f in Definition 11.1 is measurable, we know that $F_\alpha = \{x | f(x) \geq \alpha\} \in \mathbf{F}$ for $\alpha \in [0, \infty)$ and, therefore, $F_\alpha \cap A \in \mathbf{F}$. So, $\mu(F_\alpha \cap A)$ is well defined for all $\alpha \in [0, \infty)$. Furthermore, $\{F_\alpha\}$ is a class of sets that are nonincreasing with

Z. Wang, G.J. Klir, *Generalized Measure Theory*,
DOI: 10.1007/978-0-387-76852-6_11, © Springer Science+Business Media, LLC 2009

respect to α and so are sets in $\{F_\alpha \cap A\}$. Since monotone measure μ is a nondecreasing set function, we know that $\mu(F_\alpha \cap A)$ is a nondecreasing function of α and, therefore, the above Riemann integral makes sense. Thus, the Choquet integral of a nonnegative measurable function with respect to a monotone measure on a measurable set is well defined.

The following theorem establishes an equivalent form for the definition of the Choquet integral with respect to finite monotone measures.

Theorem 11.1. *Let $\mu(A)$ be finite. Then,*

$$(\mathrm{C})\int_A f d\mu = \int_0^\infty \mu(F_{\alpha+} \cap A) d\alpha,$$

where $F_{\alpha+} = \{x | f(x) > \alpha\}$ for $\alpha \in [0, \infty)$.

Proof. For any given $\varepsilon > 0$, we have

$$\begin{aligned}(\mathrm{C})\int_A f d\mu &= \int_0^\infty \mu(F_\alpha \cap A) d\alpha = \int_0^\infty \mu(\{x | f(x) \geq \alpha\} \cap A) d\alpha \\ &\geq \int_0^\infty \mu(\{x | f(x) > \alpha\} \cap A) d\alpha \geq \int_0^\infty \mu(\{x | f(x) \geq \alpha + \varepsilon\} \cap A) d\alpha \\ &= \int_0^\infty \mu(\{x | f(x) \geq \alpha + \varepsilon\} \cap A) d(\alpha + \varepsilon) = \int_\varepsilon^\infty \mu(\{x | f(x) \geq \alpha\} \cap A) d\alpha \\ &\geq \int_0^\infty \mu(\{x | f(x) \geq \alpha\} \cap A) d\alpha - \varepsilon \cdot \mu(A) \\ &= (\mathrm{C})\int_A f d\mu - \varepsilon \cdot \mu(A).\end{aligned}$$

Since $\mu(A) < \infty$, letting $\varepsilon \to 0$, we get

$$(\mathrm{C})\int_A f d\mu = \int_0^\infty \mu(\{x | f(x) > \alpha\} \cap A) d\alpha = \int_0^\infty \mu(F_{\alpha+} \cap A) d\alpha. \qquad \square$$

In the special case when the monotone measure is σ-additive, the Choquet integral coincides with the Lebesgue integral since the definition of the Choquet integral is just an equivalent definition of the Lebesgue integral. So, the Choquet integral is a real generalization of the Lebesgue integral.

Example 11.1. Let $X = [0, 1]$, $f(x) = x$ for $x \in X$, $\mathbf{F}$ be the class of all Borel sets in $[0, 1]$, and $\mu(B) = [m(B)]^2$ for $B \in \mathbf{F}$, where m is the Lebesgue measure. We know that μ is a monotone measure on σ-algebra $\mathbf{F}$ and f is a nonnegative measurable function on X. According to Definition 11.1, the Choquet integral of f with respect to μ is

$$\begin{aligned}(C)\int f d\mu &= \int_0^\infty \mu(\{x|f(x) \geq \alpha\})d\alpha = \int_0^\infty \mu(\{x|x \geq \alpha\})d\alpha \\ &= \int_0^1 \mu([\alpha,\ 1])d\alpha = \int_0^1 [m([\alpha,\ 1])]^2 d\alpha = \int_0^1 (1-\alpha)^2 d\alpha \\ &= \int_0^1 (1 - 2\alpha + \alpha^2)\ d\alpha = \alpha|_0^1 - \alpha^2|_0^1 + \frac{1}{3}\alpha^3|_0^1 \\ &= 1 - 1 + \frac{1}{3} = \frac{1}{3}.\end{aligned}$$

When the integrand $\mu(F_\alpha)$ of the above Riemann integral cannot be expressed by an explicit algebraic expression of α, or the expression is too complex, the value of the Choquet integral has to be approximately calculated by using some numerical method (e.g., the Simpson method).

11.2 Properties of the Choquet Integral

Unlike the Lebesgue integral, the Choquet integral is generally nonlinear with respect to its integrand due to the nonadditivity of μ. That is, we may have

$$(C)\int (f+g)\ d\mu \neq (C)\int f\ d\mu + (C)\int g\ d\mu$$

for some nonnegative measurable functions f and g.

Example 11.2. Let $X = \{a,\ b\}$, $\mathbf{F} = \mathbf{P}(X)$, and

$$\mu(A) = \begin{cases} 0 & \text{if } A = \emptyset \\ 1 & \text{otherwise.} \end{cases}$$

In this case, any function on X is measurable. Considering two functions,

$$f(x) = \begin{cases} 0 & \text{if } x = a \\ 1 & \text{if } x = b \end{cases}$$

and

$$g(x) = \begin{cases} 0 & \text{if } x = b \\ 1 & \text{if } x = a, \end{cases}$$

we have

$$(C)\int f d\mu = \int_0^\infty \mu(\{x|f(x) \geq \alpha\})d\alpha = \int_0^1 \mu(\{b\})d\alpha = 1 \times 1 = 1$$

and

$$(\text{C})\int g\,d\mu = \int_0^\infty \mu(\{x|g(x) \geq \alpha\})d\alpha = \int_0^1 \mu(\{a\})d\alpha = 1 \times 1 = 1.$$

Since $f+g \equiv 1$, we obtain

$$(\text{C})\int (f+g)d\mu = (\text{C})\int 1\,d\mu = \int_0^\infty \mu(\{x|1 \geq \alpha\})d\alpha = \int_0^1 1\,d\alpha = 1.$$

Thus, $(\text{C})\int (f+g)d\mu \neq (\text{C})\int f\,d\mu + (\text{C})\int g\,d\mu$. This shows that the Choquet integral is not linear with respect to its integrand in general.

However, the Choquet integral has some properties of the Lebesgue integral. These properties are listed in the following theorems.

Theorem 11.2. *Let f and g be nonnegative measurable functions on $(X, \mathbf{F}, \mu)$, A and B be measurable sets, and a be a nonnegative real constant. Then,*

(1) $(\text{C})\int_A 1\,d\mu = \mu(A)$;
(2) $(\text{C})\int_A f\,d\mu = (\text{C})\int f \cdot \chi_A d\mu$;
(3) *If $f \leq g$ on A, then* $(\text{C})\int_A f\,d\mu \leq (\text{C})\int_A g\,d\mu$;
(4) *If $A \subset B$ then,* $(\text{C})\int_A f\,d\mu \leq (\text{C})\int_B f\,d\mu$;
(5) $(\text{C})\int_A af\,d\mu = a \cdot (\text{C})\int_A f\,d\mu$.

Proof. These results follow directly from Definition 11.1. We leave the detailed proofs to the reader. □

Theorem 11.3. $(\text{C})\int_A f\,d\mu = 0$ if $\mu(\{x|f(x)>0\} \cap A) = 0$, *i.e., $f = 0$ on A almost everywhere; conversely, if monotone measure μ is continuous from below and* $(\text{C})\int_A f\,d\mu = 0$, then $\mu(\{x|f(x)>0\} \cap A) = 0$.

Proof. For the first conclusion, from $\mu(\{x|f(x)>0\} \cap A) = 0$ we know that

$$\mu(\{x|f(x) \geq \alpha\} \cap A) \leq \mu(\{x|f(x)>0\} \cap A) = 0$$

for every $\alpha > 0$. Since μ is nonnegative we have

$$\mu(\{x|f(x) \geq \alpha\} \cap A) = 0$$

for every $\alpha > 0$. Thus,

$$(\text{C})\int_A f\,d\mu = \int_0^\infty \mu(F_\alpha \cap A)d\alpha = \int_0^\infty \mu(\{x|f(x) \geq \alpha\} \cap A)d\alpha = 0.$$

Now we turn to prove the second conclusion. We first use a proof by contradiction to show that $\mu(\{x|f(x) \geq \alpha\} \cap A) = 0$ for any $\alpha > 0$ if $(\text{C})\int_A f\,d\mu = 0$.

In fact, $\mu(\{x|f(x) \geq \alpha\} \cap A) = c > 0$ for some $\alpha_0 > 0$ implies $\mu(\{x|f(x) \geq \alpha\} \cap A) = c$ for all $\alpha \in (0, \alpha_0]$ since μ is nondecreasing. Thus,

$$(C)\int_A f d\mu = \int_0^\infty \mu(\{x|f(x) \geq \alpha\} \cap A)d\mu \geq \int_0^{\alpha_0} \mu(\{x|f(x) \geq \alpha\} \cap A)d\mu$$
$$\geq \int_0^{\alpha_0} c\, d\mu = c \cdot \alpha_0 > 0.$$

This contradicts $(C)\int_A f d\mu = 0$. Secondly,

$$\{x|f(x) > 0\} \cap A = \bigcup_{j=1}^{\infty}\left(\left\{x \middle| f(x) \geq \frac{1}{j}\right\} \cap A\right).$$

That is, $\{x|f(x) > 0\} \cap A$ is the limit of nondecreasing set sequence $\{\{x|f(x) \geq \frac{1}{j}\} \cap A\}$. By using the continuity from below of μ, we have

$$\mu(\{x|f(x) > 0\} \cap A) = \lim_{j\to\infty} \mu\left(\left\{x \middle| f(x) \geq \frac{1}{j}\right\} \cap A\right) = 0. \qquad \square$$

The property given in the next theorem is called the *translatability of the Choquet integral.* It is important for defining the Choquet integral with real-valued integrand shown in Section 11.3.

Theorem 11.4. *For any constant c satisfying $f + c \geq 0$, we have*

$$(C)\int_A (f + c)d\mu = (C)\int_A f d\mu + c \cdot \mu(A).$$

Proof. From the definition of the Choquet integral directly, noticing that $f(x) + c \geq \alpha$ for every $x \in X$ when α is between 0 and c, we have

$$\begin{aligned}
(C)\int_A (f + c)d\mu &= \int_0^\infty \mu(\{x|f(x) + c \geq \alpha\} \cap A)d\alpha \\
&= \int_c^\infty \mu(\{x|f(x) + c \geq \alpha\} \cap A)d\alpha \\
&\quad + \int_0^c \mu(\{x|f(x) + c \geq \alpha\} \cap A)d\alpha \\
&= \int_c^\infty \mu(\{x|f(x) \geq \alpha - c\} \cap A)\, d(\alpha - c) + \int_0^c \mu(X \cap A)d\alpha \\
&= \int_0^\infty \mu(\{x|f(x) \geq \alpha\} \cap A)d\alpha + \int_0^c \mu(A)d\alpha \\
&= (C)\int_A f d\mu + c \cdot \mu(A).
\end{aligned} \qquad \square$$

11.3 Translatable and Symmetric Choquet Integrals

The definition of the Choquet integral given in Section 11.1 is restricted to nonnegative measurable functions. In this section, we extend it to real-valued measurable functions. A direct idea from Section 8.1 for the Lebesgue integral may be used, that is, decomposing a real-valued measurable function to be the difference of two nonnegative measurable functions.

Let $(X, \mathbf{F}, \mu)$ be a monotone measure space, A be a measurable set, and f be a real-valued measurable function on A. To simplify the discussion, without any loss of generality, we assume that $A = X$.

Similarly as in Section 8.1, let

$$f^+(x) = \begin{cases} f(x) & \text{if } f(x) \geq 0 \\ 0 & \text{if } f(x) < 0 \end{cases}$$

and

$$f^-(x) = \begin{cases} -f(x) & \text{if } f(x) \leq 0 \\ 0 & \text{if } f(x) > 0. \end{cases}$$

Both f^+ and f^- are nonnegative measurable functions. Then, we can give a definition of the Choquet integral for a real-valued measurable function with respect to a monotone measure.

Definition 11.2. The symmetric Choquet integral of real-valued measurable function f with respect to monotone measure μ, denoted by $(\mathrm{C_s})\int f\,d\mu$, is defined by the difference

$$(\mathrm{C_s})\int f\,d\mu = (\mathrm{C})\int f^+ d\mu - (\mathrm{C})\int f^- d\mu,$$

if not both terms on the right-hand side are infinite.

It is evident that the symmetric Choquet integral is symmetric. That is,

$$(\mathrm{C_s})\int (-f)d\mu = -(\mathrm{C_s})\int f\,d\mu$$

for any real-valued measurable function f. Unfortunately, such an integral loses an important and very useful property, translatability, which the Choquet integral with nonnegative integrand has. It means that equality

$$(\mathrm{C_s})\int (f + c)d\mu = (\mathrm{C_s})\int f\,d\mu + c \cdot \mu(X),$$

where c is a real number, may not be always true.

There is an alternative way to extend the Choquet integral for real-valued measurable functions, in which, the translatability is used directly. First, we define the translatable Choquet integral for lower-bounded measurable functions.

Definition 11.3. Let $\mu(X) < \infty$ and f be a measurable function on $(X, \mathbf{F}, \mu)$ with a lower bound b; that is, $f(x) \geq b$ for all $x \in X$. The translatable Choquet integral of f with respect to μ, denoted by $(\mathrm{C_t})\int f\,d\mu$, is

$$(\mathrm{C_t})\int f\,d\mu = (\mathrm{C_t})\int (f - b)d\mu + b \cdot \mu(X). \tag{11.1}$$

This definition is unambiguous. That is,

$$(\mathrm{C_t})\int (f - b_1)d\mu + b_1 \cdot \mu(X) = (\mathrm{C_t})\int (f - b_2)d\mu + b_2 \cdot \mu(X)$$

if both b_1 and b_2 are lower bounds of function f.

Theorem 11.5. *The translatable Choquet integral with lower-bounded integrand is nondecreasing with respect to its integrand. That is, for two lower-bounded measurable function f and g,*

$$(\mathrm{C_t})\int f\,d\mu \leq (\mathrm{C_t})\int g\,d\mu$$

if $f \leq g$.

Proof. Let $b \leq f \leq g$. Then $0 \leq f - b \leq g - b$. By using the result shown in (3) of Theorem 11.2, we know that

$$(\mathrm{C_t})\int (f - b)d\mu \leq (\mathrm{C_t})\int (g - b)d\mu.$$

Hence, from the definition of the translatable Choquet integral for nonnegative measurable functions, we have

$$(\mathrm{C_t})\int f\,d\mu = (\mathrm{C_t})\int (f - b)d\mu + b \cdot \mu(X) \leq (\mathrm{C_t})\int (g - b)d\mu + b \cdot \mu(X)$$
$$= (\mathrm{C_t})\int g\,d\mu.$$
□

Now, by using the translatable Choquet integral with lower-bounded integrand, we can define the translatable Choquet integral for real-valued measurable functions that are not necessarily lower bounded.

Definition 11.4. Let $\mu(X) < \infty$ and f be a real-valued measurable function on $(X, \mathbf{F}, \mu)$. The translatable Choquet integral of f with respect to μ, denoted by $(\mathrm{C_t})\int f\,d\mu$, is

$$(\mathrm{C_t})\int f\,d\mu = \lim_{b \to -\infty} (\mathrm{C_t})\int f_b\,d\mu,$$

where function f_b is defined as

$$f_b(x) = \begin{cases} f(x) & \text{if } f(x) \geq b \\ b & \text{otherwise} \end{cases} \quad , \forall b \in (-\infty, \; 0]$$

The limit in Definition 11.4 exists (including negative or positive infinities) due to Theorem 11.5 and the fact that $\{f_b\}$ is nondecreasing with respect to b. Furthermore, it is not difficult to show that the translatable Choquet integral with real-valued integrand is also nondecreasing, that is, for any two real-valued measurable function f and g satisfying $f \leq g$, we have

$$(\mathrm{C_t})\int f \, d\mu \leq (\mathrm{C_t})\int g \, d\mu.$$

Theorem 11.6. *Let f be a measurable function on $(X, \mathbf{F}, \mu)$. Then,*

$$(\mathrm{C_t})\int (f + c) d\mu = (\mathrm{C_t})\int f \, d\mu + c \cdot \mu(X)$$

for any real number c.

Proof. Let $g = f + c$ and

$$g_b(x) = \begin{cases} g(x) & \text{if } g(x) \geq b \\ b & \text{otherwise} \end{cases} \quad , \forall b \in (-\infty, \; 0]$$

We have $g_b = f_{b-c} + c$. Thus,

$$\begin{aligned}
(\mathrm{C_t})\int (f + c) d\mu &= (\mathrm{C_t})\int g \, d\mu \\
&= \lim_{b \to -\infty} (\mathrm{C_t})\int g_b d\mu \\
&= \lim_{b \to -\infty} (\mathrm{C_t})\int (f_{b-c} + c) d\mu \\
&= \lim_{b \to -\infty} [(\mathrm{C_t})\int f_{b-c} d\mu + c \cdot \mu(X)] \\
&= \lim_{b \to -\infty} (\mathrm{C_t})\int f_{b-c} d\mu + c \cdot \mu(X) \\
&= (\mathrm{C_t})\int f \, d\mu + c \cdot \mu(X).
\end{aligned}$$ □

This theorem shows that the translatable Choquet integral introduced in Definition 11.4 for real-valued measurable functions keeps the translatability.

Indeed, an alternative definition of the translatable Choquet integral for real-valued measurable functions can be also obtained by using the corresponding equivalent definition of the Lebesgue integral for real-valued measurable functions shown in Section 8.1. That is,

$$(\mathrm{C_t})\int f d\mu = \int_{-\infty}^{0} [\mu(F_\alpha) - \mu(X)]d\alpha + \int_{0}^{\infty} \mu(F_\alpha)d\alpha. \tag{11.2}$$

The proof of the equivalence of these two definitions for the translatable Choquet integral is left to the reader as an exercise. Of course, similar to Theorem 11.1, this equivalent definition for the translatable Choquet integral can also be expressed as

$$(\mathrm{C_t})\int f d\mu = \int_{-\infty}^{0} [\mu(F_{\alpha+}) - \mu(X)]d\alpha + \int_{0}^{\infty} \mu(F_{\alpha+})d\alpha.$$

However, this form is not symmetric. That is, $(\mathrm{C_t})\int(-f)d\mu$ may not be $-(\mathrm{C_t})\int f d\mu$. In fact, we have the following theorem.

Theorem 11.7. *Let $\mu(X)<\infty$ and f be a real-valued measurable function on (X, **F**, μ). Then*

$$(\mathrm{C_t})\int(-f)d\mu = -(\mathrm{C_t})\int f d\bar{\mu}$$

where $\bar{\mu}$ is the dual of μ defined by

$$\bar{\mu}(A) = \mu(X) - \mu(\bar{A}) \quad \forall A \in \mathbf{F}.$$

Proof. We use Theorem 11.1 where an equivalent definition of the Choquet integral with real-valued integrand is given.

$$\begin{aligned}
(\mathrm{C_t})\int(-f)d\mu &= \int_{-\infty}^{0} [\mu(\{x| -f(x) \geq \alpha\}) - \mu(X)]d\alpha + \int_{0}^{\infty} \mu(\{x| -f(x) \geq \alpha\})d\alpha \\
&= \int_{-\infty}^{0} [\mu(\{x| -f(x) > \alpha\}) - \mu(X)]d\alpha + \int_{0}^{\infty} \mu(\{x| -f(x) > \alpha\})d\alpha \\
&= \int_{-\infty}^{0} [\mu(\{x|f(x) < -\alpha\}) - \mu(X)]d\alpha + \int_{0}^{\infty} \mu(\{x|f(x) < -\alpha\})d\alpha \\
&= \int_{0}^{-\infty} [\mu(\{x|f(x) < -\alpha\}) - \mu(X)]d(-\alpha) \\
&\quad + \int_{\infty}^{0} \mu(\{x|f(x) < -\alpha\})d(-\alpha)
\end{aligned}$$

$$= \int_0^{\infty} [\mu(\{x|f(x) < \alpha\}) - \mu(X)]d(\alpha) + \int_{-\infty}^{0} \mu(\{x|f(x) < \alpha\})d(\alpha)$$

$$= \int_0^{\infty} \bar{\mu}(\{x|f(x) \geq \alpha\})d(\alpha) + \int_{-\infty}^{0} [\bar{\mu}(\{x|f(x) \geq \alpha\}) - \bar{\mu}(X)]d(\alpha)$$

$$= -(\mathrm{C_t})\int f d\bar{\mu}. \qquad \square$$

In comparison with the symmetric Choquet integral, the translatable Choquet integral is more natural since the symmetric one violates the original translatability, though it has a new property of symmetry. Moreover, the translatable Choquet integral for real-valued measurable functions is more reasonable than the symmetric Choquet integral in real problems as we can see in the next section. Thus, from this point, we simply call it the Choquet integral and omit the subscript t from the integral's type indicator $(\mathrm{C_t})$. That is, we use the same symbol as we did for the Choquet integral with nonnegative integrand introduced in Section 11.1. Also, we use only the translatable one in the applications discussed in Chapter 15.

11.4 Convergence Theorems

Let the set of all real-valued measurable functions for which the Choquet integral is well defined be denoted here by **G** and let its subset consisting of all measurable functions having finite value of the Choquet integral be denoted by $\mathbf{G}_0$. Furthermore, we assume in this section that monotone measure μ on $(X, \mathbf{F})$ is finite and continuous.

Let $f \in \mathbf{G}$ and $\{f_n\} \subset \mathbf{G}$. To investigate the convergence theorems of sequences of Choquet integrals, besides the convergences discussed in Chapter 7, we also need a convergence concept of measurable function sequence based on the Choquet integral.

Definition 11.5. $\{f_n\}$ is mean convergent (with respect to the Choquet integral) to f, denoted by $f_n \xrightarrow{\text{m.c.}} f$, if

$$(\mathrm{C})\int |f_n - f| d\mu \to 0.$$

The new concept of convergence in Definition 11.5 is related to other concepts of convergence introduced in Chapter 7, as expressed in the following theorem.

Theorem 11.8. *Uniform convergence implies mean convergence, while the latter implies convergence in measure.*

Proof. The first implication follows from (1), (3), and (5) of Theorem 11.2. To prove the second implication, we use a proof by contradiction. Suppose that $\{f_n\}$ does not converge in measure to f, that is, there exist $\alpha_0 > 0$, $\varepsilon > 0$, and a sequence $\{n_i\}$ such that

$$\mu(|f_{n_i} - f| \geq \alpha_0) \geq \varepsilon \qquad i = 1, 2, \ldots .$$

Since $\mu(|f_{n_i} - f| \geq \alpha)$ is nonincreasing with respect to α,

$$\int_0^\infty \mu(|f_{n_i} - f| \geq \alpha)d\alpha \geq \int_0^{\alpha_0} \mu(|f_{n_i} - f| \geq \alpha)d\alpha \geq \alpha_0 \cdot \varepsilon > 0$$

for $i = 1, 2, \ldots$. We obtain a contradiction with $f_n \xrightarrow{m.c.} f$. □

By using an approach similar to the one taken in Theorem 9.6 for the Sugeno integral, we obtain the following generalization of classical Fatou's lemma.

Lemma 11.1. *Let $A \in \mathbf{F}$. If there exists $g \in \mathbf{G}_0$ such that $f_n \geq g$ for $n = 1, 2, \ldots$, then*

$$(\mathrm{C})\int_A (\underline{\lim}_n f_n) d\mu \leq \underline{\lim}_n (\mathrm{C})\int_A f_n d\mu.$$

The condition that μ be a continuous monotone measure in Lemma 11.1 can be reduced to the condition that μ be a lower semicontinuous monotone measure. Furthermore, a generalization of classical bounded convergence theorem can be established. We omit the proof here since it is similar to the one established for Sugeno integral sequences in Theorem 9.7.

Theorem 11.9. *Let $A \in \mathbf{F}$. If $f_n \to f$ and there exists $g \in \mathbf{G}_0$ such that $|f_n| \leq g$, for $n = 1, 2, \ldots$, on A, then*

$$(\mathrm{C})\int_A f_n d\mu \to (\mathrm{C})\int_A f d\mu.$$

Lemma 11.2. *Let $A \in \mathbf{F}$ and $f, g \in \mathbf{G}$. If μ is a null-additive and $f = g$ a. e., then*

$$(\mathrm{C})\int_A f d\mu = (\mathrm{C})\int_A g d\mu.$$

Proof. Since $\{x|f(x) \geq \alpha\} \subset \{x|g(x) \geq \alpha\} \cup \{x|f(x) \neq g(x)\}$ and $\mu(\{x|f(x) \neq g(x)\}) = 0$, by the monotonicity and the null-additivity of μ, we have

$$\begin{aligned}\mu(\{x|f(x) \geq \alpha\} \cap A) &\leq \mu((\{x|g(x) \geq \alpha\} \cap A) \cup \{x|f(x) \neq g(x)\}) \\ &= \mu(\{x|g(x) \geq \alpha\} \cap A).\end{aligned}$$

From the definition of the Choquet integral, we directly obtain

$$(C)\int_A f d\mu \leq (C)\int_A g d\mu.$$

The inverse inequality can be obtained in a similar way. □

The following theorem is called an *almost everywhere convergence theorem.*

Theorem 11.10. *Let μ be a null-additive and $A \in \mathbf{F}$. If $f_n \xrightarrow{a.e.} f$ and there exists $g \in \mathbf{G}_0$ such that $|f_n| \leq g$ a. e., for $n = 1, 2, \ldots$, on A, then*

$$(C)\int_A f_n d\mu \to (C)\int_A f d\mu.$$

Proof. The result follows directly from Theorem 11.9, Lemma 11.2, and the fact that a countable union of μ-null sets is still a μ-null set when μ is a null-additive monotone measure. □

To formulate and prove an important theorem regarding the convergence for a sequence of Choquet integrals, we need a lemma and a new concept of equi-integrability for a sequence of measurable functions as follows.

Lemma 11.3. *If μ is autocontinuous, then for any class $\mathbf{C} \subset \mathbf{F}$ that is totally ordered according to set inclusion and any $\varepsilon > 0$, there exists $\delta > 0$ such that*

$$\mu(A \cup B) \leq \mu(A) + \varepsilon$$

whenever $A \in \mathbf{C}$, $B \in \mathbf{F}$, and $\mu(B) < \delta$; similarly, for any totally ordered class $\mathbf{C}$ of sets in $\mathbf{F}$ and any $\varepsilon > 0$, there exists $\delta > 0$ such that

$$\mu(A - B) \geq \mu(A) - \varepsilon$$

whenever $A \in \mathbf{C}$, $B \in \mathbf{F}$, and $\mu(B) < \delta$.

The proof of Lemma 11.3 is omitted here. The reader may consult [Wang, 1984] for details. The first implication in Lemma 11.3 is called *local-uniform autocontinuity from above* of μ, while the second one is called *local-uniform autocontinuity from below* of μ. When μ satisfies both these implications, it is called *local-uniformly autocontinuous.*

Definition 11.6. Let $A \in \mathbf{F}$. Sequence $\{f_n\}$ is called *equi-integrable* on A if, for any given $\varepsilon > 0$, there exists $N(\varepsilon) > 0$ such that

$$(C)\int_A f_n^+ d\mu + (C)\int_A f_n^- d\bar{\mu} \le \int_0^N \mu(F_\alpha^n \cap A)d\alpha - \int_{-N}^0 [\mu(F_\alpha^n \cap A) - \mu(X)]d\alpha + \varepsilon$$

for all $n = 1, 2, \ldots$, where $\bar{\mu}$ is the dual of μ.

It is obvious that $\{f_n\}$ is equi-integrable if there exists $g \in \mathbf{G}_0$ such that $|f_n| \le g$ for all $n = 1, 2, \ldots$.

Theorem 11.11. *Let $A \in \mathbf{F}$, $\{f_n\}$ be equi-integrable, and $f \in \mathbf{G}_0$. If $f_n \xrightarrow{\mu} f$ and μ is autocontinuous, then*

$$(C)\int_A f_n d\mu \to (C)\int_A f d\mu.$$

Proof. There is no loss of generality in assuming $A = X$, $f \ge 0$, and $f_n \ge 0$, $n = 1, 2, \ldots$. Denote $\mu(X) = c$. For any given $\varepsilon > 0$, by using the equi-integrability of $\{f_n\}$, we can find $N > 0$ such that

$$(C)\int f_n \, d\mu \le \int_0^N \mu(F_\alpha^n)d\alpha + \varepsilon$$

for any $n = 1, 2, \ldots$. Since $\{F_\alpha | \alpha \ge 0\}$ is totally ordered according to set inclusion and, by Lemma 11.3, μ is local-uniformly autocontinuous from above, we know from above, we know from

$$F_{\alpha+\varepsilon}^n \subset F_\alpha \cup \{x | \, |f_n(x) - f(x)| \ge \varepsilon\}$$

and $f_n \xrightarrow{\mu} f$ that there exists $n_1 > 0$ for the given ε such that

$$\mu(F_{\alpha+\varepsilon}^n) \le \mu(F_\alpha) + \varepsilon/N$$

whenever $n > n_1$ and $\alpha \ge 0$. Thus,

$$\begin{aligned}(C)\int f_n d\mu &\le \int_0^N \mu(F_\alpha^n)d\alpha + \varepsilon \le \int_\varepsilon^N \mu(F_\alpha^n)d\alpha + c\cdot\varepsilon + \varepsilon = \int_0^{N-\varepsilon} \mu(F_{\alpha+\varepsilon}^n)d\alpha + c\cdot\varepsilon + \varepsilon \\ &\le \int_0^{N-\varepsilon} [\mu(F_\alpha) + \varepsilon/N]d\alpha + c\cdot\varepsilon + \varepsilon \le \int_0^{N-\varepsilon} \mu(F_\alpha)d\alpha + c\cdot\varepsilon + 2\varepsilon \\ &\le (C)\int f d\mu + c\cdot\varepsilon + 2\varepsilon\end{aligned}$$

for any $n > n_1$.

Conversely, for any given $\varepsilon > 0$, by using the integrability of f, we can find $N > 0$ such that

$$(C)\int f d\mu - \varepsilon \le \int_0^N \mu(F_\alpha)d\alpha.$$

Since μ is local-uniformly autocontinuous from below by Lemma 11.3, we know from

$$F^n_{\alpha-\varepsilon} \supset F_\alpha - \{x|\ |f_n(x) - f(x)| \geq \varepsilon\}$$

and $f_n \xrightarrow{\mu} f$ that there exists $n_2 > 0$ for the given ε such that

$$\mu(F^n_{\alpha-\varepsilon}) \geq \mu(F_\alpha) - \varepsilon/N$$

whenever $n > n_2$ and $\alpha \geq \varepsilon$. Thus,

$$\begin{aligned}(C)\int f_n d\mu &\geq \int_0^N \mu(F^n_\alpha)d\alpha \geq \int_\varepsilon^N \mu(F^n_{\alpha-\varepsilon})d\alpha \geq \int_\varepsilon^N [\mu(F_\alpha) - \varepsilon/N]d\alpha \\ &\geq \int_0^N \mu(F_\alpha)d\alpha - c\cdot\varepsilon - \varepsilon \geq (C)\int fd\mu - c\cdot\varepsilon - 2\varepsilon\end{aligned}$$

for any $n > n_2$.

Combining these two inequalities, we obtain

$$(C)\int f_n d\mu \to (C)\int f d\mu.$$

□

From Theorems 11.8 and 11.11, we obtain the following corollary.

Corollary 11.1. *Let $A \in \mathbf{F}$, $\{f_n\}$ be equi-integrable, and $f \in \mathbf{G}_o$. If $f_n \xrightarrow{\text{m.c.}} f$ and μ is autocontinuous, then*

$$(C)\int_A f_n d\mu \to (C)\int_A f d\mu.$$

The following theorem establishes the inverse result of Theorem 11.11. Thus, the autocontinuity of finite monotone measure μ plays a necessary and sufficient condition for the convergence in monotone measure theorem of the Choquet integral sequence.

Theorem 11.12. *If $(C)\int_A f_n d\mu \to (C)\int_A f d\mu$, whenever $A \in \mathbf{F}, \{f_n\}$ is equi-integrable, $f \in \mathbf{G}_0$, and $f_n \xrightarrow{\mu} f$, then μ is autocontinuous.*

Proof. For any given $A \in \mathbf{F}$ and $\{B_n\} \subset \mathbf{F}$ with $\mu(B_n) \to 0$, taking $f = \chi_A$ and $f_n = \chi_{A\cup B_n}$, $n = 1, 2, \ldots$, we have

$$\mu(|f_n - f| \geq \varepsilon) = \mu(B_n) \to 0$$

for any $\varepsilon \in (0, 1)$. This means that $f_n \xrightarrow{\mu} f$. Since

$$(C)\int fd\mu = (C)\int \chi_A d\mu = \mu(A),$$

and

$$(\mathrm{C})\int f_n d\mu = (\mathrm{C})\int \chi_{A\cup B_n} d\mu = \mu(A \cup B_n),$$

we obtain $\mu(A \cup B_n) \to \mu(A)$ from the assumption that $(\mathrm{C})\int f_n d\mu \to (\mathrm{C})\int f d\mu$. This shows that μ is autocontinuous. □

The following theorem is called a *uniform convergence theorem.*

Theorem 11.13. *Let* $A \in \mathbf{F}$. *If* $f_n \xrightarrow{\text{u.}} f$ *on* A, *then* $(\mathrm{C})\int_A f_n d\mu \to (\mathrm{C})\int_A f d\mu$.

Proof. For any given $\varepsilon > 0$, there exists n_0 such that

$$f(x) - \varepsilon \le f_n(x) \le f(x) + \varepsilon$$

for any $n > n_0$ and $x \in A$. Using Theorem 11.2.(3) and Theorem 11.4, we have

$$(\mathrm{C})\int_A f d\mu - \varepsilon \cdot \mu(A) \le (\mathrm{C})\int_A f_n d\mu \le (\mathrm{C})\int_A f d\mu + \varepsilon \cdot \mu(A)$$

for any $n > n_0$. This means that $(\mathrm{C})\int_A f_n d\mu \to (\mathrm{C})\int_A f d\mu$ since $\mu(A) < \infty$. □

11.5 Choquet Integrals on Finite Sets

In any database, the number of attributes is always finite. Let $X = \{x_1, x_2, \ldots, x_n\}$ denote a finite set of attributes. Then, $(X, \mathbf{P}(X))$ is a measurable space. Each record (or, observation) of $x_1, x_2, \ldots, x_n$, denoted by $f(x_1), f(x_2), \ldots, f(x_n)$ respectively, is just a real-valued function f on X. Since the power set of X is taken as the σ-algebra, any real-valued function on X is measurable. A monotone measure μ defined on $\mathbf{P}(X)$ is usually used to describe the joint importance as well as the individual importance of attributes in X towards a certain target. To obtain a global contribution from f towards the target, we need an aggregation tool. Due to the nonadditivity of μ, as a common aggregation tool the Lebesgue integral fails. In this case, we saw in Section 11.1 that the Choquet integral can replace the Lebesgue integral.

The Choquet integral of function f with respect to monotone measure μ is defined via a Riemann integral as

$$(\mathrm{C})\int f d\mu = \int_{-\infty}^{0} [\mu(F_\alpha) - \mu(X)] d\alpha + \int_0^{\infty} \mu(F_\alpha) d\alpha.$$

Since X is a finite set we may develop a simple formula for calculating the value of $(\mathrm{C})\int f d\mu$ once f and μ are given.

Let $b_1 = \min_{1 \le i \le n} f(x_i)$ and $b_2 = \max_{1 \le i \le n} f(x_i)$. Since the values of f is between b_1 and b_2, we have $F_\alpha = X$ when $\alpha \le b_1$ and $F_\alpha = \emptyset$ when $\alpha > b_2$. Hence, by using the translatability of the Choquet integral, we have

$$(C)\int f d\mu = (C)\int (f - b_1)d\mu + b_1 \cdot \mu(X)$$

$$= \int_0^{b_2 - b_1} \mu(\{x | f(x) - b_1 \ge \alpha\})d\alpha + b_1 \cdot \mu(X).$$

If the values of function f, $\{f(x_1), f(x_2), \ldots, f(x_n)\}$, are rearranged into a nondecreasing order as,

$$b_1 = f(x_1^*) \le f(x_2^*) \le \cdots \le f(x_n^*) = b_2$$

where $(x_1^*, x_2^*, \ldots, x_n^*)$ is a permutation of $\{x_1, x_2, \ldots, x_n\}$, then set $\{x | f(x) - b_1 \ge \alpha\}$ is always $\{x_i^*, x_{i+1}^*, \ldots, x_n^*\}$ when $\alpha \in [f(x_{i-1}^*) - b_1, f(x_i^*) - b_1]$ for $i = 2, 3, \ldots, n$. Thus, we have

$$(C)\int f d\mu = \sum_{i=2}^{n} [f(x_i^*) - f(x_{i-1}^*)] \cdot \mu(\{x_i^*, x_{i+1}^*, \ldots, x_n^*\}) + b_1 \cdot \mu(X)$$

$$= \sum_{i=1}^{n} [f(x_i^*) - f(x_{i-1}^*)] \cdot \mu(\{x_i^*, x_{i+1}^*, \ldots, x_n^*\}) \tag{11.3}$$

with a convention $f(x_0^*) = 0$.

When μ is a general measure, that is, when μ is not necessarily nondecreasing, though $\mu(F_\alpha)$ may not be monotonic with respect to α, it is still with bounded variation. So, the Riemann integrals $\int_{-\infty}^{0} [\mu(F_\alpha) - \mu(X)]d\alpha$ and $\int_0^\infty \mu(F_\alpha)d\alpha$ exist. If these integrals are not both infinite, the Choquet integral $(C)\int f d\mu$ is also well defined. The calculation formula given above is also available for the Choquet integral with respect to general measures.

Recall (see Section 3.4) that any signed general measure μ can be decomposed as the difference of two general measures: $\mu = \mu^+ - \mu^-$, where

$$\mu^+(A) = \begin{cases} \mu(A) & \text{if } \mu(A) \ge 0 \\ 0 & \text{otherwise} \end{cases}$$

and

$$\mu^-(A) = \begin{cases} -\mu(A) & \text{if } \mu(A) < 0 \\ 0 & \text{otherwise.} \end{cases}$$

Thus, we may define the Choquet integral of function f with respect to signed general measure μ by

$$(\mathrm{C})\int f d\mu = (\mathrm{C})\int f d\mu^+ - (\mathrm{C})\int f d\mu^-.$$

Hence,

$$\begin{aligned}(\mathrm{C})\int f d\mu &= \left(\int_{-\infty}^{0} [\mu^+(F_\alpha) - \mu^+(X)]d\alpha + \int_0^{\infty} \mu^+(F_\alpha)d\alpha\right) \\ &\quad - \left(\int_{-\infty}^{0} [\mu^-(F_\alpha) - \mu^-(X)]d\alpha + \int_0^{\infty} \mu^-(F_\alpha)d\alpha\right) \\ &= \int_{-\infty}^{0} [(\mu^+(F_\alpha) - \mu^-(F_\alpha)) - (\mu^+(X) - \mu^-(X))]d\alpha \\ &\quad + \int_0^{\infty} [\mu^+(F_\alpha) - \mu^-(F_\alpha)]d\alpha \\ &= \left(\int_{-\infty}^{0} [\mu(F_\alpha) - \mu(X)]d\alpha + \int_0^{\infty} \mu(F_\alpha)d\alpha\right).\end{aligned}$$

That is, we have the same expression as before. Moreover, the calculation formula for the Choquet integral with respect to a signed general measure is the same as the one for the Choquet integral with respect to a general measure.

Now, we can also see that using the translatable Choquet integral for real-valued functions is rather convenient for generalizing it and its calculation.

Example 11.3. Recall Example 8.2 where three workers, x_1, x_2, and x_3, work separately for 6, 3, and 4 days, respectively. If they work together sometimes, we must consider their joint efficiencies to calculate the total number of the manufactured toys in the given period of time. We may use $\mu(\{x_1, x_2\})$ to denote the joint efficiency of x_1 and x_2. Similarly, $\mu(\{x_1, x_3\})$, $\mu(\{x_2, x_3\})$, and $\mu(X)$ are joint efficiencies of x_1 and x_3, x_2, and x_3, all respectively. Assume that $\mu(\{x_1, x_2\}) = 14$, $\mu(\{x_1, x_3\}) = 13$, $\mu(\{x_2, x_3\}) = 9$, and $\mu(X) = 17$. Then, with $\mu(\{x_1\}) = 5$, $\mu(\{x_2\}) = 6$, $\mu(\{x_3\}) = 7$, and $\mu(\emptyset) = 0$ in Example 8.2, $\mu : \mathbf{P}(X) \to (-\infty, \infty)$ is a signed general measure (indeed, is a general measure, since it is nonnegative). It is nonadditive. For instance, $\mu(\{x_1, x_2\}) > \mu(\{x_1\}) + \mu(\{x_2\})$. This inequality means that workers x_1 and x_2 cooperate well. The nonadditivity of μ describes the interaction among the contribution rates of these three workers towards the total amount of their manufactured toys. Thus, the total number of manufactured toys by these three workers in the given period depends on their cooperation. Assume that they work in such a manner: they start the work together on the first day of the week and continue their work until their respective terminal day. That is, all of them work together for the first 3 days; then x_1 and x_3 work together for one more day; finally x_1

works alone for two other days. Thus, the total amount of toys manufactured by these three workers in this week is

$$3 \times 17 + 1 \times 13 + 2 \times 5 = 74.$$

Regarding their efficiencies as a signed general measure μ and the number of working days as a function f on the set of these three workers, the above total amount of the manufactured toys is exactly the value of the Choquet integral of f with respect to μ. In fact, from $f(x_1) = 6$, $f(x_2) = 3$, and $f(x_3) = 4$, we have $x_1^* = x_2$, $x_2^* = x_3$, and $x_3^* = x_1$. According to the calculation formula of the Choquet integral, we have

$$\begin{aligned}(\mathrm{C})\int f d\mu &= \sum_{i=1}^{3}[f(x_i^*) - f(x_{i-1}^*)] \cdot \mu(\{x_i^*, x_{i+1}^*, \ldots, x_n^*\}) \\ &= [f(x_2) - 0] \cdot \mu(X) + [f(x_3) - f(x_2)] \cdot \mu(\{x_3, x_1\}) \\ &\quad + [f(x_1) - f(x_3)] \cdot \mu(\{x_1\}) \\ &= (3 - 0) \times 17 + (4 - 3) \times 13 + (6 - 4) \times 5 \\ &= 74.\end{aligned}$$

We obtain the same total number of manufactured toys.

According to the general view of integration on finite sets expressed in Section 8.4, the Choquet integral with nonnegative integrand with respect to a signed monotone measure on a finite set facilitates a special type of integration. The partitioning rule corresponding to the Choquet integral can be described as follows: for any given nonnegative function $f: X \to [0, \infty)$, partition $\pi: \mathbf{P}(X) - \{\emptyset\} \to [0, \infty)$ is obtained by

$$\pi(A) = \begin{cases} f(x_i^*) - f(x_{i-1}^*) & \text{if } A = \{x_i^*, x_{i+1}^*, \ldots, x_n^*\} \text{ for some } i = 1, 2, \ldots, n \\ 0 & \text{otherwise} \end{cases}$$

for every $A \in \mathbf{P}(X) - \{\emptyset\}$, where $(x_1^*, x_2^*, \ldots, x_n^*)$ is a permutation of $\{x_1, x_2, \ldots, x_n\}$ such that $f(x_1^*) \leq f(x_2^*) \leq \cdots \leq f(x_n^*)$ and $f(x_0^*) = 0$ as the convention made above. It is easy to verify that

$$\sum_{A: x \in A \subset X} \pi(A) = f(x) \qquad \forall x \in X.$$

So, π is a partition of f. In such a partition there are only at most n sets A with $\pi(A) > 0$. Regarding both set functions μ and π as $(2^n - 1)$-dimensional vectors, we have $(\mathrm{C})\int f d\mu = \pi \cdot \mu$.

Example 11.4. Let us consider again the data specified in Example 11.3. The partition of f corresponding to the Choquet integral is illustrated in Fig. 11.1 where the black part, grey part, and the light part show $\pi(X) = 3$,

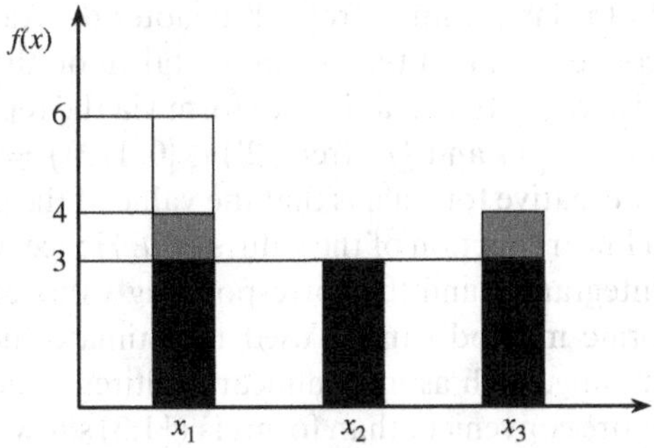

Fig. 11.1 The partition of *f* corresponding to the Choquet integral

$\pi(\{x_1, x_3\}) = 1$, and $\pi(\{x_1\}) = 2$, respectively. The values of π at other sets are zeros. Geometrically, this partitioning rule divides function *f* horizontally.

In comparison with the Lebesgue integral, the horizontal partitioning rule corresponding to the Choquet integral takes the coordination of the attributes into account maximally, that is, the manner of the partition is to make the coordination among the attributes as much as possible, while the vertical partitioning rule corresponding to the latter takes the coordination into account minimally (zero coordination). These are two extreme cases regarding the coordination among the attributes.

11.6 An Alternative Calculation Formula

In data mining, a learning data set is given while the values of signed general measure μ need to be optimally estimated. This is an inverse problem of the information fusion. In such a problem, the calculation formula shown in the last section is not convenient since the value of the Choquet integral is not expressed as an explicit linear function of the values of μ. Fortunately, the Choquet integral of real-valued function *f* with respect to a signed general measure μ can be expressed by the alternative formula

$$(C)\int f d\mu = \sum_{j=1}^{2^n-1} z_j \mu_j, \tag{11.4}$$

where $\mu_j = \mu(\bigcup_{j_i=1} \{x_i\})$ if j is expressed in terms of binary digits $j_n j_{n-1} \cdots j_1$ for every $j = 1, 2, \ldots, 2^n - 1$ and

$$z_j = \begin{cases} \min\limits_{i:\mathrm{frc}(j/2^i)\in[1/2,\,1)} f(x_i) - \max\limits_{i:\mathrm{frc}(j/2^i)\in[0,\,1/2)} f(x_i), & \text{if it is} > 0 \text{ or } j = 2^n - 1 \\ 0, & \text{otherwise} \end{cases} \tag{11.5}$$

for $j = 1, 2, \ldots, 2^n - 1$.

In the last formula,frc($j/2^i$)denotes the fractional part of $j/2^i$, and we need the convention that the maximum taken on the empty set is zero. The formula can also be written in a simpler form via the replacement $\{i \mid \text{frc}(j/2^i) \in [1/2, 1)\} = \{i \mid j_i = 1\}$ and $\{i \mid \text{frc}(j/2^i) \in [0, 1/2)\} = \{i \mid j_i = 0\}$. The significance of this alternative formula is that the value of the Choquet integral is now expressed as a linear function of the values of μ. Hence, when the data set of the values of the integrand f and the corresponding value of the integral are available, an algebraic method can be used to estimate the optimal values of μ. So, in data mining, such as in nonlinear multiregressions, this new calculation formula is more convenient than formula (11.3) shown in the last section, though the latter is convenient in information fusion.

As for the validation of this new formula, rewriting the old formula as

$$(\text{C})\int f d\mu = \sum_{i=1}^{n} [f(x_i^*) - f(x_{i-1}^*)] \cdot \mu(\{x_i^*, x_{i+1}^*, \ldots, x_n^*\}) = \sum_{j=1}^{2^n-1} a_{A_j} \cdot \mu(A_j),$$

where

$$A_j = \bigcup_{j_i=1} \{x_i\}$$

and

$$a_{A_j} = \begin{cases} f(x_i^*) - f(x_{i-1}^*) & \text{if } A_j = \{x_i^*, x_{i+1}^*, \ldots, x_n^*\} \text{ for some } i = 1, 2, \ldots, n \\ 0 & \text{otherwise} \end{cases}$$

$$= \begin{cases} \min_{x \in A_j} f(x) - \max_{x \notin A_j} f(x), & \text{if } A_j = \{x_i^*, x_{i+1}^*, \ldots, x_n^*\} \text{ for some } i = 1, 2, \ldots, n \\ 0, & \text{otherwise} \end{cases}$$

$$= \begin{cases} \min_{x \in A_j} f(x) - \max_{x \notin A_j} f(x), & \text{if } A_j = \{x_i^*, x_{i+1}^*, \ldots, x_n^*\} \text{ for some } i = 1, 2, \ldots, n \\ 0, & \min_{x \in A_j} f(x) - \max_{x \notin A_j} f(x) \le 0 \end{cases}$$

for $j = 1, 2, \ldots, 2^{n-1}$, and noticing that $j_i = 1$ if and only if $x_i \in A_j$, we can see that the new formula is equivalent to the old one. In the above expression for a_{A_j}, we need the convention that

$$\max_{x \notin X} f(x) = \max_{\emptyset} f(x) = 0.$$

Also we should note that in the above expression the function is defined in two parts. They overlap when $x_{i-1}^* = x_i^*$ for some $i = 1, 2, \ldots, n$. Fortunately, they are both zero at the overlapped j and, therefore, these two parts are consistent.

Notes

11.1. It seems that the name "Choquet integral" was coined by Schmeidler [1986] to give credit to Gustave Choquet, who introduced the integral in his seminal work [Choquet, 1953–54]. It turns out, however, that the integral was introduced first by Giuseppe Vitali in a paper published in Italian in 1925, which was only recently translated into English [Vitali, 1997]. Since the late 1980s, the Choquet integral has been discussed in the literature quite extensively. A few representative references are [Benvenuti and Mesiar, 2000; De Campos and Bolaños, 1992; Denneberg, 1994a, 2000a; Grabisch et al., 2000; Grabisch and Labreuche, 2005; Krätschmer, 2003a; Morufushi and Sugeno, 1989, 1991a,b, 1993; Pap, 1995; Wang et al., 1996b, 2000b; Wang, 1997; Wang and Klir, 1997a].

11.2. The symmetric Choquet integral was introduced by Šipoš [1979a,b] and is often referred to in the literature as the Šipoš integral. Properties of this integral are thoroughly examined by Denneberg [1994a], Mesiar and Šipoš [1994], and Pap [1995].

11.3. The Choquet integral is usually discussed in the context of monotone measures. Murofushi et al. [1994] discuss the meaning of the Choquet integral with respect to general measures. Wang and Ha [2006] investigate Choquet integrals of fuzzy-valued functions, and Wang et al. [2006a] investigate Choquet integrals with respect to monotone measures defined on L-fuzzy sets.

Exercises

11.1. Let $X = [0,\ 1]$, $f(x) = x$ for $x \in X$, $\mathbf{F}$ be the class of all Borel sets in $[0,\ 1]$, and $\mu(B) = [m(B)]^{1/2}$ for $B \in \mathbf{F}$, where m is the Lebesgue measure. Calculate $(\mathrm{C})\int f\, d\mu$.

11.2. Prove Theorem 11.2.

11.3. Let $(X, \mathbf{F}, \mu)$ be a monotone measure space, and let f and g be real-valued measurable functions. Prove that $(\mathrm{C_s})\int f\, d\mu \le (\mathrm{C_s})\int g\, d\mu$ if $f \le g$.

11.4. Find a counterexample to show that the symmetric Choquet integral with real-valued integrand is not translatable in general.

11.5. Prove the unambiguity of Definition 11.3, that is, when μ is a finite monotone measure,

$$(\mathrm{C_t})\int (f - b_1)d\mu + b_1 \cdot \mu(X) = (\mathrm{C_t})\int (f - b_2)d\mu + b_2 \cdot \mu(X)$$

if both b_1 and b_2 are lower bounds of function f.

11.6. If we use

$$(\mathrm{C_t})\int f\, d\mu = \int_{-\infty}^{0} [\mu(F_\alpha) - \mu(X)]d\alpha + \int_{0}^{\infty} \mu(F_\alpha)d\alpha$$

to define the Choquet integral with real-valued integrand, show that it is translatable and, therefore, this definition is equivalent to Definitions 11.3 and 11.4.

11.7. The way to express a signed general measure as the difference of two general measures is not unique. Show that if a signed general measure μ can also be expressed as $\mu = \mu_1 - \mu_2$, where both μ_1 and μ_2 are general measure, then

$$(\mathrm{C})\int f d\mu = (\mathrm{C})\int f d\mu^+ - (\mathrm{C})\int f d\mu^- = (\mathrm{C})\int f d\mu_1 - (\mathrm{C})\int f d\mu_2.$$

11.8. Let $X = \{x_1, x_2, \ldots, x_5\}$. The values of μ and f are shown in Table 11.1. Find the value of $(\mathrm{C})\int f d\mu$.

Table 11.1. Given functions in Exercise 11.8

Set A	$\mu(A)$	Set A	$\mu(A)$	x_i	$f(x_i)$
$\emptyset$	0	$\{x_5\}$	4	x_1	3
$\{x_1\}$	2	$\{x_1, x_5\}$	7	x_2	7
$\{x_2\}$	3	$\{x_2, x_5\}$	3	x_3	5
$\{x_1, x_2\}$	4	$\{x_1, x_2, x_5\}$	6	x_4	2
$\{x_3\}$	1	$\{x_3, x_5\}$	6	x_5	1
$\{x_1, x_3\}$	5	$\{x_1, x_3, x_5\}$	8		
$\{x_2, x_3\}$	4	$\{x_2, x_3, x_5\}$	5		
$\{x_1, x_2, x_3\}$	6	$\{x_1, x_2, x_3, x_5\}$	9		
$\{x_4\}$	5	$\{x_4, x_5\}$	7		
$\{x_1, x_4\}$	3	$\{x_1, x_4, x_5\}$	7		
$\{x_2, x_4\}$	4	$\{x_2, x_4, x_5\}$	6		
$\{x_1, x_2, x_4\}$	7	$\{x_1, x_2, x_4, x_5\}$	8		
$\{x_3, x_4\}$	6	$\{x_3, x_4, x_5\}$	5		
$\{x_1, x_3, x_4\}$	8	$\{x_1, x_3, x_4, x_5\}$	6		
$\{x_2, x_3, x_4\}$	8	$\{x_2, x_3, x_4, x_5\}$	9		
$\{x_1, x_2, x_3, x_4\}$	9	X	8		

11.9. The properties of the Choquet integral presented in Section 11.2 are given under the assumption that the involved set function μ is a monotone measure. When the set function is a general measure or a signed general measure, identify those properties which no longer hold.

Chapter 12
Upper and Lower Integrals

12.1 Definitions

Throughout this chapter, unless specified otherwise, we assume that $(X, \mathbf{F}, \mu)$ is a general measure space. That is, X is a nonempty set, $\mathbf{F}$ is a σ-algebra of subsets of X, and $\mu : \mathbf{F} \to [0, \infty)$ is a general measure.

Definition 12.1. Given a measurable function f: $X \to [0, \infty)$ and a set $A \in \mathbf{F}$, the upper integral of f with respect to μ on A, in symbol $(\mathrm{U})\int_A f\,d\mu$, is defined as

$$(\mathrm{U})\int_A f\,d\mu = \lim_{\varepsilon \to 0+} U_\varepsilon,$$

where

$$U_\varepsilon = \sup\left\{ \sum_{j=1}^{\infty} \lambda_j \cdot \mu(E_j) \,\middle|\, f \geq \sum_{j=1}^{\infty} \lambda_j \cdot \chi_{E_j} \geq f - \varepsilon, E_j \in \mathbf{F} \cap A, \lambda_j \geq 0,\ j = 1, 2, \ldots \right\} \tag{12.1}$$

for $\varepsilon > 0$, where χ_{E_j} denotes the characteristic function of E_j and $\mathbf{F} \cap A = \{B \cap A | B \in \mathbf{F}\}$. Similarly, the lower integral of f with respect to μ on A, $(\mathrm{L})\int_A f\,d\mu$, is defined as

$$(\mathrm{L})\int_A f\,d\mu = \lim_{\varepsilon \to 0+} L_\varepsilon,$$

where

$$L_\varepsilon = \inf\left\{ \sum_{j=1}^{\infty} \lambda_j \cdot \mu(E_j) \,\middle|\, f \leq \sum_{j=1}^{\infty} \lambda_j \cdot \chi_{E_j} \leq f + \varepsilon, E_j \in \mathbf{F} \cap A, \lambda_j \geq 0,\ j = 1, 2, \ldots \right\} \tag{12.2}$$

for $\varepsilon > 0$.

Z. Wang, G.J. Klir, *Generalized Measure Theory*,
DOI: 10.1007/978-0-387-76852-6_12, © Springer Science+Business Media, LLC 2009

In the above definition, functions expressed in the form $\sum_{j=i}^{\infty} \lambda_j \cdot \chi_{E_j}$, where $E_j \in \mathbf{F}$ and $\lambda_j \geq 0$ for $j = 1, 2, \ldots$, are called elementary functions. In Eq. (12.1) and Eq. (12.2), the requirement of the measurability of function f is necessary to guarantee the existence of some elementary functions (but not simple functions since f may not be upper bounded!) between f and $f + \varepsilon$. When the given function is allowed to be non-measurable, we may use simple functions to give relatively looser concepts of widened-upper integral and widened-lower integral as follows.

Definition 12.2. Let f be a nonnegative function on X and set $A \in \mathbf{F}$. The widened-upper integral, denoted by $(\overline{\mathrm{W}})\int_A f\,d\mu$, is defined as

$$(\overline{\mathrm{W}})\int_A f d\mu = \sup\left\{ \sum_{j=1}^{k} \lambda_j \cdot \mu(E_j) \,\middle|\, f \geq \sum_{j=1}^{k} \lambda_j \cdot \chi_{E_j}, k \geq 0, E_j \in \mathbf{F} \cap A, \lambda_j \geq 0,\ j = 1, 2, \ldots, k \right\};$$

while the widened-lower integral of f with respect to μ, denoted by $(\underline{\mathrm{W}})\int_A f\,d\mu$, is defined as

$$(\underline{\mathrm{W}})\int_A f\,d\mu = \lim_{N \to \infty} (\underline{\mathrm{W}})\int_A f_N\,d\mu,$$

where $f_N = \min(N, f)$ and

$$(\underline{\mathrm{W}})\int_A f_N\,d\mu = \inf\left\{ \sum_{j=1}^{k} \lambda_j \cdot \mu(E_j) \,\middle|\, f_N \leq \sum_{j=1}^{k} \lambda_j \cdot \chi_{E_j}, k \geq 0, E_j \in \mathbf{F} \cap A, \lambda_j \geq 0,\ j = 1, 2, \ldots, k \right\}.$$

When f is bounded, we may use f to replace f_N in the above definition. Similar to the Lebesgue integral and the Choquet integral, we omit the subscript A in the symbol of the integral when $A = X$. When X contains only a few attributes, such as 2 or 3, it is not difficult to find the maximum and the minimum in above definitions by hand.

Example 12.1. We use the data given in Example 11.3 for three workers, x_1, x_2, and x_3, who manufacture toys. Assume that the workers' individual and group efficiencies are expressed by the following general measure: $\mu(\{x_1\}) = 5$, $\mu(\{x_2\}) = 6$, $\mu(\{x_3\}) = 7$, $\mu(\{x_1, x_2\}) = 14$, $\mu(\{x_1, x_3\}) = 13$, $\mu(\{x_2, x_3\}) = 9$, $\mu(\{x_1, x_2, x_3\}) = 17$. Assume also that their working days in a specified week are expressed by the function

$$f(x) = \begin{cases} 6 & \text{if } x = x_1 \\ 3 & \text{if } x = x_2. \\ 4 & \text{if } x = x_3 \end{cases}$$

From Definitions 12.1 and 12.2, we have $(\mathrm{U})\int f d\mu = (\overline{\mathrm{W}})\int f d\mu = 88$ and $(\mathrm{L})\int f d\mu = (\underline{\mathrm{W}})\int f d\mu = 64$. This means that these three workers, in any cooperative manner, can manufacture at most 88 but not less than 64 toys in the considered week.

A general method for calculating the value of the upper integral and the lower integral (as well as the widened-upper integral and the widened-lower integral) on finite sets is given in Section 12.5.

12.2 Properties

In general, either the upper integral or the lower integral is not linear. In fact, we may even have both

$$(\mathrm{U})\int (f+g)\, d\mu \neq (\mathrm{U})\int f\, d\mu + (\mathrm{U})\int g\, d\mu$$

and

$$(\mathrm{L})\int (f+g)\, d\mu \neq (\mathrm{L})\int f\, d\mu + (\mathrm{L})\int g\, d\mu$$

for some monotone measure μ and nonnegative measurable functions f and g. Similar to the Choquet integral, the nonlinearity of the upper integral and the lower integral comes from the nonadditivity of the monotone measure.

Example 12.2. Let $X = \{x_1, x_2, x_3\}$ and $\mathbf{F} = \mathbf{P}(X)$. Monotone measure μ is defined as follows: $\mu(\{x_1\}) = 3$, $\mu(\{x_2\}) = 3$, $\mu(\{x_3\}) = 1$, $\mu(\{x_1, x_2\}) = 5$, $\mu(\{x_1, x_3\}) = 5$, $\mu(\{x_2, x_3\}) = 5$, $\mu(\{x_1, x_2, x_3\}) = 5$. Considering functions

$$f(x) = \begin{cases} 1 & \text{if } x = x_1 \\ 1 & \text{if } x = x_2 \\ 0 & \text{if } x = x_3 \end{cases}$$

and

$$g(x) = \begin{cases} 0 & \text{if } x = x_1 \\ 0 & \text{if } x = x_2, \\ 1 & \text{if } x = x_3 \end{cases}$$

we obtain

$$(\mathrm{U})\int f\,d\mu = 1\cdot\mu(x_1) + 1\cdot\mu(x_2) = 1\times 3 + 1\times 3 = 6,$$

$$(\mathrm{U})\int g\,d\mu = 1\cdot\mu(x_3) = 1\times 1 = 1,$$

and

$$(\mathrm{U})\int (f+g)\,d\mu = 1\cdot\mu(x_1) + 1\cdot\mu(\{x_2,\ x_3\}) = 1\times 3 + 1\times 5 = 8.$$

That is, we have

$$(\mathrm{U})\int (f+g)\,d\mu > (\mathrm{U})\int f\,d\mu + (\mathrm{U})\int g\,d\mu.$$

Similarly,

$$(\mathrm{L})\int f\,d\mu = 1\cdot\mu(\{x_1,\ x_2\}) = 1\times 5 = 5,$$

$$(\mathrm{L})\int g\,d\mu = 1\cdot\mu(x_3) = 1\times 1 = 1,$$

and

$$(\mathrm{L})\int (f+g)\,d\mu = 1\cdot\mu(\{x_1,\ x_2,\ x_3\}) = 1\times 5 = 5.$$

That is,

$$(\mathrm{L})\int (f+g)\,d\mu < (\mathrm{L})\int f\,d\mu + (\mathrm{L})\int g\,d\mu.$$

These results suggest the following general inequalities of the upper and lower integrals.

Theorem 12.1. *Let f and g be nonnegative measurable functions on $(X, \mathbf{F})$. Then,*

$$(\mathrm{U})\int (f+g)\,d\mu \geq (\mathrm{U})\int f\,d\mu + (\mathrm{U})\int g\,d\mu$$

and

$$(\mathrm{L})\int (f+g)\,d\mu \leq (\mathrm{L})\int f\,d\mu + (\mathrm{L})\int g\,d\mu.$$

Proof. We only prove the first inequality since the proof for the second one is similar. For any given $\varepsilon > 0$, let

$$U_{\varepsilon/2}^{(f)} = \sup\left\{\sum_{j=1}^{\infty} \lambda_j \cdot \mu(E_j) \middle| f \geq \sum_{j=1}^{\infty} \lambda_j \cdot \chi_{E_j} \geq f - \varepsilon/2, E_j \in \mathbf{F} \cap A, \lambda_j \geq 0,\ j = 1, 2, \ldots\right\},$$

$$U_{\varepsilon/2}^{(g)} = \sup\left\{\sum_{j=1}^{\infty} \lambda_j \cdot \mu(E_j) \middle| g \geq \sum_{j=1}^{\infty} \lambda_j \cdot \chi_{E_j} \geq g - \varepsilon/2, E_j \in \mathbf{F} \cap A, \lambda_j \geq 0,\ j = 1, 2, \ldots\right\},$$

and

$$U_{\varepsilon}^{(f+g)} = \sup\left\{\sum_{j=1}^{\infty} \lambda_j \cdot \mu(E_j) \middle| f + g \geq \sum_{j=1}^{\infty} \lambda_j \cdot \chi_{E_j} \geq f + g - \varepsilon, E_j \in \mathbf{F} \cap A, \lambda_j \geq 0, j = 1, 2, \ldots\right\}.$$

Then, for any

$$\sum_{j=1}^{\infty} \lambda_j' \cdot \chi_{E_j'} \in \left\{ f \geq \sum_{j=1}^{\infty} \lambda_j' \cdot \chi_{E_j'} \geq f - \varepsilon/2, E_j' \in \mathbf{F} \cap A, \lambda_j' \geq 0, j = 1, 2, \ldots \right\}$$

and

$$\sum_{j=1}^{\infty} \lambda''_j \cdot \chi_{E''_j} \in \left\{ g \geq \sum_{j=1}^{\infty} \lambda''_j \cdot \chi_{E''_j} \geq g - \varepsilon/2, E''_j \in \mathbf{F} \cap A, \lambda''_j \geq 0, j = 1, 2, \ldots \right\},$$

we have

$$\sum_{j=1}^{\infty} \lambda_j' \cdot \chi_{E_j'} + \sum_{j=1}^{\infty} \lambda''_j \cdot \chi_{E''_j} \in [f + g - \varepsilon,\ f + g].$$

This means that

$$U_{\varepsilon}^{(f+g)} \geq \sum_{j=1}^{\infty} \lambda_j' \cdot \mu(E_j') + \sum_{j=1}^{\infty} \lambda''_j \cdot \mu(E''_j)$$

and, therefore,

$$U_{\varepsilon}^{(f+g)} \geq U_{\varepsilon/2}^{(f)} + U_{\varepsilon/2}^{(f)}.$$

Letting $\varepsilon \to 0$, we obtain

$$(\mathrm{U})\int (f+g)\, d\mu \geq (\mathrm{U})\int f\, d\mu + (\mathrm{U})\int g\, d\mu.$$

□

A property similar to that of Theorem 11.2(1) does not exist for the upper and lower integrals. We can see it from the following example.

Example 12.3. Let $X = \{x_1, x_2\}$ and $\mathbf{F} = \mathbf{P}(X)$. Set function μ is defined as

$$\mu(A) = \begin{cases} 0 & \text{if } A = \emptyset \\ 1 & \text{otherwise} . \end{cases}$$

Clearly, μ is a monotone measure. We have

$$(\mathrm{U})\int 1\, d\mu = 1 \cdot \mu(\{x_1\}) + 1 \cdot \mu(\{x_2\}) = 1 \times 1 + 1 \times 1 = 2 \neq \mu(X).$$

Finding a similar counterexample for the lower integral is left to the reader. Though the equalities do not hold, we still have the inequalities expressed in the following theorem.

Theorem 12.2. $0 \leq (\mathrm{L})\int_A 1\, d\mu \leq \mu(A) \leq (\mathrm{U})\int_A 1\, d\mu.$

Proof. These inequalities follow directly from Definition 12.1. Details of the proof are left to the reader as an exercise. □

In addition to the above inequalities, the upper and lower integrals possess most of the common properties of the Lebesgue integral and the Choquet integral.

Theorem 12.3. *Let f and g be nonnegative measurable functions on* $(X, \mathbf{F}, \mu)$, *A and B be measurable sets, and a be a nonnegative real constant. Then,*

(1) $(\mathrm{U})\int_A f\, d\mu = (\mathrm{U})\int f \cdot \chi_A\, d\mu$ *and* $(\mathrm{L})\int_A f\, d\mu = (\mathrm{L})\int f \cdot \chi_A\, d\mu$;

(2) *if* $f \leq g$ *on A, then* $(\mathrm{U})\int_A f\, d\mu \leq (\mathrm{U})\int_A g\, d\mu$;

(3) *if* $A \subset B$, *then* $(\mathrm{U})\int_A f\, d\mu \leq (\mathrm{U})\int_B f\, d\mu$;

(4) $(\mathrm{U})\int_A af\, d\mu = a \cdot (\mathrm{U})\int_A f\, d\mu$ *and* $(\mathrm{L})\int_A af\, d\mu = a \cdot (\mathrm{L})\int_A f\, d\mu$.

Proof. These properties follow directly from Definition 12.1. The proofs are left to the reader as an exercise. □

Note that there are no counterparts for the lower integral in (2) and (3) of Theorem 12.3. These counterparts hold only under additional conditions, as expressed by the next theorem.

Theorem 12.4. *Let f and g be nonnegative measurable functions on monotone measure space $(X, \mathbf{F}, \mu)$, and let A and B be measurable sets. Then,*

(1) *if $f \le g$ on A, then* $(\mathrm{L})\int_A f\,d\mu \le (\mathrm{L})\int_A g\,d\mu$;

(2) *if $A \subset B$, then* $(\mathrm{L})\int_A f\,d\mu \le (\mathrm{L})\int_B f\,d\mu$.

Proof. We only need to prove (1), since (2) is a direct consequence of (1). Without any loss of generality, we may assume that $A = X$. Let f and g be measurable functions satisfying $f \le g$. For any fixed $\varepsilon > 0$, and for any elementary function $h_{g\varepsilon}$ with expression $h_{g\varepsilon} = \sum_j \lambda_j \cdot \chi_{E_j}$ satisfying $g \le h_{g\varepsilon} \le g + \varepsilon$, where $E_j \in \mathbf{F}, \lambda_j > 0$ for all j (finitely or countable-infinitely many, no loss of generality in assuming that $j = 0, 1, 2, \ldots$), from $f \le h_{g\varepsilon}$ and the measurability of f, we know that there exists elementary function $h_{f\varepsilon}$ with expression $h_{f\varepsilon} = \sum_j \sum_l \lambda_{jl} \cdot \chi_{F_{jl}}$, where $F_{jl} \in \mathbf{F}$, $F_{jl} \subseteq E_j$, $\lambda_{jl} \ge 0$, $j, l = 0, 1, 2, \ldots$, and $\sum_{l: F_{jl} \ne \emptyset} \lambda_{jl} \le \lambda_j$ for all j such that $f \le h_{f\varepsilon} \le f + \varepsilon$. In fact, on E_0, we may construct elementary function $d_0 = \lambda_0 \wedge \sum_l \frac{\varepsilon}{2^1} \chi_{F_{0l}}$, where $F_{0l} = \{x | \frac{l\varepsilon}{2^1} < f(x)\} \cap E_0$, $l = 0, 1, 2, \ldots$; for E_1, let $d_1 = \lambda_1 \wedge \sum_l \frac{\varepsilon}{2^2} \chi_{F_{1l}}$, where $F_{1l} = \{x | \frac{l\varepsilon}{2^2} < f(x) - d_0(x)\} \cap E_1$, $l = 0, 1, 2, \ldots$; generally, let $d_j = \lambda_j \wedge \sum_l \frac{\varepsilon}{2^{j+1}} \chi_{F_{jl}}$, where $F_{jl} = \{x | \frac{l\varepsilon}{2^{j+1}} < f(x) - \sum_{i=1}^{j-1} d_i(x)\} \cap E_j$, $l = 0, 1, 2, \ldots$. Continuing this procedure for j going to infinity, we obtain the required elementary function $h_{f\varepsilon} = \sum_j \sum_l \lambda_{jl} \cdot \chi_{F_{jl}}$, where $\lambda_{jl} = \frac{\varepsilon}{2^{j+1}}$, $j, l = 0, 1, 2, \ldots$ except $l = l_j = \lfloor 2^{j+1} \lambda_j / \varepsilon \rfloor$, for which $0 \le \lambda_{jl_j} = \lambda_j - l_j \frac{\varepsilon}{2^{j+1}} \le \frac{\varepsilon}{2^{j+1}}$. From the fact that $\sum_{l \le 2^{j+1} \lambda_j / \varepsilon} \lambda_{jl} = \lambda_j$ and $F_{jl} = \emptyset$ when $l > 2^{j+1} \lambda_j / \varepsilon$ for every j, and the monotonicity of μ, we have

$$\sum_j \sum_l \lambda_{jl}\, \mu(F_{jl}) \le \sum_j \lambda_j \mu(E_j),$$

where $\sum_j \sum_l \lambda_{jl}\, \mu(F_{jl})$ can be still expressed as a countable summation $\sum_k \lambda_k^* \mu(E_k^*)$. On the one hand, from the construction of its corresponding elementary function $\sum_k \lambda_k^* \chi_{E_k^*}$, since pairs (λ_{jl}, F_{jl}) are exhaustively defined, we may also see that $f \le \sum_k \lambda_k^* \chi_{E_k^*}$. On the other hand, we have $\sum_k \lambda_k^* \chi_{E_k^*} - f \le \sum_{j=0}^{\infty} \frac{\varepsilon}{2^{j+1}} = \varepsilon$. Hence,

$$\sum_k \lambda_k^* \chi_{E_k^*} \in \left\{ \sum_{k=1}^{\infty} \lambda_k \cdot \chi_{E_k} \middle| f \le \sum_{k=1}^{\infty} \lambda_k \cdot \chi_{E_k} \le f + \varepsilon, E_k \in \mathbf{F}, \lambda_k \ge 0, \text{ for all } k \right\}$$

and, therefore,

$$L_{f\varepsilon} = \inf \left\{ \sum_{k=1} \lambda_k \cdot \mu(E_k) \middle| f \le \sum_{k=1} \lambda_k \cdot \chi_{E_k} \le f + \varepsilon, E_k \in \mathbf{F}, \lambda_k \ge 0, \text{ for all } k \right\}$$
$$\le \sum_j \lambda_j \mu(E_j).$$

From the arbitrariness of $h_{g\varepsilon}$ and its expression, we have

$$L_{f\varepsilon} \le \inf \left\{ \sum_{j=1}^{\infty} \lambda_j \cdot \mu(E_j) \middle| g \le \sum_{j=1}^{\infty} \lambda_j \cdot \chi_{E_j} \le g + \varepsilon, E_j \in \mathbf{F} \cap A, \lambda_j \ge 0, j = 1, 2, \ldots \right\} = L_{g\varepsilon}.$$

Let $\varepsilon \to 0$. We get

$$(\mathrm{L}) \int f \, d\mu \le (\mathrm{L}) \int g \, d\mu.$$ □

Similarly, for the widened-lower integral, we have the following theorem.

Theorem 12.5. *Let f and g be nonnegative measurable functions on monotone measure space* $(X, \mathbf{F}, \mu)$, *and let A and B be measurable sets.*

(1) *If* $f \le g$ *on* A, then $(\underline{\mathrm{W}}) \int_A f \, d\mu \le (\underline{\mathrm{W}}) \int_A g \, d\mu$.

(2) *If* $A \subset B$, *then* $(\underline{\mathrm{W}}) \int_A f \, d\mu \le (\underline{\mathrm{W}}) \int_B f \, d\mu$.

Proof. If $f \le g$, then $f_N \le g_N$ and, therefore, for any simple function

$$h \in \left\{ \sum_{j=1}^{k} \lambda_j \cdot \chi_{E_j} \middle| g_N \le \sum_{j=1}^{k} \lambda_j \cdot \chi_{E_j}, k \ge 0, E_j \in \mathbf{F} \cap A, \lambda_j \ge 0, j = 1, 2, \ldots, k \right\},$$

we have

$$h \in \left\{ \sum_{j=1}^{k} \lambda_j \cdot \chi_{E_j} \middle| f_N \le \sum_{j=1}^{k} \lambda_j \cdot \chi_{E_j}, k \ge 0, E_j \in \mathbf{F} \cap A, \lambda_j \ge 0, j = 1, 2, \ldots, k \right\},$$

where $g_N = \min(N, g)$ and $f_N = \min(N, f)$ for $N > 0$. This means that

$$\left\{g_N \leq \sum_{j=1}^{k} \lambda_j \cdot \chi_{E_j}, k \geq 0, E_j \in \mathbf{F} \cap A, \lambda_j \geq 0,\ j = 1, 2, \ldots, k\right\}$$

$$\subset \left\{f_N \leq \sum_{j=1}^{k} \lambda_j \cdot \chi_{E_j}, k \geq 0, E_j \in \mathbf{F} \cap A, \lambda_j \geq 0,\ j = 1, 2, \ldots, k\right\}.$$

Hence,

$$\inf\left\{g_N \leq \sum_{j=1}^{k} \lambda_j \cdot \chi_{E_j}, k \geq 0, E_j \in \mathbf{F} \cap A, \lambda_j \geq 0,\ j = 1, 2, \ldots, k\right\}$$

$$\geq \inf\left\{f_N \leq \sum_{j=1}^{k} \lambda_j \cdot \chi_{E_j}, k \geq 0, E_j \in \mathbf{F} \cap A, \lambda_j \geq 0,\ j = 1, 2, \ldots, k\right\}.$$

That is,

$$(\underline{\mathrm{W}})\int_A f_N d\mu \leq (\underline{\mathrm{W}})\int_A g_N d\mu.$$

Letting $N \to \infty$, we obtain

$$(\underline{\mathrm{W}})\int_A f\, d\mu \leq (\underline{\mathrm{W}})\int_A g\, d\mu.$$

The proof of (1) now is complete, and (2) follows directly from (1). □

Due to Theorem 12.7 given in the next section, we need not list the inequalities similar to those in above theorem for widened-upper integral here.

Finally, the following theorem shows that the upper integral and the Lebesgue integral share a particular property.

Theorem 12.6. *Let f be a nonnegative measurable function and μ be a monotone measure on $(X, \mathbf{F})$. If $\mu(\{x|f(x) > 0\} \cap A) = 0$, i.e., $f = 0$ on A almost everywhere, then $(\mathrm{U})\int_A f\, d\mu = 0$. Conversely, if $(\mathrm{U})\int_A f\, d\mu = 0$ and μ is continuous from below, then $\mu(\{x|f(x) > 0\} \cap A) = 0$.*

Proof. There is no loss of generality in assuming $A = X$. First, we prove the necessary part of the theorem. If $\mu(\{x|f(x) > 0\}) = 0$, we know , by the monotonicity of μ, that each term $\lambda_j \cdot \mu(E_j)$ of the summation in the expression of U_ε in Definition 12.1 is zero for every $\varepsilon > 0$. So,

$$(\mathrm{U})\int f\, d\mu = \lim_{\varepsilon \to 0} U_\varepsilon = 0.$$

Next, we prove the sufficiency. A proof by contradiction is used here. Assume that $\mu(\{x|f(x) > 0\}) > 0$. From

$$\{x|f(x) > 0\} = \lim_{n \to \infty}\left\{x\middle|f(x) \geq \frac{1}{n}\right\},$$

we know, by the continuity from below of μ, that there is a positive integer n_0 such that

$$\mu(E_{n_0}) = \mu(\{x | f(x) \geq \frac{1}{n_0}\}) > 0.$$

Since $\chi_{E_{n_0}} \leq f$, we have

$$(\mathrm{U})\int f\,d\mu = (\overline{\mathrm{W}})\int f\,d\mu \geq \frac{1}{n_0} \cdot \mu(E_{n_0}) > 0.$$

This contradicts the fact that $(\mathrm{U})\int_A f\,d\mu = 0$. □

The condition of the lower continuity of μ in the second part of Theorem 12.6 is essential. This is shown by the following counterexample, in which the condition is violated.

Example 12.4. Let $X = (0, 1]$, $\mathbf{F}$ be the class of all Borel sets in X, and monotone measure μ on $\mathbf{F}$ be defined as

$$\mu(A) = \begin{cases} 1 & \text{if } A = X \\ 0 & \text{otherwise} \end{cases} \quad \forall A \in \mathbf{F}.$$

This monotone measure is clearly not continuous from below. Taking $f(x) = x$ for $x \in X$, we have $\mu(E) = 0$ for every $E \in \mathbf{F}$ satisfying $\lambda \cdot \chi_E \leq f$ with $\lambda > 0$. So,

$$(\mathrm{U})\int f\,d\mu = (\overline{\mathrm{W}})\int f\,d\mu = 0.$$

However,

$$\mu(\{x | f(x) > 0\}) = 1 \neq 0.$$

12.3 Relations Between Integrals

When function f is measurable, we may compare the four integrals defined in Section 12.1. The principal results of this comparison are expressed by Theorems 12.7–12.9.

Theorem 12.7. *If f: $X \to [0, \infty)$ is a measurable function on $(X, \mathbf{F})$ and $A \in \mathbf{F}$, then*

$$(\overline{\mathrm{W}})\int_A f\,d\mu = (\mathrm{U})\int_A f\,d\mu.$$

Proof. Since

$$\left\{ \sum_{j=1}^{\infty} \lambda_j \cdot \mu(E_j) \middle| f \geq \sum_{j=1}^{\infty} \lambda_j \cdot \chi_{E_j}, E_j \in \mathbf{F} \cap A, \lambda_j \geq 0, j = 1, 2, \ldots \right\}$$

$$\supset \left\{ \sum_{j=1}^{\infty} \lambda_j \cdot \mu(E_j) \middle| f \geq \sum_{j=1}^{\infty} \lambda_j \cdot \chi_{E_j} \geq f - \varepsilon, E_j \in \mathbf{F} \cap A, \lambda_j \geq 0, j = 1, 2, \ldots \right\},$$

we have

$$\sup \left\{ \sum_{j=1}^{k} \lambda_j \cdot \mu(E_j) \middle| f \geq \sum_{j=1}^{k} \lambda_j \cdot \chi_{E_j}, k \geq 0, E_j \in \mathrm{F} \cap A, \lambda_j \geq 0, j = 1, 2, \ldots, k \right\}$$

$$= \sup \left\{ \sum_{j=1}^{\infty} \lambda_j \cdot \mu(E_j) \middle| f \geq \sum_{j=1}^{\infty} \lambda_j \cdot \chi_{E_j}, E_j \in \mathrm{F} \cap A, \lambda_j \geq 0, j = 1, 2, \ldots \right\}$$

$$\geq \sup \left\{ \sum_{j=1}^{\infty} \lambda_j \cdot \mu(E_j) \middle| f \geq \sum_{j=1}^{\infty} \lambda_j \cdot \chi_{E_j} \geq f - \varepsilon, E_j \in \mathrm{F} \cap A, \lambda_j \geq 0, j = 1, 2, \ldots \right\}$$

$$= U_\varepsilon$$

for any $\varepsilon > 0$. Hence,

$$(\overline{\mathrm{W}}) \int_A f d\mu \geq \lim_{\varepsilon \to 0+} U_\varepsilon = (\mathrm{U}) \int_A f d\mu.$$

To show the inverse inequality, consider any given function having a form

$$g = \sum_{j=1}^{k} \lambda_j \cdot \chi_{E_j}$$

and satisfying $g \leq f$, where $k \geq 1, E_j \in \mathbf{F} \cap A, \lambda_j \geq 0, j = 1, 2, \ldots, k$. Since $f - g$ is measurable, for any given $\varepsilon > 0$, there exists function

$$h = \sum_{j=k+1}^{\infty} \lambda_j \cdot \chi_{E_j},$$

where $E_j \in \mathbf{F} \cap A, \lambda_j \geq 0, j = k+1, k+2, \ldots$ (some or all λ_j may be zeros), such that $f - g \geq h \geq f - g - \varepsilon$. Thus,

$$g + h \in \{ \sum_{j=1}^{\infty} \lambda_j \cdot \chi_{E_j} | f \geq \sum_{j=1}^{\infty} \lambda_j \cdot \chi_{E_j} \geq f - \varepsilon, E_j \in \mathbf{F} \cap A, \lambda_j \geq 0, j = 1, 2, \ldots \}$$

and $U_\varepsilon \geq \sum_{j=1}^{\infty} \lambda_j \cdot \mu(E_j) \geq \sum_{j=1}^{k} \lambda_j \cdot \mu(E_j)$. Letting ε go to zero, we get

$$(\mathrm{U}) \int_A f d\mu \geq \sum_{j=1}^{k} \lambda_j \cdot \mu(E_j).$$

Consequently,

$$(\mathrm{U})\int_A f\,d\mu \geq \sup\left\{\sum_{j=1}^{k}\lambda_j\cdot\mu(E_j)\,\middle|\, f\geq\sum_{j=1}^{k}\lambda_j\cdot\chi_{E_j}, k\geq 1, E_j\in\mathbf{F}\cap A, \lambda_j\geq 0,\ j=1,2,\ldots,k\right\}$$

$$= (\overline{\mathrm{W}})\int_A f\,d\mu. \qquad \square$$

From the proof of Theorem 12.7, we know that U_ε is independent of ε. It should be noted that there is no similar good property for L_ε. Furthermore, between $(\underline{\mathrm{W}})\int_A f\,d\mu$ and $(\mathrm{L})\int_A f\,d\mu$, there is no similar equality (or even only inequality) as the one shown in Theorem 12.7 for $(\overline{\mathrm{W}})\int_A f\,d\mu$ and $(\mathrm{U})\int_A f\,d\mu$. We can see this from the following counterexamples.

Example 12.5. Let $X = \{a, b\}$, $\mathbf{F} = \mathbf{P}(X)$, function $f = \chi_{\{a\}}$, and $\mu(A) = |A| (\mathrm{mod}\ 2)$ for $\forall A \in \mathbf{F}$, where $|A|$ is the cardinality of A. For $\varepsilon \geq 1$, we have $L_\varepsilon = 0$, with $k = 1$, $\lambda_1 = 1$ and $E_1 = X$ reaching the infimum; while when $0 < \varepsilon < 1$, we have $L_\varepsilon = 1 - \varepsilon$ with $k = 2$, $\lambda_1 = 1 - \varepsilon$, $\lambda_2 = \varepsilon$, $E_1 = \{a\}$, and $E_2 = X$ reaching the infimum. So, we have $(\mathrm{L})\int f\,d\mu = 1$. However, in this example, $(\underline{\mathrm{W}})\int f\,d\mu = 0$ with $k = 1$, $\lambda_1 = 1$ (or larger), and $E_1 = X$ reaching the infimum. This shows that $(\underline{\mathrm{W}})\int f\,d\mu \geq (\mathrm{L})\int f\,d\mu$ may not be true, though function f is measurable.

Example 12.6. Let $X = \{1, 2, \ldots\}$, $\mathbf{F} = \mathbf{P}(X)$, and

$$\mu(A) = \begin{cases} 1 & \text{if } |A| = \infty \\ 0 & \text{otherwise.} \end{cases}$$

Clearly, μ is a monotone measure. For function $f = \chi_X = 1$, we have

$$(\underline{\mathrm{W}})\int f\,d\mu = 1.$$

However, for any $\varepsilon > 0$, $L_\varepsilon = 0$ with expression $f = \sum_{j=1}^{\infty}\chi_{\{j\}}$ reaching the infimum and, therefore,

$$(\mathrm{L})\int f\,d\mu = 0.$$

This shows that $(\underline{\mathrm{W}})\int f\,d\mu \leq (\mathrm{L})\int f\,d\mu$ may not be true, though function f is measurable.

However, when μ is a monotne measure, we have a result for the lower integral and the widened-lower integral that is similar to, but weaker than, the one stated in Theorem 12.7 for the upper integral and widened-upper integral.

Theorem 12.8. *Let μ be a monotone measure. If $f\colon X \to [0, \infty)$ is a measurable function on $(X, \mathbf{F})$ and $A \in \mathbf{F}$, then*

$$\lim_{N\to\infty} (\mathrm{L})\int_A f_N \, d\mu \le (\underline{\mathrm{W}})\int_A f \, d\mu,$$

where $f_N = \min(N, f)$ for $N > 0$.

Proof. Similarly as in Theorem 12.7, we assume that $A = X$. For any given $N > 0$ and $\varepsilon > 0$, and for any given simple function g satisfying the inequality

$$g = \sum_{j=1}^{k} \lambda_j \cdot \chi_{E_j} \ge f_N,$$

where $k \ge 1, E_j \in \mathbf{F}, \lambda_j \ge 0, j = 1, 2, \ldots, k$, there exists simple function

$$h = \sum_{j=1}^{k} \sum_{l=1}^{k_j} \lambda_{jl} \cdot \chi_{F_{jl}},$$

where $k_j \ge 1, F_{jl} \in \mathbf{F}, \lambda_{jl} \ge 0, l = 1, 2, \ldots, k_j, E_j \supset F_{j1} \supset F_{j2} \supset \ldots \supset F_{jk_j}$ with $\sum_{l=1}^{k_j} \lambda_{jl} \le \lambda_j$ for $j = 1, 2, \ldots, k$, that satisfies the inequality

$$f_N \le h \le \min(f_N + \varepsilon, \ g).$$

In fact, for function

$$g_1 = f_N - \sum_{j=2}^{k} \lambda_j \cdot \chi_{E_j},$$

its positive part,

$$g_1^+ = \begin{cases} g_1 & \text{if } g_1 \ge 0 \\ 0 & \text{otherwise.} \end{cases}$$

is a measurable function satisfying $g_1^+ \le \lambda_1 \cdot \chi_{E_1}$. Let k_1 be a positive integer large enough so that $\frac{\lambda_1}{k_1} \le \varepsilon$. Denote $\frac{\lambda_1}{k_1}$ by ε_1, and take

$$F_{1l} = \{x | g_1^+ > (l-1)\varepsilon_1, \ x \in X\}$$

and $\lambda_{1l} = \varepsilon_1$ for $l = 1, 2, \ldots, k_1$. We know that $F_{1l} \in \mathbf{F}$ for $l = 1, 2, \ldots, k_1$ and

$$E_1 \supset F_{11} \supset F_{12} \supset \ldots \supset F_{1k_1}.$$

Therefore,

$$g_1^+ \leq \sum_{l=1}^{k_1} \lambda_{1l} \cdot \chi_{F_{1l}} \leq \min(g_1^+ + \varepsilon,\ \lambda_1 \cdot \chi_{E_1}).$$

Similarly, for measurable function

$$g_2 = f_N - \sum_{l=1}^{k_1} \lambda_{1l} \cdot \chi_{F_{1l}} - \sum_{j=3}^{k} \lambda_j \cdot \chi_{E_j},$$

take k_2 large enough so that $\varepsilon_2 = \frac{\lambda_2}{k_2} \leq \varepsilon$. We can find

$$F_{2l} = \{x | g_2^+ > (l-1)\varepsilon_2,\ x \in X\},$$

satisfying $F_{2l} \in \mathbf{F}$ for $l = 1, 2, \ldots, k_2$ and

$$E_2 \supset F_{21} \supset F_{22} \supset \ldots \supset F_{2k_2},$$

and corresponding $\lambda_{2l} = \varepsilon_2$ for $l = 1, 2, \ldots, k_2$ such that

$$g_2^+ \leq \sum_{l=1}^{k_2} \lambda_{2l} \cdot \chi_{F_{2l}} \leq \min(g_2^+ + \varepsilon,\ \lambda_2 \cdot \chi_{E_2}),$$

where g_2^+ is the positive part of g_2. Recursively, for function

$$g_s = f_N - \sum_{j=1}^{s-1} \sum_{l=1}^{k_j} \lambda_{jl} \cdot \chi_{F_{jl}} - \sum_{j=s+1}^{k} \lambda_j \cdot \chi_{E_j},$$

$s = 3,\ 4, \ldots,\ k$, we can find

$$F_{sl} = \{x | g_s^+ > (l-1)\varepsilon_s,\ x \in X\},$$

satisfying $F_{sl} \in \mathbf{F}$ for $l = 1, 2, \ldots, k_s$ and

$$E_s \supset F_{s1} \supset F_{s2} \supset \ldots \supset F_{sk_s},$$

and corresponding $\lambda_{sl} = \varepsilon_s \leq \varepsilon$ for $l = 1, 2, \ldots, k_s = \frac{\lambda_s}{\varepsilon_s}$ such that

$$g_s^+ \leq \sum_{l=1}^{k_s} \lambda_{sl} \cdot \chi_{F_{sl}} \leq \min(g_s^+ + \varepsilon,\ \lambda_s \cdot \chi_{E_s}),$$

where g_s^+ is the positive part of g_s. Now, we take

$$h = \sum_{j=1}^{k} \sum_{l=1}^{k_j} \lambda_{jl} \cdot \chi_{F_{jl}}.$$

From

$$f_N - \sum_{j=1}^{k-1} \sum_{l=1}^{k_j} \lambda_{jl} \cdot \chi_{F_{jl}} = g_k \le g_k^+ \le \sum_{l=1}^{k_k} \lambda_{kl} \cdot \chi_{F_{kl}},$$

we know, on the one hand, that

$$f_N \le \sum_{l=1}^{k_k} \lambda_{kl} \cdot \chi_{F_{kl}} + \sum_{j=1}^{k-1} \sum_{l=1}^{k_j} \lambda_{jl} \cdot \chi_{F_{jl}} = \sum_{j=1}^{k} \sum_{l=1}^{k_j} \lambda_{jl} \cdot \chi_{F_{jl}} = h.$$

On the other hand, for any given $x \in X$, if $g_s(x) = 0$ for all $s = 1, 2, \ldots, k$, then $x \notin F_{s1}$ for $s = 1, 2, \ldots, k$ and, therefore,

$$h(x) = 0 \le f_N(x) + \varepsilon;$$

otherwise, let

$$s(x) = \max\{s | g_s(x) > 0\}.$$

From

$$\sum_{l=1}^{k_{s(x)}} \lambda_{s(x)l} \cdot \chi_{F_{s(x)l}}(x) \le g_{s(x)}^+(x) + \varepsilon = g_{s(x)}(x) + \varepsilon,$$

$$\begin{aligned} g_{s(x)}(x) &= f_N(x) - \sum_{j=1}^{s(x)-1} \sum_{l=1}^{k_j} \lambda_{jl} \cdot \chi_{F_{jl}}(x) - \sum_{j=s(x)+1}^{k} \lambda_j \cdot \chi_{E_j}(x) \\ &\le f_N(x) - \sum_{j=1}^{s(x)-1} \sum_{l=1}^{k_j} \lambda_{jl} \cdot \chi_{F_{jl}}(x), \end{aligned}$$

and $x \notin F_{jl}$ for all $j > s(x)$ and all $l = 1, 2, \ldots, k_j$, we have

$$\begin{aligned} h(x) &= \sum_{j=1}^{s(x)-1} \sum_{l=1}^{k_j} \lambda_{jl} \cdot \chi_{F_{jl}}(x) + \sum_{l=1}^{k_{s(x)}} \lambda_{jl} \cdot \chi_{F_{jl}}(x) + \sum_{j=s(x)+1}^{k} \sum_{l=1}^{k_j} \lambda_{jl} \cdot \chi_{F_{jl}}(x) \\ &= \sum_{j=1}^{s(x)-1} \sum_{l=1}^{k_j} \lambda_{jl} \cdot \chi_{F_{jl}}(x) + \sum_{l=1}^{k_{s(x)}} \lambda_{jl} \cdot \chi_{F_{jl}}(x) \\ &\le f_N(x) + \varepsilon. \end{aligned}$$

From the construction of h directly, we also know that $h \le g$. So, we have $f_N \le h \le \min(f_N + \varepsilon,\ g)$. Thus, by using the nonnegativity and the monotonicity of μ, it is not difficult to show that

$$\sum_{j=1}^{k}\sum_{l=1}^{k_j}\lambda_{jl}\cdot\mu(F_{jl}) \le \sum_{j=1}^{k}\lambda_j\cdot\mu(E_j).$$

This means that

$$L_\varepsilon^{(FN)} \le \sum_{j=1}^{k}\lambda_j\cdot\mu(E_j),$$

where

$$L_\varepsilon^{(FN)} = \inf\left\{\sum_{j=1}^{k}\lambda_j\cdot\mu(E_j)\,\middle|\, f_N \le \sum_{j=1}^{k}\lambda_j\cdot\chi_{E_j} \le f_N+\varepsilon, k\ge 1,\right.$$
$$\left. E_j\in\mathbf{F}, \lambda_j\ge 0, j=1,2,\ldots,k\right\}.$$

Denoting

$$L_\varepsilon^{(N)} = \inf\left\{\sum_{j=1}^{\infty}\lambda_j\cdot\mu(E_j)\,\middle|\, f_N \le \sum_{j=1}^{\infty}\lambda_j\cdot\chi_{E_j} \le f_N+\varepsilon, E_j\in\mathbf{F}, \lambda_j\ge 0, j=1,2,\ \ldots\right\},$$

we have

$$L_\varepsilon^{(N)} \le L_\varepsilon^{(FN)}$$

for every $N>0$ and $\varepsilon>0$ because some (even infinitely many) λ_j in the expression of $L_\varepsilon^{(N)}$ are allowed to be zero. Since ε may be any small positive number, we have

$$(\mathrm{L})\int f_N d\mu = \lim_{\varepsilon\to+0} L_\varepsilon^{(N)} \le \lim_{\varepsilon\to+0} L_\varepsilon^{(FN)} \le \sum_{j=1}^{k}\lambda_j\cdot\mu(E_j)$$

and, therefore,

$$(\mathrm{L})\int f_N d\mu \le \inf\left\{\sum_{j=1}^{k}\lambda_j\cdot\mu(E_j)\,\middle|\, f_N \le \sum_{j=1}^{k}\lambda_j\cdot\chi_{E_j}, k\ge 1,\right.$$
$$\left. E_j\in\mathbf{F}, \lambda_j\ge 0, j=1,2,\ldots,k\right\}$$
$$= (\underline{\mathrm{W}})\int f_N d\mu.$$

Guaranteed by the monotonicity of (L)$\int f_N\,d\mu$ with respect to N shown in Theorem 12.4, $\lim_{N\to\infty}$ (L)$\int f_N\,d\mu$ exists. Letting $N \to \infty$, we obtain

$$\lim_{N\to\infty} (\text{L})\int f_N\,d\mu \le (\underline{\text{W}})\int f\,d\mu.$$

The proof of this theorem is now complete. □

When $f: X \to [0, \infty)$ is upper-bounded, we may readily obtain the following corollary of Theorem 12.8 since $\lim_{N\to\infty}$ (L)$\int_A f_N\,d\mu =$(L)$\int_A f\,d\mu$.

Corollary 12.1. *Let μ be a monotone measure. if $f: X \to [0, \infty)$ is an upper-bounded measurable function on $(X, \mathbf{F})$ and $A \in \mathbf{F}$, then*

$$(\text{L})\int_A f\,d\mu \le (\underline{\text{W}})\int_A f\,d\mu.$$

Comparing the lower integral, the upper integral, and the Choquet integral discussed in Chapter 11, we have the following result.

Theorem 12.9. *Let μ be a general measure. For any given measurable function $f: X \to [0, \infty)$ and any set $A \in \mathbf{F}$,*

$$(\text{L})\int_A f\,d\mu \le (\text{C})\int_A f\,d\mu \le (\text{U})\int_A f\,d\mu,$$

provided the involved Choquet integral exists.

Proof. Without any loss of generality, we may assume that $A = X$.

(1) When (C)$\int f\,d\mu = \infty$, based on the result in Theorem 12.7, we just need to show that

$$(\overline{\text{W}})\int f\,d\mu = \infty.$$

In fact, from the definition of the Choquet integral (as a Riemann integral), we know that for any given large number $N > 0$, there exists $\delta(N) > 0$ such that

$$\delta(N)\sum_{i=1}^{\infty} \mu(F_{i\delta(N)}) > 2N,$$

where $F_{i\delta(N)} = \{x | f(x) \ge i\delta(N)\}$ for $i = 1, 2, \ldots$. So, there exists a positive integer k such that

$$\delta(N)\sum_{i=1}^{k} \mu(F_{i\delta(N)}) > N.$$

Since f is a measurable function and, therefore,

$$\delta(N)\sum_{i=1}^{k}\chi_{F_{i\delta(N)}} \in \left\{\sum_{j=1}^{k}\lambda_j\cdot\chi_{E_j}\,\middle|\, f\geq\sum_{j=1}^{k}\lambda_j\cdot\chi_{E_j}, k\geq 1,\right.$$
$$\left. E_j\in\mathbf{F}, \lambda_j\geq 0, j=1,2,\ldots k\right\},$$

we have

$$(\overline{\mathrm{W}})\int f\,d\mu \geq \delta(N)\sum_{i=1}^{k}\mu(F_{i\delta(N)}) > N.$$

This means that $(\overline{\mathrm{W}})\int f\,d\mu = \infty$.

(2) When $(\mathrm{C})\int f\,d\mu < \infty$, we may use a similar idea as above. For any given $\varepsilon > 0$, by using the measurability of f, there exists $\delta(\varepsilon) \in (0, \varepsilon]$ such that

$$f-\varepsilon \leq f-\delta(\varepsilon) \leq \delta(\varepsilon)\sum_{i=1}^{\infty}\chi_{F_{i\delta(\varepsilon)}} \leq f \leq \delta(\varepsilon)\sum_{i=0}^{\infty}\chi_{F_{i\delta(\varepsilon)}} \leq f+\delta(\varepsilon) \leq f+\varepsilon$$

with

$$(\mathrm{C})\int f\,d\mu - \varepsilon \leq \delta(\varepsilon)\sum_{i=1}^{\infty}\mu(F_{i\delta(\varepsilon)}) \leq \delta(\varepsilon)\sum_{i=0}^{\infty}\mu(F_{i\delta(\varepsilon)}) \leq (\mathrm{C})\int f\,d\mu + \varepsilon.$$

Thus,

$$(\mathrm{L})\int f\,d\mu \leq (\mathrm{C})\int f\,d\mu + \varepsilon$$

and

$$(\mathrm{U})\int f\,d\mu \geq (\mathrm{C})\int f\,d\mu - \varepsilon.$$

By the arbitrariness of ε, we obtain that

$$(\mathrm{L})\int f\,d\mu \leq (\mathrm{C})\int f\,d\mu \leq (\mathrm{U})\int f\,d\mu.$$

□

Example 12.7. Recall Examples 11.3 and 12.1, where $(\mathrm{C})\int f\,d\mu = 74$, $(\mathrm{U})\int f\,d\mu = 88$, and $(\mathrm{L})\int \mathrm{f}\,\mathrm{d}\mu = 64$. This exemplifies the inequalities in Theorem 12.9.

12.4 Lower and Upper Integrals on Finite Sets

In this section we restrict our discussion to a finite set $X = \{x_1, x_2, \ldots, x_n\}$. We also assume that f is a nonnegative function on X, and μ is a general measure on $\mathbf{P}(X)$. Recalling the concept of r-integral introduced in Section 8.4, we may regard the upper and lower integrals on finite sets as special r-integrals. They are a pair of extreme cases in regard to the integration value. In fact, since X is finite, the supremum and the infimum in Definition 12.1 are accessible. Hence, the upper integral of f with respect to μ, $(\mathrm{U})\int f\,d\mu$, can be expressed as

$$(\mathrm{U})\int f\,d\mu = \sup\left\{\sum_{j=1}^{2^n-1} \lambda_j \cdot \mu(A_j)\middle|\sum_{j=1}^{2^n-1} \lambda_j \chi_{A_j} = f\right\},$$

where $\lambda_j \geq 0$ and $A_j = \bigcup_{i:j_i=1} \{x_i\}$ if j is expressed in binary digits as $j_n\, j_{n-1} \cdots j_1$ for every $j = 1, 2, \ldots, 2^n - 1$. The value of $(\mathrm{U})\int f\,d\mu$ then is just the solution of the following linear programming problem, where $\lambda_1, \lambda_2, \ldots, \lambda_{2^n-1}$ are unknown parameters:

$$\text{maximize} \qquad z = \sum_{j=1}^{2^n-1} \lambda_j \cdot \mu_j$$

$$\text{subject to} \qquad \sum_{j=1}^{2^n-1} \lambda_j \chi_{A_j}(x_i) = f(x_i), i = 1, 2, \ldots, n$$

$$\lambda_j \geq 0,\ j = 1, 2, \ldots, 2^n - 1,$$

where $\mu_j = \mu(A_j)$ for $j = 1, 2, \ldots, 2^n - 1$. The above n constraints can be also rewritten as

$$\sum_{j:x\in A_j\subseteq X} \lambda_j = f(x) \quad \forall x \in X.$$

Defining set function $\lambda:\mathbf{P}(X) \to [0, \infty)$ by $\lambda(A_j) = \lambda_j$ for $j = 1, 2, \ldots, 2^n - 1$, we may see that λ is a partition of f. So, the upper integral is just a special r-integral. Its corresponding partitioning rule is "dividing integrand such that the integration value is maximized."

By a knowledge on the linear programming, the above maximum can be accessed by at most n nonzero-valued λ_j, that is, the solution can be expressed as

$$\sum_{i=1}^{n} \lambda_{j_i} \mu_{j_i},$$

where $\{j_1, j_2, \ldots, j_n\}$ is a subset of $\{1, 2, \ldots, 2^n - 1\}$.

Example 12.8. We still use the data given in Example 11.3. Now the question is: how to arrange these workers such that the total amount of the manufactured toys during this week is as large as possible. This is just a linear programming problem:

$$\text{maximize} \quad z = 5\lambda_1 + 6\lambda_2 + 14\lambda_3 + 7\lambda_4 + 13\lambda_5 + 9\lambda_6 + 17\lambda_7$$

$$\text{subject to} \quad \lambda_1 + \lambda_3 + \lambda_5 + \lambda_7 = 6$$

$$\lambda_2 + \lambda_3 + \lambda_6 + \lambda_7 = 3$$

$$\lambda_4 + \lambda_5 + \lambda_6 + \lambda_7 = 4$$

$$\lambda_j \geq 0, \ j = 1, 2, \ldots, 7$$

Using the simplex method, a solution of this linear programming problem can be obtained as $\lambda_3 = 3$, $\lambda_4 = 1$, and $\lambda_5 = 3$ with $z = 88$. That is, we should arrange x_1 and x_2 to work together for 3 days, x_1 and x_3 to work together for 3 days, and x_3 works alone for one day. Then, the total number of manufactured toys will be maximized. This maximum solution is illustrated in Fig. 12.1. From it, we can see that function f is partitioned into three parts, where 3 is just the number of workers. The maximized value $z = 88$ is just the upper integral $(\mathrm{U})\int f\, d\mu$.

Similarly, we can express the lower integral of f with respect to μ, $(\mathrm{U})\int f\, d\mu$, as

$$(\mathrm{L})\int f\, d\mu = \inf\left\{ \sum_{j=1}^{2^n-1} \lambda_j \cdot \mu(A_j) \,\middle|\, \sum_{j=1}^{2^n-1} \lambda_j \chi_{A_j} = f \right\},$$

where $\lambda_j \geq 0$ and $A_j = \bigcup_{i:j_i=1} \{x_i\}$ if j is expressed in binary digits as $j_n j_{n-1} \cdots j_1$ for every $j = 1, 2, \ldots, 2^n - 1$. The value of $(\mathrm{L})\int f\, d\mu$ then is the solution of the linear programming problem

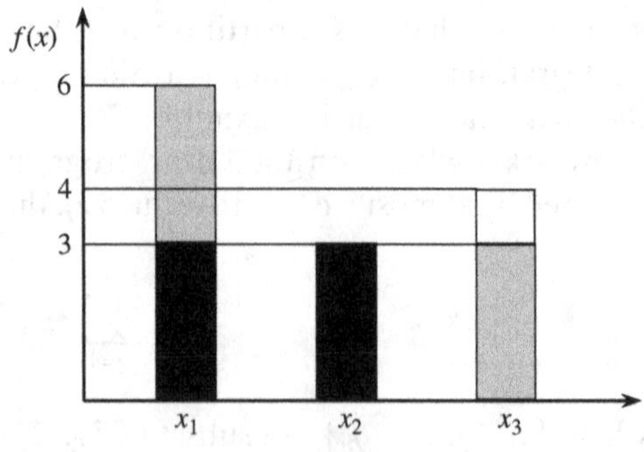

Fig. 12.1 The partition of f corresponding to the upper integral

$$\text{minimize} \qquad z = \sum_{j=1}^{2^n-1} \lambda_j \cdot \mu_j$$

$$\text{subject to} \qquad \sum_{j=1}^{2^n-1} \lambda_j \chi_{A_j}(x_i) = f(x_i), i = 1, 2, \ldots, n$$

$$\lambda_j \geq 0, \ j = 1, 2, \ldots, 2^n - 1,$$

where $\mu_j = \mu(A_j)$ for $j = 1, 2, \ldots, 2^n - 1$, and $\lambda_1, \lambda_2, \ldots, \lambda_{2^n-1}$ are unknown parameters. The corresponding partitioning rule to the lower integral is "dividing integrand such that the integration value is minimized." The above minimum can be accessed by at most n nonzero-valued a_j as well.

Example 12.9. Continuing Example 12.8, now the question is: What is the most conservative estimation for the total number of the toys that can be produced by these workers in this week? This is another linear programming problem:

$$\text{minimize} \qquad z = 5\lambda_1 + 6\lambda_2 + 14\lambda_3 + 7\lambda_4 + 13\lambda_5 + 9\lambda_6 + 17\lambda_7$$

$$\text{subject to} \qquad \lambda_1 + \lambda_3 + \lambda_5 + \lambda_7 = 6$$

$$\lambda_2 + \lambda_3 + \lambda_6 + \lambda_7 = 3$$

$$\lambda_4 + \lambda_5 + \lambda_6 + \lambda_7 = 4$$

$$\lambda_j \geq 0, \ \ j = 1, 2, \ldots, 7$$

The solution of this linear programming problem is $\lambda_1 = 6$, $\lambda_4 = 1$, and $\lambda_6 = 3$ with $z = 64$. That is, when x_2 and x_3 work together for 3 days, x_1 works alone for 6 days, and x_3 works alone for one day, the total amount of manufactured toys will be at least 64. It is just the lower integral $(\text{L})\int f\, d\mu$. This minimum solution is illustrated in Fig. 12.2.

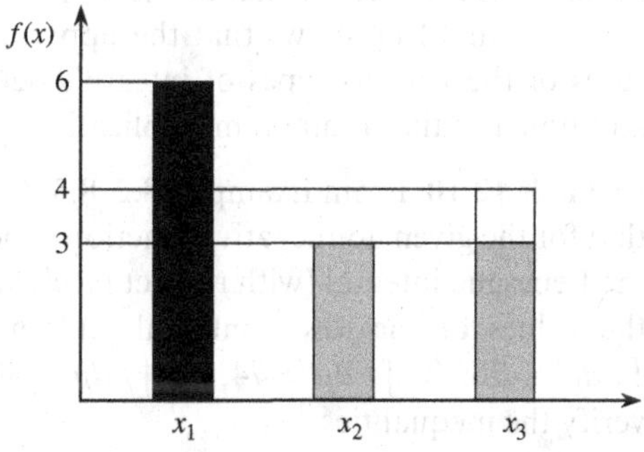

Fig. 12.2 The partition of f corresponding to the lower integral

Now, as is shown in the next theorem, we can strengthen the property of the upper integral established in Theorem 12.9.

Theorem 12.10. *Let f be a nonnegative function on X, and let μ be a monotone measure on* $\mathbf{P}(X)$. *Then,* $(\mathrm{U})\int f\,d\mu = 0$ *if and only if for every set A with* $\mu(A) > 0$ *there exists* $x \in A$ *such that* $f(x) = 0$.

Proof. First, we prove the "if" part. Suppose that for every set A with $\mu(A) > 0$, there exists $x \in A$ such that $f(x) = 0$. Thus, for each A_j with $\mu(A_j) > 0$, there exists some $x_{i(j)} \in A_j$ such that $f(x_{i(j)}) = 0$. So, we have $\lambda_j = 0$ since $0 \le \lambda_j = \lambda_j \cdot \chi_{A_j}(x_{i(j)}) \le f(x_{i(j)}) = 0$. Hence,

$$\sum_{j=1}^{2^n-1} \lambda_j \cdot \mu(A_j) = 0$$

for any λ_j and A_j, $j = 1, 2, \ldots, 2^n - 1$, satisfying $\sum_{j=1}^{2^n-1} \lambda_j \chi_{A_j} = f$. Consequently, according to the definition $(\mathrm{U})\int f\,d\mu = 0$.

Next, we prove the "only if" part. A proof by contradiction is used. Suppose that there exists some A_j with $\mu(A_j) > 0$ such that no x in A_j has $f(x) = 0$. Denote $\min_{x \in A_j} f(x)$ by λ_j. Then $\lambda_j > 0$ and, therefore, $\lambda_j \cdot \mu(A_j) > 0$. Since inequality $\lambda_j \cdot \chi_{A_j} \le f$ holds, we know that $(\mathrm{U})\int f\,d\mu \ge \lambda_j \cdot \mu(A_j) > 0$. This contradicts to the fact $(\mathrm{U})\int f\,d\mu = 0$. □

Theorem 12.10 can be expressed in an alternative but equivalent way as follows.

Theorem 12.10′. *Let* f *be a nonnegative function on X, and let μ be a monotone measure on* $\mathbf{P}(X)$. *Then,* $(\mathrm{U})\int f\,d\mu = 0$ *if and only if* $\mu(\{x \mid f(x) > 0\}) = 0$.

For comparing Lebesque integrals to Choquet integrals, or more generally, to r-integrals, we have the following theorem.

Theorem 12.11. *Let f be a nonnegative function on X, and let μ be a monotone measure on* $\mathbf{P}(X)$. *Then,* $0 \le (\mathrm{L})\int f\,d\mu \le (\mathrm{r})\int f\,d\mu \le (\mathrm{U})\int f\,d\mu$ *for any partitioning rule r.*

Proof. This is a direct result of the definitions of these integrals. □

Theorem 12.11 shows that the upper and lower integrals are two extreme cases of the various types of integrals defined on finite sets via the common addition and the common multiplication.

Example 12.10. From Examples 8.2, 8.5, 11.3, 11.4, 12.8, and 12.9, we have seen that for the given nonnegative function f and monotone measure μ, the values of the Lebesgue integral (with respect to μ') and the Choquet integral are between the values of the lower integral and the upper integral. In fact, we have $\int f\,d\mu' = 76$, $(\mathrm{C})\int f\,d\mu = 74$, $(\mathrm{U})\int f\,d\mu = 88$, and $(\mathrm{L})\int f\,d\mu = 64$. These results verify the inequalities

$$(\mathrm{L})\int f\,d\mu \le \int f\,d\mu' \le (\mathrm{U})\int f\,d\mu$$

and

$$(\mathrm{L})\int f\,d\mu \le (\mathrm{C})\int f\,d\mu \le (\mathrm{U})\int f\,d\mu$$

assessed in Theorem 12.11.

Theorem 12.12. *Let* μ *be monotone measures on* **P**(*X*). *Then,* $(\mathrm{U})\int 1\,d\mu \le n\cdot\mu(X)$.

Proof. Consider each $\sum_{j=1}^{2^n-1}\lambda_j\cdot\mu(A_j)$ satisfying $\sum_{j=1}^{2^n-1}\lambda_j\chi_{A_j}(x)=1$ for every $x\in X$. Since $\sum_{j=1}^{2^n-1}\lambda_j\chi_{A_j}(x)=1$ means $\sum_{x_i\in A_j}\lambda_j=1$ for every x_i, $i=1,2,\ldots,n$, we have

$$\begin{aligned}\sum_{j=1}^{2^n-1}\lambda_j\cdot\mu(A_j)\le\sum_{j=1}^{2^n-1}\lambda_j\cdot\mu(X)&=\mu(X)\cdot\sum_{j=1}^{2^n-1}\lambda_j\le\mu(X)\cdot\sum_{i=1}^{n}\Big(\sum_{x_i\in A_j}\lambda_j\Big)\\&=\mu(X)\cdot\sum_{i=1}^{n}1=n\cdot\mu(X).\end{aligned}$$

Hence,

$$(\mathrm{U})\int 1\,d\mu=\sup\left\{\sum_{j=1}^{2^n-1}\lambda_j\cdot\mu(A_j)\,\Big|\,\sum_{j=1}^{2^n-1}\lambda_j\chi_{A_j}=1\right\}\le n\cdot\mu(X).$$ □

12.5 Uncertainty Carried by Monotone Measures

Let $X=\{x_1,x_2,\ldots,x_n\}$. In this section, we assume that set function μ is a nontrivial monotone measure on (*X*, **P**(*X*)). Here, the word "nontrivial" means that there exists at least one set $A\subset X$ such that $\mu(A)>0$. We have seen that, due to the nonadditivity of μ, for a given nonnegative function *f*, different types of integrals may result in different integration values. This may be viewed as the uncertainty carried by monotone measure μ. Since the upper integral and the lower integral are too extremes in regard to the integration value, we may estimate the uncertainty by their difference.

Definition 12.3. Given a monotone measure μ on (*X*, **P**(*X*)), the degree of the uncertainty carried by μ is defined by

$$\gamma_\mu=\frac{(\mathrm{U})\int 1\,d\mu-(\mathrm{L})\int 1\,d\mu}{\mu(X)}.$$

It is evident that when μ is a classical measure, the upper integral coincides with the lower integral and, therefore, $\gamma_\mu=0$.

Theorem 12.13. *For any monotone measure μ on $(X, \mathbf{P}(X))$, $0 \le \gamma_\mu \le n$.*

Proof. On the one hand, from

$$(\mathrm{U})\int 1\, d\mu \ge (\mathrm{L})\int 1\, d\mu,$$

we obtain

$$\gamma_\mu = \frac{(\mathrm{U})\int 1\, d\mu - (\mathrm{L})\int 1\, d\mu}{\mu(X)} \ge 0.$$

On the other hand, from Theorem 12.12 and Definition 12.3, since $(\mathrm{L})\int 1\, d\mu \ge 0$, we have

$$\gamma_\mu = \frac{(\mathrm{U})\int 1\, d\mu - (\mathrm{L})\int 1\, d\mu}{\mu(X)} \le \frac{(\mathrm{U})\int 1\, d\mu}{\mu(X)} \le \frac{n \cdot \mu(X)}{\mu(X)}$$
$$= n$$

□

To present an estimate formula for the difference between the upper integral and the lower integral of a given nonnegative function, we need the following lemma.

Lemma 12.1. *For any given monotone measure μ and a bounded nonnegative function f,*

$$(\mathrm{U})\int f\, d\mu - (\mathrm{L})\int f\, d\mu \le (\mathrm{U})\int c\, d\mu - (\mathrm{L})\int c\, d\mu,$$

where c may be any upper bound of f.

Proof. From the expressions of the upper integral and the lower integral on a finite set given in Section 12.4, we know that there are $\lambda_j \ge 0$ and $\nu_j \ge 0$, $j = 1, 2, \ldots, 2^n - 1$, satisfying $\sum_{j: x \in A_j \subset X} \lambda_j = f(x)$ and $\sum_{j: x \in A_j \subset X} \nu_j = f(x)$ for every $x \in X$, such that

$$(\mathrm{U})\int f\, d\mu = \sum_{j=1}^{2^n - 1} \lambda_j \cdot \mu(A_j)$$

and

$$(\mathrm{L})\int f\, d\mu = \sum_{j=1}^{2^n - 1} \nu_j \cdot \mu(A_j).$$

For the nonnegative function $c - f$, we can find $\lambda'_j \ge 0$ and $\nu'_j \ge 0$, $j = 1, 2, \ldots, 2^n - 1$, satisfying $\sum_{j: x \in A_j \subset X} \lambda'_j = c - f(x)$ and $\sum_{j: x \in A_j \subset X} \nu_j{}' = c - f(x)$ for every $x \in X$, such that

$$(\mathrm{U})\int (c-f)\,d\mu = \sum_{j=1}^{2^n-1} \lambda'_j \cdot \mu(A_j)$$

and

$$(\mathrm{U})\int (c-f)\,d\mu = \sum_{j=1}^{2^n-1} \nu'_j \cdot \mu(A_j).$$

Since

$$\sum_{j:x\in A_j\subset X} \lambda_j + \sum_{j:x\in A_j\subset X} \lambda'_j = \sum_{j:x\in A_j\subset X} (\lambda_j + \lambda'_j) = c$$

and

$$\sum_{j:x\in A_j\subset X} \nu_j + \sum_{j:x\in A_j\subset X} \nu'_j = \sum_{j:x\in A_j\subset X} (\nu_j + \nu'_j) = c,$$

we have

$$\sum_{j=1}^{2^n-1} (\lambda_j + \lambda'_j) \cdot \mu(A_j) \le (\mathrm{U})\int c\,d\mu$$

and

$$\sum_{j=1}^{2^n-1} (\nu_j + \nu'_j) \cdot \mu(A_j) \ge (\mathrm{L})\int c\,d\mu.$$

Thus, from

$$(\mathrm{L})\int (c-f)\,d\mu \le (\mathrm{U})\int (c-f)\,d\mu,$$

we obtain

$$\begin{aligned}
(\mathbf{U})\int f\,d\mu - (\mathbf{L})\int f\,d\mu &= \sum_{j=1}^{2^n-1} \lambda_j \cdot \mu(A_j) - \sum_{j=1}^{2^n-1} \nu_j \cdot \mu(A_j) \le \sum_{j=1}^{2^n-1} \lambda_j \cdot \mu(A_j) \\
&\quad + \sum_{j=1}^{2^n-1} \lambda'_j \cdot \mu(A_j) - \sum_{j=1}^{2^n-1} \nu_j \cdot \mu(A_j) - \sum_{j=1}^{2^n-1} \nu'_j \cdot \mu(A_j) \\
&= \sum_{j=1}^{2^n-1} (\lambda_j + \lambda'_j) \cdot \mu(A_j) - \sum_{j=1}^{2^n-1} (\nu_j + \nu'_j) \cdot \mu(A_j) \\
&\le (\mathbf{U})\int c\,d\mu - (\mathbf{L})\int c\,d\mu.
\end{aligned}$$

□

Theorem 12.14. *Given a monotone measure μ on $(X, \mathbf{P}(X))$ and any nonnegative function f on X, we have*

$$0 \le (\mathrm{U})\int f\,d\mu - (\mathrm{L})\int f\,d\mu \le \gamma_\mu \cdot \mu(X) \cdot \max_{x\in X} f(x).$$

Proof. Let $c = \max_{x\in X} f(x)$. From the definition of γ_μ, Corollary 12.1 , Theorem 12.3(4), and Lemma 12.1, we have

$$\begin{aligned} 0 \le &(\mathbf{U})\int f\,d\mu - (\mathbf{L})\int f\,d\mu \le (\mathbf{U})\int c\,d\mu - (\mathbf{L})\int c\,d\mu \\ \le &c\cdot[(\mathbf{U})\int 1\,d\mu - (\mathbf{L})\int 1\,d\mu] = \gamma_\mu \cdot \mu(X) \cdot \max_{x\in X} f(x). \end{aligned}$$

□

Example 12.11. The data and some results in Examples 12.1, 12.8, and 12.9 are used here. Since $(\mathrm{U})\int 1\,d\mu = 21$ and $(\mathrm{L})\int 1\,d\mu = 14$, we have

$$\gamma_\mu = \frac{21-14}{17} = \frac{7}{17}.$$

From $(\mathrm{U})\int f\,d\mu - (\mathrm{U})\int f\,d\mu = 88 - 64 = 22$, $\max_{x\in X} f(x) = 6$, and $\mu(X) = 17$, Theorem 12.14 is verified: $22 \le \frac{7}{17} \times 6 \times 17 = 42$.

Theorem 12.14 can be used to estimate the uncertainty carried by the monotone measure in an aggregation process if the coordination manner is unknown.

Notes

12.1. The concept of an upper integral on a finite set and an algorithm for its computation were first introduced in [Wang and Xu, 1998] and further discussed in [Wang et al., 2000a]. This concept is similar to the concept of natural extension in [Walley, 1991], but it is more restricted.

12.2. A general form of integrals on finite sets with respect to signed general measures is discussed in [Wang et al., 2006a]. In this context, the two extreme types of integrals — the upper and lower integrals — were introduced.

12.3. The uncertainty carried by monotone measures was first introduced in [Wang and Klir, 2007].

Exercises

12.1. We use monotone measure space $(X, \mathbf{F}, \mu)$ and function f given in Example 11.1. That is, $X = [0, 1]$, $f(x) = x$ for $x \in X$, $\mathbf{F}$ is the class of all Borel sets in $[0, 1]$, and $\mu(B) = [m(B)]^2$ for $B \in \mathbf{F}$, where m is the Lebesgue measure. Calculate $(\mathrm{U})\int f\,d\mu$ and $(\mathrm{L})\int f\,d\mu$. Compare this result with $(\mathrm{C})\int f\,d\mu$ obtained in Example 11.1 to verify Theorem 12.9.

12.2. In Exercise 12.1, replacing $\mu(B) = [m(B)]^2$ by $\mu(B) = [m(B)]^{1/2}$, find $(\mathrm{U})\int f\,d\mu$ and $(\mathrm{L})\int f\,d\mu$. Compare this result with $(\mathrm{C})\int f\,d\mu$ obtained in Exercise 11.1 to verify Theorem 12.9.

12.3. Prove Theorem 12.3.

12.4. Find a counterexample to show that the conclusion in Theorem 12.4 may not be true if set function μ is a general measure.

12.5. Find a counterexample to show that, when μ is a general measure, the first part of Theorem 12.6 may not be true. That is, show that

$$\mu(\{x | f(x) > 0\} \cap A) = 0 \Longrightarrow (\mathrm{U})\int_A f\,d\mu = 0$$

may not be true.

12.6. Cite a counterexample to show that, when μ is a monotone measure that is not continuous from below, the second part of Theorem 12.6 may not be true, that is,

$$(\mathrm{U})\int_A f\,d\mu = 0 \Longrightarrow \mu(\{x | f(x) > 0\} \cap A) = 0$$

may not be true.

12.7. Three workers x_1, x_2, and x_3 manufacture toys. Their efficiencies μ can be regarded as a general measure:

Set	Value of μ
$\{x_1\}$	9
$\{x_2\}$	6
$\{x_1, x_2\}$	5
$\{x_3\}$	7
$\{x_1, x_3\}$	13
$\{x_2, x_3\}$	19
$\{x_1, x_2, x_3\}$	16

The numbers of their working days in this week is a function

$$f(x) = \begin{cases} 5 & \text{if } x = x_1 \\ 3 & \text{if } x = x_2 \\ 2 & \text{if } x = x_3. \end{cases}$$

Find (U)$\int f d\mu$, (L)$\int f d\mu$, (C)$\int f d\mu$, and the Lebesgue integral $\int f d\mu'$, where μ' is a classical measure satisfying $\mu'(\{x_i\}) = \mu(\{x_i\}), i = 1, 2, 3$.

12.8. Find the degree of uncertainty of the general measure μ given in Exercise 12.7.

Chapter 13
Constructing General Measures

13.1 An Overview

The problem of constructing general measures in various application contexts is not one of generalized measure theory per se. It is rather a problem of knowledge acquisition. Generalized measure theory provides in this case a framework within which the process of knowledge acquisition takes place and in which the elicited knowledge is represented (Fig. 13.1). Developing methods for knowledge acquisition have been the subject of knowledge engineering, an area of engineering that emerged in the 1970s. Although these methods are beyond the scope of this book, we want to illustrate in this chapter some of the main issues involved in constructing general measures.

General measures are considerably more expressive than classical measures, and this added expressiveness is essential in some applications. However, constructing general measures is a more difficult problem than constructing classical measures. There are two primary reasons for this increased difficulty. One of them is the substantially larger number of parameters that must be determined. While each classical measure is fully characterized by its values on singletons, a general measure on a measurable space $(X, \mathbf{F})$ is fully characterized by its values on all sets in $\mathbf{F}$. This implies an exponential increase in computational complexity. The second reason is that in most applications it is required that the measure involved be monotone. This requirement must be checked during the construction process, which further increases computational complexity.

To construct a general measure on a given a measurable space $(X, \mathbf{F})$ requires, in general, that $|\mathbf{F}| - 1$ unknown parameter be determined, provided that $\mathbf{F}$ contains a finite number of sets and the measure is normalized. When this requirement leads to prohibitively high computational complexity, it is often desirable to consider only some special class of general measures, which are characterized by a smaller number of unknown parameters. By considering only measures of a special type, we inevitably lose some expressiveness, but the construction process becomes more tractable. This may be in some applications a reasonable trade-off. Classes of lambda measures and possibility measures are examples of special classes of measures that are suitable for this purpose. When X is finite, each lambda measure is uniquely characterized by determining $|X|$

Z. Wang, G.J. Klir, *Generalized Measure Theory*,
DOI: 10.1007/978-0-387-76852-6_13,

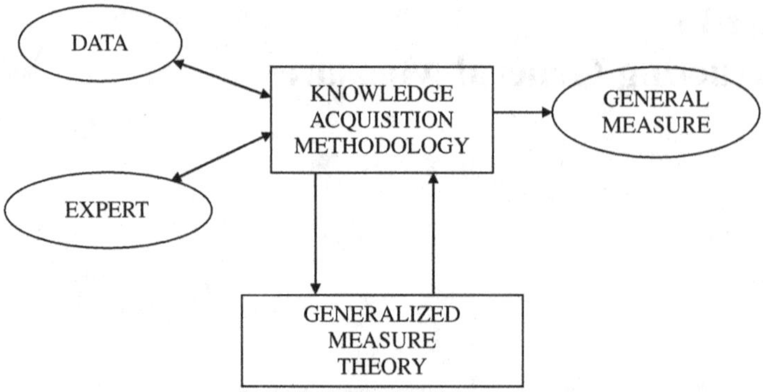

Fig. 13.1 Knowledge acquisition problem within generalized measure theory

parameters, and each possibility measure is uniquely characterized by determining $|X| - 1$ parameters.

When constructing general measures, additional properties may be required. Monotonicity, for example, is usually required, but some stronger properties such as maxitivity, superadditivity, 2-monotonicity, and the like are often required as well. When dealing with infinite sets, it is almost always required that the measure be continuous or semicontinuous.

Measures that satisfy some required properties are sometimes conveniently constructed from given measures. This can be done, for example, by integrating some nonnegative measurable function with respect to the given measure. This method is discussed for the Sugeno and Choquet integrals in Section 13.2. New measures can also be constructed from given measures by suitable transformations, as is discussed in Section 13.3. Construction methods whose goal is to determine a measure of a specified type that is as close as possible to a given measure are usually referred to as *identification methods*. These methods, which are sometimes combined with extensions, are discussed in Section 13.4.

Constructing measures of specified types from data or eliciting them from experts by knowledge-engineering methods represent perhaps the most important construction methods in many applications. Data-driven methods are surveyed in Section 13.5, but the many knowledge-engineering methods, notwithstanding their significance, are far beyond the scope of this book. A few additional methods for constructing measures of various types are examined in Section 13.6.

13.2 Constructing New Measures via Integration

As already mentioned, $|\mathbf{F}| - 1$ unknown parameter must be determined to construct a general measure on $(X, \mathbf{F})$ when $\mathbf{F}$ is finite. Since $2 \le |\mathbf{F}| \le 2^{|X|}$, it is clear that the problem of constructing general measures is computationally highly demanding, especially when $|\mathbf{F}|$ is close to its upper bound.

In some applications we can utilize a given general measure on $(X, \mathbf{F})$ to reduce the number of parameters to be determined to construct a measure with desirable properties. To explain this possibility, we let μ denote a given general measure on $(X, \mathbf{F})$. Then, by choosing a nonnegative measurable function f on X, a new measure υ on $(X, \mathbf{F})$ can be obtained for all $A \in \mathbf{F}$ by

$$\upsilon(A) = \int_A f \, d\mu,$$

where the integration may be done via the Sugeno integral, Choquet integral, or an integral of some other type. That is, given μ and some properties that υ is required to posses, we can construct an acceptable measure υ by determining an appropriate function f. Since the number of unknown parameters of f is $|X|$, this indirect way of determining υ is simpler than a direct way whenever $|X| < |\mathbf{F}| - 1$.

There are various methods for determining the function f in this problem. Whether a particular method is applicable or not depends on the requirements imposed on the constructed measure υ. In some applications the function f is determined from given data by methods analogous to those discussed in Section 13.5.

The choice of the type of integral to be used in this problem (only the Sugeno and Choquet integrals are considered here) depends, by and large, on the structural properties of μ that it is required υ preserve. The following properties are particularly important: monotonicity, continuity from below and from above, subadditivity and superadditivity, null-additivity and converse null-additivity, autocontinuity and converse autocontinuity from below and from above, uniform autocontinuity and uniform converse autocontinuity, and maxitivity. It is known that the Choquet integral preserves all these properties except maxitivity. The Sugeno integral preserves maxitivity, but does not preserve superadditivity, converse null-additivity, converse autocontinuity from below and from above, and uniform converse autocontinuity. The Choquet integral is thus generally preferable. Only when maxitivity of υ is required do we need to use the Sugeno integral.

13.3 Constructing New Measures by Transformations

Another way of constructing new measures from given ones is to use suitable transformations. These are generalizations of transformations introduced in Section 4.4 for constructing quasi-measures from classical measures. Constructing measures by transformations has the advantage that the number of parameters to be determined is very small, usually one or two.

A transformation employed in constructing a new normalized measure from a given one is a function of the form $\theta : [0, 1] \to [0, 1]$ that is continuous, strictly monotone, and such that $\theta(0) = 0$ and $\theta(1) = 1$. Applying a chosen transformation θ to any given measure μ on a measurable space $(X, \mathbf{F})$ results in a new measure v, which is obtained by composing μ with v. That is,

$$v(A) = \theta(\mu(A))$$

for all $A \in \mathbf{F}$. It is known that any measure v obtained from measure μ by transformation θ preserves all the structural properties of μ that are listed in Section 13.2, except subadditivity and superadditivity. Moreover, μ and v are order-isomorphic in the sense that

$$\mu(A) < \mu(B) \Leftrightarrow v(A) < v(B)$$

for all $A, B \in \mathbf{F}$, which is a desirable property in most applications. Moreover, given any pair μ_1 and μ_2 of general measures on $(X, \mathbf{F})$, these measures are order-isomorphic if and only if there exists a transformation θ such that $\mu_2(A) = \theta(\mu_1(A))$ for all $A \in \mathbf{F}$.

Using transformations to obtain new measures from given measures is one of the most effective ways of constructing measures. The advantage of transformations is that they preserve order and virtually all the desirable structural characteristics of the given measures, with the exception of subadditivity and superadditivity. It makes them suitable for revising given measures in face of new evidence. The two exceptions are not necessarily a disadvantage of transformations. They allow us, in some cases, to use an appropriate transformation to obtain a desired superadditive measure from a given subadditive measure, and vice versa.

Some common types of transformations with one or two parameters, which have been introduced and investigated in the literature, are defined in Table 13.1. To guarantee that measures obtained by the listed transformations are monotone, which is almost always required in applications, the parameters involved in each of these transformations must satisfy certain specific restrictions, as shown in the table.

13.4 Constructing New Measures by Identification and Extension

Identification is a problem of converting a given monotone measure on a finite class of sets to a monotone measure of a specified type, such as λ-measure or belief measure. Generally, the identification problem can be described as follows. Let X be a finite set and $\mu : \mathbf{P}(X) \to [0, \infty)$ be a given monotone measure. We want to find a monotone measure $\nu : \mathbf{P}(X) \to [0, \infty)$ with the specified type such that $\sum_{A \subset X} (\mu(A) - \nu(A))^2$ is minimized. Mathematically, this is an

Table 13.1 Some common transformations of measures

	$\theta(x)$	Parameter(s)	Restrictions on parameters
Quadratic transformations	$ax + (1-a)x^2$	a	$0 \le a \le 2$
Cubic transformations	$(1-a-b)x + ax^2 + bx^3$	a, b	$a+b \le 1, 1+a+2b>0$, and $\frac{-a \pm \sqrt{a^2 - 3b + 3ab + 3b^3}}{3b} \notin (0,1)$ unless $a^2 - 3b + 3ab + 3b^3 = 0$
Simple rational transformations	$\frac{(1+a)x}{1+ax}$	a	$a > -1$
Quadratic/linear transformations	$\frac{x+ax^2}{b+(1+a-b)x}$	a, b	$a < -1,\ 0<b \le 1$, *or* $a=-1,\ b=1$, *or* $-1<a<0,\ b \ge -a$, *or* $a \ge 0,\ b>0$
Linear/quadratic transformations	$\frac{x}{1-a-b+ax+bx^2}$	a, b	$b \le 0,\ a<1-b$, *or* $b>0,\ a \le 2b$

optimization problem, which can be solved by using a suitable analytical or numerical method.

When the domain of μ in the identification problem is smaller than $\mathbf{P}(X)$, a similar problem is called an *extension* for the specified type of monotone measure. It is a generalization (but restricted to the case of a finite universal set) of the extension problem discussed in Chapter 5. It can be solved in a similar way as the identification problem.

The use of genetic algorithms has been proven particularly suitable for dealing with these optimization problems. See Note 13.4 for an overview of relevant literature.

13.5 Data-Driven Construction Methods

Constructing monotone measures from data is perhaps the most important way of obtaining desired monotone measures in practical applications. Methods for constructing monotone measures from data are usually referred to as *data-driven methods*. Unlike methods discussed in the Section 13.4, where a given set function is converted in an optimal way to another set function of a desired type, the data-driven methods construct a set function of a desired type from given input-output data under the assumption that each output value is obtained by aggregating associated input values by a nonlinear integral of some specific type. It is thus an inverse problem of information fusion, which is discussed in Section 15.5. From a set of observed input data and the associated output observations, each of which is assumed to be obtained by aggregating the input values by a nonlinear integral of a particular type with respect to an unknown measure of some type, we want to determine the measure. The following is a more specific description of this inverse problem.

Given a set of attributes, $X = \{x_1, x_2, \ldots, x_n\}$, consider an input-output system whose inputs are values of the attributes and whose output y for each observation is the fused value of the input values, which is assumed to be obtained by a nonlinear integral of some type. Then, considering the observation of attributes $x_1, x_2, \ldots, x_n$ as a function defined on X, $f: X \to (-\infty, \infty)$, the input-output relation of the system can be expressed as $y = (\cdot)\int f\,d\mu$, where $(\cdot)$ indicates the assumed type of the integral and μ is an unknown monotone measure (or a signed general measure) defined on $\mathbf{P}(X)$. The problem is to determine μ on the basis of l input-output observations, as illustrated in general terms in Table 13.2a. The jth row in this table ($j = 1, 2, \ldots, l$) denotes the jth observation of the inputs (attributes $x_1, x_2, \ldots, x_n$) and the aggregated output. The positive integer l characterizes the number of observations in the given data, and it is usually much larger than the number of attributes. We write $f_{ji} = f_j(x_i), i = 1, 2, \ldots, n$ for $j = 1, 2, \ldots, l$, to denote conveniently the jth observation of attribute $x_i (i = 1, 2, \ldots, n, \text{and } j = 1, 2, \ldots, l)$. Using the data, our aim is to determine a monotone measure (or a signed general measure) μ (if it exists) so that $y_j = (\cdot)\int f_j\,d\mu, j = 1, 2, \ldots, l$, for some specified type of integral. Except for some contradictory cases, such a monotone measure (or signed general measure) usually exists when $l \le 2^n - 1$. If the specified type of integral is the Choquet integral, an algebraic method can be applied to obtain the values of μ since the value of the Choquet integral is a linear function of μ's values as shown in Section 11.6. The method is illustrated by the following example.

Table 13.2 Input-output observations
(a) general scheme

x_1	x_2	$\ldots$	x_n	y
f_{11}	f_{12}	$\ldots$	f_{1n}	y_1
f_{21}	f_{22}	$\ldots$	f_{2n}	y_2
$\vdots$				
f_{l1}	f_{l2}	$\ldots$	f_{ln}	y_l

(b) observations in Example 13.1

	x_1	x_2	x_3	y
week 1	5	3	1	55
week 2	3	5	2	60
week 3	4	1	5	63
week 4	2	4	4	52
week 5	1	2	3	33
week 6	5	4	3	70
week 7	2	5	4	38

Example 13.1. Let $X = \{x_1, x_2, x_3\}$ be a set of three workers. They are hired for producing toys and work together each week in the manner described in Example 11.3. Now, we only have the data consisting of the record of their attendance (days) and the total numbers of manufactured toys each week for seven weeks, as shown in Table 13.2b. The individual and joint efficiencies of these workers are not known. We want to use the data to determine the efficiencies. If the data in the ith row are denoted by f_j and y_j, $j = 1, 2, \ldots, 7$, then the relation among efficiencies μ, f_j, and y_j can be expressed, assuming that the inputs are aggregated by the Choquet integral, as

$$y_j = (\mathrm{C})\int f_j \, d\mu, \; j = 1, 2, \ldots, 7.$$

Thus, by using the calculation formula of the Choquet integral, we obtain a system of linear algebraic equations with unknown variables μ_j ($j = 1, 2, \ldots, 7,$), where $\mu_1 = \mu(\{x_1\})$, $\mu_2 = \mu(\{x_2\})$, $\mu_3 = \mu(\{x_1, x_2\})$, $\mu_4 = \mu(\{x_3\})$, $\mu_5 = \mu(\{x_1, x_3\})$, $\mu_6 = \mu(\{x_2, x_3\})$, and $\mu_7 = \mu(\{x_1, x_2, x_3\})$, as follows:

$$\begin{aligned} 2\mu_1 + 2\mu_3 + \mu_7 &= 55 \\ 2\mu_2 + \mu_3 + 2\mu_7 &= 60 \\ \mu_4 + 3\mu_5 + \mu_7 &= 63 \\ 2\mu_6 + 2\mu_7 &= 52 \\ \mu_4 + \mu_6 + \mu_7 &= 33 \\ \mu_1 + \mu_3 + 3\mu_7 &= 70 \\ \mu_2 + 2\mu_6 + 2\mu_7 &= 58 \end{aligned} \tag{13.1}$$

Solving system (13.1), we obtain the following unique solution: $\mu_1 = 5$, $\mu_2 = 6$, $\mu_3 = 14$, $\mu_4 = 7$, $\mu_5 = 13$, $\mu_6 = 9$, and $\mu_7 = 17$.

The data size l is often much larger than $2^n - 1$, where n is the number of attributes. In such cases, systems consisting of l linear equations have generally no precise solutions. However, an optimal approximate solution can be found by minimizing the total squared error

$$e^2 = \sum_{j=1}^{l} (y_j - (\mathrm{C})\int f_j d\mu)^2.$$

When integrals of the other types are considered, such as Sugeno integrals, upper integrals, or lower integrals, the algebraic method may fail. However, based on the observed input-output data of the system, some soft computing techniques, such as genetic algorithms or neural networks, can be used to search the above-mentioned optimal approximate solution and obtain an estimation of the values of μ.

13.6 Other Construction Methods

Methods for constructing general measures that are discussed in Sections 13.2–13.5 do not cover all conceivable types of methods for this purpose. Although our aim in this chapter is not to give a comprehensive overview of all types of construction methods, we consider it desirable to introduce in this section two additional types of construction methods: (i) methods based on the usual semantics of propositional modal logic; and (ii) methods based on suitable uncertainty principles. These types of construction methods are conceptually quite interesting and have a great potential utility. However, they are not fully developed as yet. Moreover, their full description would require in each case to introduce fairly extensive relevant preliminaries. For these reasons, we describe them only conceptually, focusing on basic ideas upon which they are based, and provide the reader with relevant references in Note 13.6.

13.6.1 Methods Based on Modal Logic

Propositional modal logic (or, simply, "modal logic") is an extension of classical propositional logic that adds to the propositional logic two unary modal operators, an operator of necessity, $\Box$, and an operator of possibility, $\Diamond$. Given a proposition p, $\Box p$ represents the proposition "it is necessary that p", while $\Diamond p$ stands for the proposition "it is possible that p." Given a universal set X, the set Q of atomic propositions for our purpose consists of all propositions of the form

$$e_A : \text{“}e \text{ is in } A\text{,”}$$

where $e \in X$ and A is a subset of X. The proposition e_A means that a given, incompletely characterized element e of X lies within se A.

Modal logic representations of some classes of general measures, including belief and plausibility measures, possibility and necessity measures, lambda measures, and additive measures, have recently been established. These representations are based on different models M of modal logic. Each model is a triple

$$M = (W, R, V),$$

where W, R, V denote, respectively, a set of possible worlds, a binary relation on W, and a value assignment function. Relation R describes accessibility between the possible worlds: $(w, w') \in R$ means that world w' is accessible to world w. This relation is usually assumed to be reflexive, which means that each world is accessible to itself. Function V assigns truth (T) or falsity (F) to each atomic proposition in each possible world. That is,

$$V: W \times Q \to \{\mathrm{T}, \mathrm{F}\},$$

where Q denotes the set of all atomic propositions. Once defined for proposition in Q, function V is inductively extended to all relevant propositions. This is done in the usual way for each propositional connective and each possible world. A proposition of the form $\Box p$ is true in a possible world w (i.e., $V(w, \Box p) = \mathrm{T}$) if the proposition p is true in all possible words that are accessible to w. Similarly, a proposition of the form $\Diamond p$ is true in a possible world w if there is at least one world accessible to w in which the proposition p is true.

The established modal logic representations of some classes of general measures can be utilized for constructing measures in these classes. Let us illustrate this utility by the following example.

Assume that the set of possible worlds W represents a group of experts in some field. Universal set X represents all possible answers to a question related to this field. Each expert is assumed to have his or her own opinion regarding the correct answer to each question of interest. The accessibility relation R may be interpreted in this context in the following way: $(w, w') \in R$ means that expert w takes into consideration the opinion (the valuations of relevant propositions) of expert w'. Naturally, every expert takes into consideration his or her own opinion, and, therefore, R is assumed to be reflexive. Assuming, for example, that W is a finite set with n possible worlds, X is a finite set, and R is an equivalence relation, it was proven that belief and plausibility measures are represented for all $A \in \mathbf{P}(X)$ by the formulas

$$\mathrm{Bel}(A) = \mathrm{T}[\Box e_A]/n, \qquad \mathrm{Pl}(A) = \mathrm{T}[\Diamond e_A]/n,$$

where $\mathrm{T}[p]$ denotes for any relevant proposition p the number of worlds in which p is true. This result has also been generalized in different ways, including the case of infinite sets. For representing and constructing other types of dual measures, we need to use appropriate types of accessibility relations.

13.6.2 Methods Based on Uncertainty Principles

New possibilities for constructing measures of various types open when the measures are used for representing uncertainty. In this case we may utilize four epistemological principles for coping with uncertainty. These principles are: a principle of minimum uncertainty, a principle of maximum uncertainty, a principle of uncertainty invariance, and a principle of requisite generalization. These four principles are applicable to four distinct classes of problems, all involving representation of uncertainty by general measures of appropriate types. Each of these principles provides guidance for dealing with the respective problems in specific ways that are epistemologically sound. When applying any of these principles, we always construct a new measure from a given measure. This new measure is an epistemological sound solution to a given problem, which is obtained by following the uncertainty principle pertaining to the problem. Depending on the problem involved, the new measure is required by the relevant principle to maximize, minimize, or preserve relevant uncertainty (measured in a justifiable way) within the constraints of the given problem.

Notes

13.1. An overview of methods for constructing measures of various types in the context of expert systems is presented in [Klir et al., 1997].

13.2. The preservation of the various structural properties of monotone measures by integrals are investigated in detail in [Wang et al., 1995a] for the Sugeno integral and in [Wang et al., 1996b] for the Choquet integral.

13.3. Constructions of new measures from given ones by transformations that are listed in Table 13.1 are investigated in papers by Klir et al. [1996] and Wang et al. [1996].

13.4. An analytical method for identifying a λ-measure on a finite universal set is presented in [Wierzchon, 1993]. The method for identifying general measures of various types by genetic algorithms is explored in papers by Wang and Wang [1996], Chen et al. [2000], and Wang and Chen [2005].

13.5. Data-driven methods for constructing general measures of various types are discussed in the literature fairly extensively. Some representative publications include methods employing genetic or evolutionary algorithms [Wang et al., 1998b; Wang et al., 1999a, b; Wang and Chen, 2005], neural networks [Wang and Wang, 1997; Wang et al., 1998a], and various other methods [Grabisch, 1995a; Klir et al., 1995; Yuan and Klir, 1996; Soria-Frisch, 2006].

13.6. A connection between modal logic [Chellas, 1980; Hughes and Cresswell, 1996] and various types of measures has been explored since the early 1990s and opened new ways for constructing measures. Some representative samples of the growing literature in this domain are [Resconi et al. 1993; Klir, 1994; Klir and Harmanec, 1994; Harmanec et al., 1994, 1996; Wang et al., 1995b; Tsiporkova et al., 1999]. The methodological principles of uncertainty mentioned in Section 13.6, which can be employed for constructing measures, are formulated in [Klir, 2006].

Chapter 14
Fuzzification of Generalized Measures and the Choquet Integral

14.1 Conventions

This chapter deals with *standard fuzzy sets*, which are introduced in Section 2.3. It is thus convenient to omit the adjective "standard." Classical sets, which are viewed in this chapter as special fuzzy sets, are called *crisp sets*. Fuzzy sets (as well as crisp sets) are denoted by capital letters printed in italics. When we refer to operations of intersection, union, and complement of fuzzy sets, it is always assumed in this chapter that they are the standard operations on fuzzy sets, as defined in Section 2.3.

14.2 Monotone Measures Defined on Fuzzy σ-Algebras

Let X be a universal set that is nonempty but may be not finite. The class of all fuzzy subsets of X, denoted by $\tilde{\mathbf{P}}(X)$, is called a *fuzzy power set* of X.

Definition 14.1. A subset of $\tilde{\mathbf{P}}(X)$ is called a *fuzzy* σ-algebra, denoted by $\tilde{\mathbf{F}}$, if it satisfies the following conditions:

(FSA1) The empty set $\emptyset$ belongs to $\tilde{\mathbf{F}}$;
(FSA2) $\tilde{\mathbf{F}}$ is closed under the formation of countable unions, i.e., $\bigcup_{i=1}^{\infty} A_i \in \tilde{\mathbf{F}}$ if each $A_i \in \tilde{\mathbf{F}}, i = 1, 2, \ldots$;
(FSA3) $\tilde{\mathbf{F}}$ is closed under the formation of complements, i.e., $\bar{A} \in \tilde{\mathbf{F}}$ if $A \in \tilde{\mathbf{F}}$.

The fuzzy power set $\tilde{\mathbf{P}}(X)$ is a fuzzy σ-algebra. The pair $(X, \tilde{\mathbf{F}})$ is called a *fuzzy measurable space* if $\tilde{\mathbf{F}}$ is a fuzzy σ-algebra of fuzzy subsets of X.

Let $(X, \mathbf{F})$ be a measurable space. The class of all fuzzy sets possessing $\mathbf{F}$-measurable membership functions, $\tilde{\mathbf{F}}(\mathbf{F}) = \{A | m_A \text{ is } \mathbf{F}\text{-measurable}\}$, forms a fuzzy σ-algebra and is called a fuzzy σ-algebra generated by $\mathbf{F}$. Such a fuzzy σ-algebra is of our primary interest in this chapter.

For a given fuzzy measurable space $(X, \tilde{\mathbf{F}})$, let $\mathbf{F}$ be the class of all crisp sets in $\tilde{\mathbf{F}}$, that is, $\mathbf{F} = \{A | A \in \tilde{\mathbf{F}}, A \text{ is crisp}\}$. Then $\mathbf{F}$ is a σ-algebra. Using σ-algebra $\mathbf{F}$, a fuzzy σ-algebra $\tilde{\mathbf{F}}(\mathbf{F})$ can be formed as mentioned above. It should be noted that $\tilde{\mathbf{F}}(\mathbf{F})$ may be different from $\tilde{\mathbf{F}}$.

Z. Wang, G.J. Klir, *Generalized Measure Theory*,
DOI: 10.1007/978-0-387-76852-6_14, © Springer Science+Business Media, LLC 2009

Example 14.1. Let $X = \{a, b\}$ and let $\tilde{\mathbf{F}}$ be the fuzzy σ-algebra consisting of all fuzzy sets whose membership function has a form of

$$m(x) = \begin{cases} c_1 & \text{if } x = a \\ c_2 & \text{if } x = b \end{cases}$$

where $c_1 \in \{0, 1\}$ and $c_2 \in [0, 1]$. Then the class of all crisp sets in $\tilde{\mathbf{F}}$, denoted by $\mathbf{F}$, is the power set of X. We can see that $\tilde{\mathbf{F}}(\mathbf{F})$ is the fuzzy power set of X, i.e., $\tilde{\mathbf{F}}(\mathbf{F}) = \tilde{\mathbf{P}}(X)$. So, we have $\tilde{\mathbf{F}} \neq \tilde{\mathbf{F}}(\mathbf{F})$.

Definition 14.2. Let $\tilde{\mu} : \tilde{\mathbf{F}} \to [0, \infty]$. Function $\tilde{\mu}$ is called a *fuzzified monotone measure* on $\tilde{\mathbf{F}}$ if:

(1) $\tilde{\mu}(\emptyset) = 0$;
(2) $\tilde{\mu}(A) \leq \tilde{\mu}(B)$ whenever $A \in \tilde{\mathbf{F}}, B \in \tilde{\mathbf{F}}$, and $A \subset B$.

A fuzzified monotone measure is also simply called monotone measure if there is no confusion. The triple $(X, \tilde{\mathbf{F}}, \tilde{\mu})$ is called a *fuzzy monotone measure space*. For any given $(X, \tilde{\mathbf{F}}, \tilde{\mu})$, restricting $\tilde{\mu}$ on σ-algebra $\mathbf{F} = \{A | A \in \tilde{\mathbf{F}}, A$ is crisp$\}$ as $\tilde{\mu}$, $(X, \mathbf{F}, \tilde{\mu})$ is a monotone measure space.

Example 14.2. The fuzzy measurable space $(X, \tilde{\mathbf{F}},)$ is given in Example 14.1. Let

$$\tilde{\mu}(A) = \begin{cases} 0.8c_2 & \text{if } c_1 = 0 \\ 0.5(c_2 + 1) & \text{if } c_1 = 1 \end{cases}$$

for fuzzy set A with membership function $m(x) = \begin{cases} c_1 & \text{if } x = a \\ c_2 & \text{if } x = b \end{cases}$. Then $\tilde{\mu}$ is a fuzzified monotone measure on $(X, \tilde{\mathbf{F}})$. Restricting $\tilde{\mu}$ on σ-algebra $\mathbf{F}$, μ is a monotone measure on $(X, \mathbf{F})$.

To simplify the notation from now on, $\tilde{\mathbf{F}}(\mathbf{F})$ is simply written as $\tilde{\mathbf{F}}$ if there is no confusion.

14.3 The Choquet Extension

Let $(X, \mathbf{F})$ be a measurable space, μ be a monotone measure on $\mathbf{F}$, and $f: X \to [0, \infty)$ be a nonnegative $\mathbf{F}$-measurable function defined on X. By using the Choquet integral, which is introduced and studied in Chapter 11, μ can be extended from σ-algebra $\mathbf{F}$ onto fuzzy σ-algebra $\tilde{\mathbf{F}}$, generated by $\mathbf{F}$.

Theorem 14.1. *For every fuzzy set* $A \in \tilde{\mathbf{F}}$, *define* $\tilde{\mu}(A) = (\mathrm{C})\int m_A d\mu = \int_0^1 \mu(A_\alpha) d\alpha$, *where* m_A *is the membership function of* A *and* $A_\alpha = \{x | m_A(x) \geq \alpha\}$ *is the* α-cut of A. Then $\tilde{\mu}$ *is a monotone measure on fuzzy* σ-algebra $\tilde{\mathbf{F}}$.

Proof. Since $A \in \tilde{\mathbf{F}}$, we know that m_A is an $\mathbf{F}$-measurable function and, therefore, $A_\alpha \in \mathbf{F}$. $\mu(A_\alpha)$ is a monotone function of α. So, $\tilde{\mu}(A)$ is well defined on $\tilde{\mathbf{F}}$.

Furthermore, $\tilde{\mu}(\emptyset) = (C)\int m_\emptyset d\mu = (C)\int 0\, d\mu = 0$. Finally, by monotonicity of the Choquet integral, $\tilde{\mu}(A) = (C)\int m_A d\mu \leq (C)\int m_B d\mu = \tilde{\mu}(B)$ whenever $A \in \tilde{\mathbf{F}}$, $B \in \tilde{\mathbf{F}}$, and $A \subset B$. Hence, $\tilde{\mu}$ is a monotone measure on $\tilde{\mathbf{F}}$. □

Since $\tilde{\mu}(A) = (C)\int \chi_A d\mu = \mu(A)$ when $A \in \mathbf{F}$, i.e., $\tilde{\mu}$ coincides with μ on $\mathbf{F}$, $\tilde{\mu}$ is an extension of μ from $\mathbf{F}$ onto $\tilde{\mathbf{F}}$, and is called a *Choquet extension* of μ.

Example 14.3. Let $X = \{x_1, x_2, x_3\}$ and a monotone measure μ be given on $\mathbf{F} = \mathbf{P}(X)$ as $\mu(\emptyset) = 0$, $\mu(\{x_1\}) = 1$, $\mu(\{x_2\}) = 2$, $\mu(\{x_1, x_2\}) = 5$, $\mu(\{x_3\}) = 3$, $\mu(\{x_1, x_3\}) = 8$, $\mu(\{x_2, x_3\}) = 4$, $\mu(\{x_1, x_2, x_3\}) = 10$. For fuzzy set A with membership function

$$m_A(x) = \begin{cases} 0.5 & \text{if } x = x_1 \\ 1 & \text{if } x = x_2, \\ 0.25 & \text{if } x = x_3 \end{cases}$$

we have

$$\begin{aligned} \tilde{\mu}(A) &= (C)\int m_A d\mu = m_A(x_3)\cdot\mu(\{x_1, x_2, x_3\}) \\ &\quad + [m_A(x_1) - m_A(x_3)]\cdot\mu(\{x_1, x_2\}) + [m_A(x_2) - m_A(x_1)]\cdot\mu(\{x_2\}) \\ &= 0.25\cdot 10 + (0.5 - 0.25)\cdot 5 + (1 - 0.5)\cdot 2 = 4.75. \end{aligned}$$

14.4 Structural Characteristics of Monotone Measures on Fuzzy σ-Algebras

In this section, we introduce some structural characteristics of monotone measures on fuzzy σ-algebras that are similar to those for monotone measures on σ-algebras in Chapter 6. Throughout this section, let $(X, \tilde{\mathbf{F}}, \tilde{\mu})$ be a fuzzy monotone measure space.

Definition 14.3. Monotone measure $\tilde{\mu}$ is continuous from below iff $\{A_i\} \subset \tilde{\mathbf{F}}$ and $A_1 \subset A_2 \subset \cdots$ imply $\lim_{i\to\infty} \tilde{\mu}(A_i) = \tilde{\mu}(\bigcup_{i=1}^{\infty} A_i)$; $\tilde{\mu}$ is continuous from above if $\{A_i\} \subset \tilde{\mathbf{F}}$, $A_1 \supset A_2 \supset \cdots$, and $\tilde{\mu}(A_1) < \infty$ imply $\lim_{i\to\infty} \tilde{\mu}(A_i) = \tilde{\mu}(\bigcap_{i=1}^{\infty} A_i)$; $\tilde{\mu}$ is continuous if it is both continuous from below and continuous from above.

Definition 14.4. Monotone measure $\tilde{\mu}$ is null-subtractive iff $\tilde{\mu}(A - B) = \tilde{\mu}(A)$ whenever $A \in \tilde{\mathbf{F}}$, $B \in \tilde{\mathbf{F}}$, and $\tilde{\mu}(B) = 0$; $\tilde{\mu}$ is null-additive iff $\tilde{\mu}(A \cup B) = \tilde{\mu}(A)$ whenever $A \in \tilde{\mathbf{F}}$, $B \in \tilde{\mathbf{F}}$, and $\tilde{\mu}(B) = 0$.

Unlike the situation where μ is defined on a crisp σ-algebra, the null-subtractivity of $\tilde{\mu}$ is not equivalent to its null-additivity. We can see this from the following example.

Example 14.4. Let $X = \{a\}$, $\tilde{\mathbf{F}}$ be the fuzzy σ-algebra consisting of all fuzzy subsets of X, and monotone measure $\tilde{\mu}$ be defined as

$$\tilde{\mu}(A) = \begin{cases} 0 & \text{if } m_A(a) \le 1/2 \\ 1 & \text{otherwise.} \end{cases}$$

Then, $\tilde{\mu}$ is null-additive. In fact, if $B \in \tilde{\mathbf{F}}$ with $\tilde{\mu}(B) = 0$, then $m_B(a) \le 1/2$. Thus, for any $A \in \tilde{\mathbf{F}}$, either $m_A(a) \le 1/2$ so that $m_{A\cup B}(a) = m_A(a) \vee m_B(a) \le 1/2$ and, therefore, $\tilde{\mu}(A \cup B) = \tilde{\mu}(A) = 0$, or $m_A(a) > 1/2$ so that $m_{A\cup B}(a) = m_A(a) \vee m_B(a) > 1/2$ and, therefore, $\tilde{\mu}(A \cup B) = \tilde{\mu}(A) = 1$. However, $\tilde{\mu}$ is not null-subtractive. We can see this as follows. Take $A = X$ and $B \in \tilde{\mathbf{F}}$ with $m_B(a) = 1/2$. Then $\tilde{\mu}(A) = 1$ and $\tilde{\mu}(B) = 0$. Noting that $m_{A-B}(a) = m_{A\cap\bar{B}}(a) = m_A(a) \wedge (1 - m_B(a)) = 1 \wedge (1/2) = 1/2$, we have $\tilde{\mu}(A - B) = 0$.

Definition 14.5. Monotone measure $\tilde{\mu}$ is autocontinuous from below iff $\lim_{i\to\infty} \tilde{\mu}(A - B_i) = \tilde{\mu}(A)$ whenever $A \in \tilde{\mathbf{F}}$, $B_i \in \tilde{\mathbf{F}}$, $i = 1, 2, \ldots$, and $\lim_{i\to\infty} \tilde{\mu}(B_i) = 0$; $\tilde{\mu}$ is autocontinuous from above iff $\lim_{i\to\infty} \tilde{\mu}(A \cup B_i) = \tilde{\mu}(A)$ whenever $A \in \tilde{\mathbf{F}}, B_i \in \tilde{\mathbf{F}}, i = 1, 2, \ldots$, and $\lim_{i\to\infty} \tilde{\mu}(B_i) = 0$; $\tilde{\mu}$ is autocontinuous iff it is both autocontinuous from below and autocontinuous from above.

Definition 14.6. Monotone measure $\tilde{\mu}$ is uniformly autocontinuous from below iff for any $\varepsilon > 0$, there exists $\delta = \delta(\varepsilon) > 0$ such that $\tilde{\mu}(A - B) \ge \tilde{\mu}(A) - \varepsilon$ whenever $A \in \tilde{\mathbf{F}}$, $B \in \tilde{\mathbf{F}}$, and $\tilde{\mu}(B) \le \delta$; $\tilde{\mu}$ is uniformly autocontinuous from above iff for any $\varepsilon > 0$, there exists $\delta = \delta(\varepsilon) > 0$ such that $\tilde{\mu}(A \cup B) \le \tilde{\mu}(A) + \varepsilon$ whenever $A \in \tilde{\mathbf{F}}$, $B \in \tilde{\mathbf{F}}$, and $\tilde{\mu}(B) \le \delta$; $\tilde{\mu}$ is uniformly autocontinuous iff it is both uniformly autocontinuous from below and uniformly autocontinuous from above.

Notice that, similarly to the situation in Definition 14.4, expression $\tilde{\mu}(A - B) \ge \tilde{\mu}(A) - \varepsilon$ cannot be replaced by expression $\tilde{\mu}(A \cup B) \le \tilde{\mu}(A) + \varepsilon$ when A and B are fuzzy sets. So, we need to define "uniform autocontinuity from below" and "uniform autocontinuity from above" in Definition 14.6 separately.

Definition 14.7. Monotone measure $\tilde{\mu}$ is subadditive iff $\tilde{\mu}(A \cup B) \le \tilde{\mu}(A) + \tilde{\mu}(B)$ whenever $A \in \tilde{\mathbf{F}}$ and $B \in \tilde{\mathbf{F}}$; $\tilde{\mu}$ is superadditive iff $\tilde{\mu}(A \cup B) \ge \tilde{\mu}(A) + \tilde{\mu}(B)$ whenever $A \in \tilde{\mathbf{F}}$, $B \in \tilde{\mathbf{F}}$, and $A \cap B = \emptyset$; $\tilde{\mu}$ is additive iff it is both subadditive and superadditive.

The concepts of structural characteristics introduced by Definitions 14.3–14.7 for monotone measures defined on fuzzy σ-algebras are fuzzy counterparts of the concepts for monotone measures defined on crisp σ-algebras shown in Chapter 6, and the former are generalizations of the latter. The relation among fuzzy counterparts of structural characteristics of monotone measure on crisp σ-algebras is summarized in Fig. 14.1.

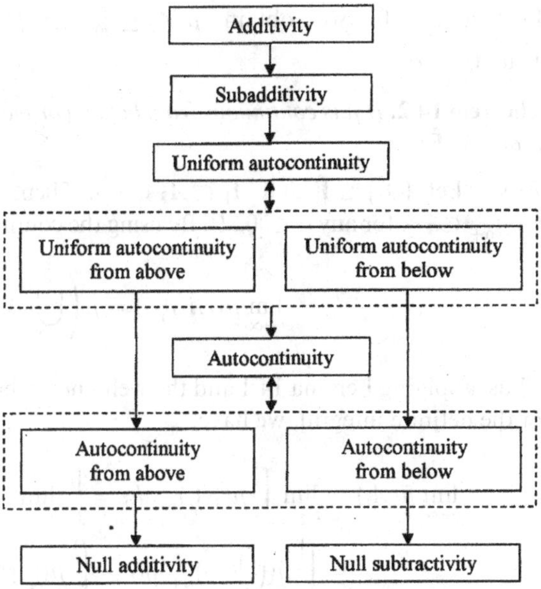

Fig. 14.1 Relation among structural characteristics of monotone measures on fuzzy σ-algebra

14.5 Hereditability of Structural Characteristics

In this section we investigate the hereditability of the various structural characteristics of monotone measures when the Choquet extension is used to establish a monotone measure on a fuzzy σ-algebra. In the rest of this section, let μ denote a monotone measure on measurable space $(X, \mathbf{F})$ and let $\tilde{\mu}$ denote its Choquet extension on fuzzy measurable space $(X, \tilde{\mathbf{F}})$.

Lemma 14.1. *For any* $A \in \tilde{\mathbf{F}}, \tilde{\mu}(A) = \int_0^1 \mu(A_{\alpha+})d\alpha$.

Proof. First, we have $\mu(A_{\alpha+}) \le \mu(A_\alpha)$ for any $\alpha \in [0, 1]$ and, therefore,

$$\int_0^1 \mu(A_{\alpha+})d\alpha \le \int_0^1 \mu(A_\alpha)d\alpha = \tilde{\mu}(A).$$

Conversely, for any $\varepsilon > 0$, since $A_\alpha \subset A_{(\alpha-\varepsilon)+}$ and μ is monotone we have $\mu(A_\alpha) \le \mu(A_{(\alpha-\varepsilon)+})$, and, therefore,

$$\begin{aligned}\tilde{\mu}(A) &= \int_0^1 \mu(A_\alpha)d\alpha \le \int_0^1 \mu(A_{(\alpha-\varepsilon)+})d\alpha \\ &= \int_{-\varepsilon}^{1-\varepsilon} \mu(A_{\alpha+})d\alpha \le \int_{-\varepsilon}^{1} \mu(A_{\alpha+})d\alpha \le \varepsilon \cdot \mu(X) + \int_0^1 \mu(A_{\alpha+})d\alpha.\end{aligned}$$

Letting $\varepsilon \to 0$, we obtain $\tilde{\mu}(A) \le \int_0^1 \mu(A_{\alpha+})d\alpha$. Consequently, $\tilde{\mu}(A) = \int_0^1 \mu(A_{\alpha+})d\alpha$. □

Theorem 14.2. *If μ is continuous from below (or from above) on $(X, \mathbf{F})$, then so is $\tilde{\mu}$ on $(X, \tilde{\mathbf{F}})$.*

Proof. Let $\{A_i\} \subset \tilde{\mathbf{F}}$ and $A_1 \subset A_2 \subset \cdots$. Then, $\{(A_i)_{\alpha+}\} \subset \mathbf{F}$ and $(A_1)_{\alpha+} \subset (A_2)_{\alpha+} \subset \cdots$ for any $\alpha \in [0, 1]$. By using the continuity from below of μ, we have

$$\lim_{i\to\infty} \mu((A_i)_{\alpha+}) = \mu\left(\bigcup_{i=1}^{\infty}(A_i)_{\alpha+}\right).$$

Thus, applying Lemma 14.1 and the well known bounded convergence theorem of the definite integral, we have

$$\begin{aligned}\lim_{i\to\infty} \tilde{\mu}(A_i) &= \lim_{i\to\infty} \int_0^1 \mu((A_i)_{\alpha+}d\alpha = \int_0^1 \lim_{i\to\infty} \mu((A_i)_{\alpha+}d\alpha \\ &= \int_0^1 \mu(\bigcup_{i=1}^{\infty}(A_i)_{\alpha+})d\alpha = \int_0^1 \mu(\bigcup_{i=1}^{\infty}(A_i)_{\alpha+})d\alpha = \tilde{\mu}(\bigcup_{i=1}^{\infty} A_i).\end{aligned}$$

Hence, $\tilde{\mu}$ is continuous from below. Similarly, noting that $\tilde{\mu}(A_1) = 0$ implies $\mu((A_1)_\alpha) = 0$ for any $\alpha \in (0, 1]$, we can prove that $\tilde{\mu}$ is continuous from above by using the equality $\bigcap_{i=1}^{\infty} (A_i)_\alpha = (\bigcap_{i=1}^{\infty} A_i)_\alpha$. The details are omitted here. □

Combining the two parts in Theorem 14.2, we obtain the following Corollary.

Corollary 14.1. *If μ is continuous on $(X, \mathbf{F})$, then so is $\tilde{\mu}$ on $(X, \tilde{\mathbf{F}})$.*

Theorem 14.3. *If μ is null-additive on $(X, \mathbf{F})$, then so is $\tilde{\mu}$ on $(X, \tilde{\mathbf{F}})$.*

Proof. Let μ be null-additive on $(X, \mathbf{F})$ and $B \in \tilde{\mathbf{F}}$ with $\tilde{\mu}(B) = 0$. From $0 = \tilde{\mu}(B) = \int_0^1 \mu(B_\alpha)d\alpha$ and the monotonicity of μ, we know that $\mu(B_\alpha) = 0$ for all $\alpha > 0$. In fact, if there exists $\alpha_0 > 0$ such that $\mu(B_{\alpha_0}) = c > 0$, then $\mu(B_\alpha) \ge \mu(B_{\alpha_0}) = c$ for every $\alpha \in (0, \alpha_0]$ and, therefore, $\int_0^1 \mu(B_\alpha)d\alpha \ge c\alpha_0 > 0$. This is a contradiction with equality $\int_0^1 \mu(B_\alpha)d\alpha = 0$. Thus, for any $A \in \tilde{\mathbf{F}}$,

$$\tilde{\mu}(A \cup B) = \int_0^1 \mu((A \cup B)_\alpha)d\alpha = \int_0^1 \mu(A_\alpha \cup B_\alpha)d\alpha.$$

By using the null-additivity of μ, we have $\mu(A_\alpha \cup B_\alpha) = \mu(A_\alpha)$ for every $\alpha > 0$. Hence, for any $A \in \tilde{\mathbf{F}}$ and any $B \in \tilde{\mathbf{F}}$ with $\tilde{\mu}(B) = 0$,

$$\tilde{\mu}(A \cup B) = \int_0^1 \mu(A_\alpha \cup B_\alpha)d\alpha = \int_0^1 \mu(A_\alpha)d\alpha = \tilde{\mu}(A).$$

This means that $\tilde{\mu}$ is null-additive on $(X, \tilde{\mathbf{F}})$. □

Theorem 14.4. *If μ is null-subtractive on $(X, \mathbf{F})$, then so is $\tilde{\mu}$ on $(X, \tilde{\mathbf{F}})$.*

Proof. Let μ be null-subtractive on $(X, \mathbf{F})$ and $B \in \tilde{\mathbf{F}}$ with $\tilde{\mu}(B) = 0$. From the partial result obtained in the proof of Theorem 14.3 that $\mu(B_\alpha) = 0$ for all $\alpha > 0$ and the fact that $B_{\alpha+} \subset B_\alpha$, we conclude that $\mu(B_{\alpha+}) = 0$ for all $\alpha \in (0, 1)$. Thus, by using the null-subtractivity of μ,

$$\begin{aligned}\tilde{\mu}(A - B) &= \int_0^1 \mu((A - B)_\alpha)d\alpha = \int_0^1 \mu((A \cap \bar{B})_\alpha)d\alpha \\ &= \int_0^1 \mu(A_\alpha \cap (\bar{B})_\alpha)d\alpha = \int_0^1 \mu(A_\alpha - B_{(1-\alpha)+})d\alpha \\ &= \int_0^1 \mu(A_\alpha)d\alpha = \tilde{\mu}(A).\end{aligned}$$

□

In Theorem 14.4, the condition "μ is null-subtractive on $(X, \mathbf{F})$" can be replaced by "μ is null-additive on $(X, \mathbf{F})$" since they are equivalent (see Theorem 6.2). A similar situation occurs in the discussion on the uniform autocontinuity.

Theorem 14.5. *If μ is autocontinuous from below (or from above) on $(X, \mathbf{F})$, then so is $\tilde{\mu}$ on $(X, \tilde{\mathbf{F}})$.*

Proof. Let $A \in \tilde{\mathbf{F}}$ and $\{B_i\} \subset \tilde{\mathbf{F}}$ with $\lim_{i\to\infty} \tilde{\mu}(B_i) = 0$. First, we know that $\lim_{i\to\infty} \tilde{\mu}(B_i) = 0$ implies $\lim_{i\to\infty} \mu((B_i)_\alpha) = 0$ for every $\alpha \in (0, 1]$. In fact, if it is not true, then there exist $\alpha_0 \in (0, 1]$ and sequence $\{i_j\}$ such that $\mu((B_{i_j})_{\alpha_0}) \geq c > 0$ for $j = 1, 2, \ldots$. Hence,

$$\int_0^1 \mu((B_{i_j})_\alpha)d\alpha \geq \int_0^{\alpha_0} \mu((B_{i_j})_\alpha)d\alpha \geq \int_0^{\alpha_0} c\, d\alpha = c\alpha_0 > 0$$

for every $j = 1, 2, \ldots$; therefore, $\lim_{i\to\infty} \int_0^1 \mu((B_i)_\alpha)d\alpha > 0$ or does not exist. This is a contradiction with $\lim_{i\to\infty} \int_0^1 \mu((B_i)_\alpha)d\alpha = \lim_{i\to\infty} \tilde{\mu}(B_i) > 0$. Furthermore, $\lim_{i\to\infty} \mu((B_i)_{(1-\alpha)+}) = 0$ for every $\alpha \in (0, 1)$ since $0 \leq \lim_{i\to\infty} \mu((B_i)_{(1-\alpha)+}) \leq \lim_{i\to\infty} \mu((B_i)_{(1-\alpha)}) = 0$. Thus,

$$\begin{aligned}\tilde{\mu}(A - B_i) &= \int_0^1 \mu((A - B_i)_\alpha)d\alpha = \int_0^1 \mu((A \cap \bar{B}_i)_\alpha)d\alpha \\ &= \int_0^1 \mu((A_\alpha \cap (\bar{B}_i)_\alpha)d\alpha = \int_0^1 \mu((A_\alpha \cap \overline{(B_i)_{(1-\alpha)+}})d\alpha \\ &= \int_0^1 \mu((A_\alpha - (B_i)_{(1-\alpha)+})d\alpha\,.\end{aligned}$$

By using the autocontinuity from below of μ and the bounded convergence theorem of the definite integral, we have

$$\lim_{i\to\infty} \tilde{\mu}(A - B_i) = \lim_{i\to\infty} \int_0^1 \mu((A_\alpha - (B_i)_{(1-\alpha)+})d\alpha$$
$$= \int_0^1 \lim_{i\to\infty} \mu((A_\alpha - (B_i)_{(1-\alpha)+})d\alpha = \int_0^1 \mu(A_\alpha)d\alpha = \tilde{\mu}(A)\,.$$

The proof for the autocontinuity from below of $\tilde{\mu}$ is now complete. Similarly, by using the autocontinuity from above of μ, we obtain

$$\lim_{i\to\infty} \tilde{\mu}(A \cup B_i) = \lim_{i\to\infty} \int_0^1 \mu((A \cup B_i)_\alpha)d\alpha = \int_0^1 \lim_{i\to\infty} \mu((A_\alpha \cup (B_i)_\alpha)d\alpha$$
$$= \int_0^1 \mu(A_\alpha)d\alpha = \tilde{\mu}(A).$$

This shows that $\tilde{\mu}$ is autocontinuous from above too. □

Corollary 14.2. *If μ is autocontinuous on $(X, \mathbf{F})$, then so is $\tilde{\mu}$ on $(X, \tilde{\mathbf{F}})$.*

Theorem 14.6. *If μ is uniformly autocontinuous on $(X, \mathbf{F})$, then $\tilde{\mu}$ is both uniformly autocontinuous from below and uniformly autocontinuous from above on $(X, \tilde{\mathbf{F}})$.*

Proof. Denote $\mu(X)$ by c. We prove the uniform autocontinuity from above of $\tilde{\mu}$ first. From the uniform autocontinuity of μ on $(X, \mathbf{F})$, for any given $\varepsilon > 0$, there exists $\delta > 0$ such that $\mu(E \cup F) \le \mu(E) + \varepsilon/2$ whenever $E \in \mathbf{F}$, $F \in \mathbf{F}$, and $\mu(F) \le 2c\delta/\varepsilon$. If $B \in \tilde{\mathbf{F}}$ with $\tilde{\mu}(B) < \delta$, we know that $\{\alpha | \mu\,(B_\alpha) > 2c\delta/\varepsilon\} \subset [0, \varepsilon/2c)$ from $\int_0^1 \mu(B_\alpha)d\alpha = \tilde{\mu}(B) < \delta$ and the fact that $\mu(B_\alpha)$ is a nonincreasing function of α. Indeed, if there exists some $\alpha \ge \varepsilon/2c$ such that $\mu(B_\alpha) > 2c\delta/\varepsilon$, then

$$\int_0^1 \mu(B_\alpha)d\alpha \ge \int_0^{\varepsilon/2c} \mu(B_\alpha)d\alpha \ge \int_0^{\varepsilon/2c} (2c\delta/\varepsilon)d\alpha = \delta.$$

This contradicts the inequality $\int_0^1 \mu(B_\alpha)d\alpha < \delta$. Thus, for any $A \in \tilde{\mathbf{F}}$ we have

$$\tilde{\mu}(A \cup B) = \int_0^1 \mu((A \cup B)_\alpha)d\alpha$$
$$= \int_0^{\varepsilon/2c} \mu((A \cup B)_\alpha)d\alpha + \int_{\varepsilon/2c}^1 \mu((A \cup B)_\alpha)d\alpha$$
$$= \int_0^{\varepsilon/2c} \mu((A \cup B)_\alpha)d\alpha + \int_{\varepsilon/2c}^1 \mu(A_\alpha \cup B_\alpha)d\alpha$$

$$\leq \int_0^{\varepsilon/2c} c\,d\alpha + \int_{\varepsilon/2c}^1 [\mu(A_\alpha) + \varepsilon/2]d\alpha \leq \varepsilon/2 + \int_0^1 [\mu(A_\alpha) + \varepsilon/2]d\alpha$$

$$= \varepsilon/2 + \int_0^1 \mu(A_\alpha)d\alpha + \varepsilon/2 = \int_0^1 \mu(A_\alpha)d\alpha + \varepsilon = \tilde{\mu}(A) + \varepsilon.$$

This means that $\tilde{\mu}$ is uniformly autocontinuous from above on $(X, \tilde{\mathbf{F}})$. Now, we use a slightly modified reasoning to prove that $\tilde{\mu}$ is uniformly autocontinuous from below on $(X, \tilde{\mathbf{F}})$. From the uniform autocontinuity of μ on $(X, \mathbf{F})$, for any given $\varepsilon > 0$, there exists $\delta > 0$ such that $\mu(E - F) \geq \mu(E) - \varepsilon/2$ whenever $E \in \mathbf{F}$, $F \in \mathbf{F}$, and $\mu(F) \leq 2c\delta/\varepsilon$. If $B \in \tilde{\mathbf{F}}$ with $\tilde{\mu}(B) < \delta$, $\int_0^1 \mu(B_{\alpha+})d\alpha = \tilde{\mu}(B) < \delta$ from Lemma 1, and from the monotonicity of $\mu(B_{\alpha+})$ with respect to α, we know that $\{\alpha | \mu(B_{(1-\alpha)+}) > 2c\delta/\varepsilon\} \subset (1 - \varepsilon/2c, 1]$. Thus, for any $A \in \tilde{\mathbf{F}}$, we have

$$\begin{aligned}
\tilde{\mu}(A - B) &= \int_0^1 \mu(A_\alpha - B_{(1-\alpha)+})d\alpha \\
&= \int_0^{1-\varepsilon/2c} \mu(A_\alpha - B_{(1-\alpha)+})d\alpha + \int_{1-\varepsilon/2c}^1 \mu(A_\alpha - B_{(1-\alpha)+})d\alpha \\
&\geq \int_0^{1-\varepsilon/2c} \mu(A_\alpha - B_{(1-\alpha)+})d\alpha \geq \int_0^{1-\varepsilon/2c} (\mu(A_\alpha) - \varepsilon/2)d\alpha \\
&= \int_0^{1-\varepsilon/2c} \mu(A_\alpha)d\alpha - \int_0^{1-\varepsilon/2c} (\varepsilon/2)d\alpha \\
&= \int_0^1 \mu(A_\alpha)d\alpha - \int_{1-\varepsilon/2c}^1 \mu(A_\alpha)d\alpha - \int_0^{1-\varepsilon/2c} (\varepsilon/2)d\alpha \\
&\geq \int_0^1 \mu(A_\alpha)d\alpha - \int_{1-\varepsilon/2c}^1 \mu(A_\alpha)d\alpha - \int_0^1 (\varepsilon/2)d\alpha \\
&\geq \int_0^1 \mu(A_\alpha)d\alpha - c \cdot \varepsilon/2c - \varepsilon/2 \\
&= \int_0^1 \mu(A_\alpha)d\alpha - \varepsilon/2 - \varepsilon/2 = \tilde{\mu}(A) - \varepsilon.
\end{aligned}$$

This means that $\tilde{\mu}$ is also uniformly autocontinuous from below on $(X, \tilde{\mathbf{F}})$. The proof of the theorem is now complete. □

Corollary 14.3. *If μ is uniformly autocontinuous on $(X, \mathbf{F})$, then so is $\tilde{\mu}$ on $(X, \tilde{\mathbf{F}})$.*

Theorem 14.7. *If μ is subadditive (or superadditive) on $(X, \mathbf{F})$, then so is $\tilde{\mu}$ on $(X, \tilde{\mathbf{F}})$.*

Proof. Let μ be subadditive on $(X, \mathbf{F})$. For any $A \in \tilde{\mathbf{F}}$ and $B \in \tilde{\mathbf{F}}$,

$$\tilde{\mu}(A \cup B) = \int_0^1 \mu((A \cup B)_\alpha)d\alpha = \int_0^1 \mu(A_\alpha \cup B_\alpha)d\alpha$$
$$\leq \int_0^1 [\mu(A_\alpha) + \mu(B_\alpha)]d\alpha$$
$$= \int_0^1 \mu(A_\alpha)d\alpha + \int_0^1 \mu(B_\alpha)d\alpha = \tilde{\mu}(A) + \tilde{\mu}(B).$$

This means that $\tilde{\mu}$ is subadditive on $(X, \tilde{\mathbf{F}})$. As for the superadditivity, let $A \in \tilde{\mathbf{F}}$ and $B \in \tilde{\mathbf{F}}$ be disjoint, i.e., $A \cap B = \emptyset$. From the superadditivity of μ on $(X, \mathbf{F})$, since $A_\alpha \cap B_\alpha = \emptyset$ for every $\alpha \in (0, 1]$, we have

$$\tilde{\mu}(A \cup B) = \int_0^1 \mu((A \cup B)_\alpha)d\alpha = \int_0^1 \mu(A_\alpha \cup B_\alpha)d\alpha \geq \int_0^1 [\mu(A_\alpha) + \mu(B_\alpha)]d\alpha$$
$$= \int_0^1 \mu(A_\alpha)d\alpha + \int_0^1 \mu(B_\alpha)d\alpha = \tilde{\mu}(A) + \tilde{\mu}(B).$$

□

Theorems in this section establish the hereditability for most structural characteristics of monotone measures extended from a σ-algebra based on crisp sets onto its generated fuzzy σ-algebra.

14.6 Real-Valued Choquet Integrals with Fuzzy-Valued Integrands

In this section, we assume that the universal set X is finite, and we use the convenient notation $X = \{x_1, x_2, \ldots, x_n\}$. We also assume that f is a fuzzy-valued function defined on X whose range is a subset of the set of all fuzzy numbers. Function f can be expressed as $(m_1, m_2, \ldots, m_n)$, where m_i is the membership function of fuzzy number $f(x_i)$, $i = 1, 2, \ldots, n$.

Example 14.5. Assume that papers submitted to a journal are evaluated by several criteria and the evaluation range for each criterion is the interval [0, 5], with 0 and 5 being the worst and best evaluations, respectively. Assume further that reviewers are asked to evaluate each paper for each specified criterion qualitatively by using the linguistic terms *bad*, *weak*, *fair*, *good*, *excellent*. The meanings of these four linguistic terms can be adequately captured by the trapezoidal fuzzy numbers $B = \langle 0, 0, 1, 1.5 \rangle$, $W = \langle 1, 1.5, 2, 2.5 \rangle$, $F = \langle 2, 2.5, 3, 3.5 \rangle$, $G = \langle 3, 3.5, 4, 4.5 \rangle$, and $E = \langle 4, 4.5, 5, 5.5 \rangle$, respectively (see Fig. 14.2). Collection $\{B, W, F, G, E\}$ is a fuzzy partition of the interval [0,5].

To consider the Choquet integral with fuzzy-valued integrand, first, we should discuss the α-level set of fuzzy-valued functions.

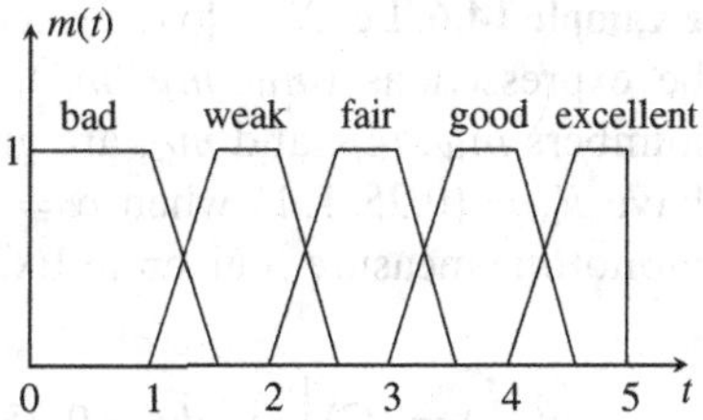

Fig. 14.2 Membership functions of fuzzy sets B, W, F, G, and E in Example 14.5

Definition 14.8. For any given $\alpha \in R$, the α-level set of fuzzy-valued function $f = (m_1, m_2, \ldots, m_n)$, denoted by F_α, is a fuzzy subset of X, whose membership function m_{F_α} has degree of membership

$$m_{F_\alpha}(x_i) = \begin{cases} \dfrac{\int_\alpha^\infty m_i(t)dt}{\int_{-\infty}^\infty m_i(t)dt} & \text{if } \int_{-\infty}^\infty m_i(t)dt \neq 0 \\ \max\limits_{t \geq \alpha} m_i(t) & \text{otherwise} \end{cases}$$

at attribute x_i, $i = 1, 2, \ldots, n$. Fuzzy set F_α can be expressed as an n-dimensional vector $(m_{F_\alpha}(x_1), m_{F_\alpha}(x_2), \ldots, m_{F_\alpha}(x_n))$.

The concept of the α-level set for a fuzzy-valued function given in Definition 14.8 is a generalization of the α-level set for a real-valued function, that is, it coincides with the α-level set of a real-valued function when the values of function f are crisp. Such a generalization is rather intuitive. It just uses the percentage of the area in the right-hand side of α under the curve of the membership function of $f(x_i)$ to define the degree of the membership for the α-level set of fuzzy-valued function f at point x_i. Of course, when the area under the curve of the membership function of $f(x_i)$ is zero (i.e., $f(x_i)$ is a crisp real number c) for some i, the above-mentioned percentage has the form of $0/0$. In this special case, we need to define the value of the membership function at point x_i separately by $\max_{t \geq \alpha} m_i(t)$, where

$$m_i(t) = \begin{cases} 1 & \text{if } t = c \\ 0 & \text{if } t \neq c \end{cases}$$

If $f(x_i)$ is an interval $[a, b]$, the degree of the membership for the α-level set of f at x_i is

$$m_{F_\alpha}(x_i) = \begin{cases} 1 & \text{if } \alpha < a \\ \dfrac{b - \alpha}{b - a} & \text{if } \alpha \in [a, b] \\ 0 & \text{if } \alpha > b. \end{cases}$$

Example 14.6. Let $X = \{x_1, x_2, x_3\}$ and let fuzzy-valued function f defined on X be expressed as (m_W, m_E, m_G), where the membership functions of fuzzy numbers m_W, m_E, and m_G, are given in Example 14.5. Then, for example, we have $F_\alpha = (0.25, 1, 1)$ when $\alpha = 2$ and $F_\alpha = (0, 1, 0.75)$ when $\alpha = 3.5$. If the monotone measure μ given in Example 14.3 is used, then

$$\mu(F_2) = (\mathrm{C})\int m_{F_2} d\mu = 0.25 \cdot 10 + (1 - 0.25) \cdot 4 + (1 - 1) \cdot 3 = 5.5$$

and

$$\mu(F_{3.5}) = (\mathrm{C})\int m_{F_{3.5}} d\mu = 0 \cdot 10 + (0.75 - 0) \cdot 4 + (1 - 0.75) \cdot 2 = 3.5.$$

To simplify the way of finding the membership function of the α-cut of a fuzzy-valued function, we first deal with only one attribute essentially. Let f be a fuzzy-valued function on $X = \{x_1, x_2, .., x_n\}$ having a form as

$$f(x) = \begin{cases} A & \text{if } x = x_{i_0} \\ 0 & \text{if } x \neq x_{i_0} \end{cases}$$

for some $i_0 \in \{1, 2, \ldots, n\}$, where A is a trapezoidal fuzzy number $\langle a_l, a_b, a_c, a_r \rangle$. Then, by calculating the quotient of two Riemann integrals shown in Definition 14.8, the degree of membership of F_α at x_{i_0} is

$$m_{F_\alpha}(x_{i_0}) = \begin{cases} 1 & \text{when } \alpha \le a_l \\ 1 - \dfrac{(\alpha - a_l)^2}{(a_r + a_c - a_l - a_b)(a_b - a_l)} & \text{when } \alpha \in (a_l, a_b] \\ \dfrac{a_r + a_c - 2\alpha}{a_r + a_c - a_l - a_b} & \text{when } \alpha \in (a_b, a_c] \\ \dfrac{(a_r - \alpha)^2}{(a_r + a_c - a_l - a_b)(a_r - a_c)} & \text{when } \alpha \in (a_c, a_r] \\ 0 & \text{when } \alpha > a_r, \end{cases}$$

(see Fig. 14.3) and at any other point $x \neq x_{i_0}$ is

$$m_{F_\alpha}(x) = \begin{cases} 1 & \text{when } \alpha \le 0 \\ 0 & \text{when } \alpha > 0. \end{cases}$$

Example 14.7. Consider a linguistic variable "the average period T between two successive pulses of the heart of a patient," denoted by x_{i_0}, whose values are *small*, *medium*, and *large*. Let the meaning of these values be represented, respectively, by trapezoidal fuzzy numbers $\langle 0, 0, 0.3, 0.4 \rangle$, $\langle 0.3, 0.4, 0.6, 0.8 \rangle$,

Fig.14.3 Illustration to Example 14.6

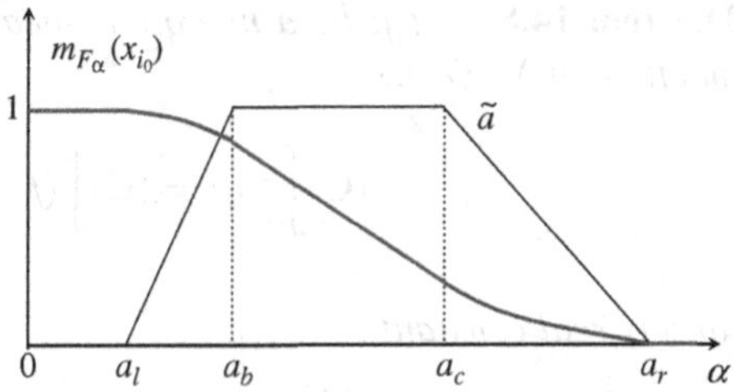

and $\langle 0.6, 0.8, 1.3, 1.3\rangle$, and let f have the value *large* for this variable. Then, for any given real number α, the fuzzy set F_α has a degree of membership

$$m_{F_\alpha}(x_{i_0}) = \begin{cases} 1 & \text{when } \alpha \leq 0.6 \\ 1 - \dfrac{(\alpha - 0.6)^2}{0.24} & \text{when } 0.6 < \alpha \leq 0.8 \\ \dfrac{1.3 - \alpha}{0.6} & \text{when } 0.8 < \alpha \leq 1.3 \\ 0 & \text{when } \alpha > 1.3 \end{cases}$$

for variable x_{i_0}.

To simplify our discussion, let the co-domain of the fuzzy-valued function employed be the set of all trapezoidal fuzzy numbers, and let μ be a monotone measure on $\mathbf{P}(X)$. As is shown in Section 14.2, monotone measure μ can be extended onto the class of all fuzzy subsets of X by using the Choquet integral. Thus, we may still use Eq. (11.2) (Section 11.3) to define the translatable Choquet integral of fuzzy-valued function f with respect to monotone measure μ:

$$(\mathrm{C})\int f d\mu = \int_{-\infty}^{0} [\mu(F_\alpha) - \mu(X)] d\alpha + \int_{0}^{\infty} \mu(F_\alpha) d\alpha.$$

Here, F_α may be a fuzzy set and it is not guaranteed that $\mu(F_\alpha)$ is nonincreasing with respect to α. However, as a function of α, $\mu(F_\alpha)$ is of bounded variation (in fact, it is finitely piecewise monotonic and bounded). So, the above Riemann integrals exist and, hence, the Choquet integral is well defined in this case.

The co-domain of a fuzzy-valued function may not be fully-ordered, and therefore the values of the function at various variables cannot be rearranged in a nondecreasing order. By the same reason, operators min and max may not be well defined on the co-domain. So, the way for calculating the value of the Choquet integral given in Section 11.5 is not directly applicable for computing the Choquet integral with a fuzzy-valued integrand. However, we may still use it for calculating $\mu(F_\alpha)$ in the above Riemann integrals since the Choquet integral is used for the extension of μ. To simplify the computation, an important property of the Choquet integral with fuzzy-valued integrand is stated in the following theorem, which is a counterpart of Theorem 11.6.

Theorem 14.8. *Let* μ *be a monotone measure on* $\tilde{\mathbf{P}}(X)$ *and* f *be a fuzzy-valued function on* X. *Then,*

$$(\mathrm{C})\int f d\mu = (\mathrm{C})\int (f - c)d\mu + c \cdot \mu(X)$$

for any real constant c.

Proof. Let $g = f - c$. Then g is also a fuzzy-valued function and its α-cut, G_α, satisfies $G_\alpha = F_{\alpha+c}$ or, equivalently, $G_{\alpha-c} = F_\alpha$, for any real number α. Thus, denoting $\alpha - c$ by β, we have

$$\begin{aligned}
(\mathrm{C})\int f d\mu &= \int_{-\infty}^{0} [\mu(F_\alpha) - \mu(X)]d\alpha + \int_{0}^{\infty} \mu(F_\alpha)d\alpha \\
&= \int_{-\infty}^{0} [\mu(G_{\alpha-c}) - \mu(X)]d\alpha + \int_{0}^{\infty} \mu(G_{\alpha-c})d\alpha \\
&= \int_{-\infty}^{0} [\mu(G_{\alpha-c}) - \mu(X)]d(\alpha - c) + \int_{0}^{\infty} \mu(G_{\alpha-c})d(\alpha - c) \\
&= \int_{-\infty}^{-c} [\mu(G_\beta) - \mu(X)]d\beta + \int_{-c}^{\infty} \mu(G_\beta)d\beta \\
&= \int_{-\infty}^{-c} [\mu(G_\beta) - \mu(X)]d\beta + \int_{-c}^{0} \mu(G_\beta)d\beta \\
&\quad + \int_{0}^{\infty} \mu(G_\beta)d\beta - \int_{-c}^{0} \mu(X)d\beta + \int_{-c}^{0} \mu(X)d\beta \\
&= \int_{-\infty}^{0} [\mu(G_\beta) - \mu(X)]d\beta + \int_{0}^{\infty} \mu(G_\beta)d\beta + \int_{-c}^{0} \mu(X)d\beta \\
&= (\mathrm{C})\int g d\mu + c \cdot \mu(X) \\
&= (\mathrm{C})\int (f - c)\, d\mu + c \cdot \mu(X).
\end{aligned}$$

□

In case the lower boundary of the support set of the integrand f at each x_i exists, denoted by a_{il} for $i = 1, 2, \ldots, n$, we may simplify the calculation of the Choquet integral by constructing a corresponding nonnegative fuzzy-valued function $g = f - c$, where $c = \min_{1 \le i \le n} a_{il}$. Due to Theorem 14.8 we can write

$$(\mathrm{C})\int f d\mu = \int_{0}^{\infty} \mu(G_\alpha)d\alpha + c \cdot \mu(X),$$

where G_α denotes the α-level set of g.

Before formulating an algorithm for calculating the crisp value of the Choquet integral of a fuzzy-valued function with respect to a given signed general measure, we first study a very special case in the following example where the integrand vanishes at all attributes except one.

Example 14.8. Let f be the fuzzy-valued function that only assigns a nonnegative trapezoidal fuzzy number $A = \langle a_l,\ a_b,\ a_c,\ a_r\rangle$ to attribute x_{i_0} and vanishes at all other attributes. Since $F_\alpha = X$ when $\alpha \le 0$, using formula (14.1) shown after Example 14.6, we have

$$(C)\int f d\mu = \int_0^\infty \mu(F_\alpha)d\alpha = \int_0^\infty \mu(\{x_{i_0}\})\cdot m_{F_\alpha}(x_{i_0})d\alpha = \mu(\{x_{i_0}\})\cdot \int_0^\infty m_{F_\alpha}(x_{i_0})d\alpha.$$

$$=\mu(\{x_{i_0}\})\cdot\Bigg[a_l + a_b - a_l - \frac{1}{3}\left[\frac{(\alpha-a_l)^3}{(a_r+a_c-a_b-a_l)(a_b-a_l)}\right]_{\alpha=a_l}^{\alpha=a_b}$$

$$-\frac{1}{4}\left[\frac{(a_r+a_c-2\alpha)^2}{a_r+a_c-a_b-a_l}\right]_{\alpha=a_b}^{\alpha=a_c} - \frac{1}{3}\left[\frac{(a_r-\alpha)^3}{(a_r+a_c-a_b-a_l)(a_r-a_c)}\right]_{\alpha=a_c}^{\alpha=a_r}\Bigg]$$

$$=\mu(\{x_{i_0}\})\Bigg[a_b - \frac{1}{3}\cdot\frac{(a_b-a_l)^2}{a_r+a_c-a_b-a_l}$$

$$-\frac{1}{4}\cdot\frac{(a_r+a_c-2a_c)^2-(a_r+a_c-2a_b)^2}{a_r+a_c-a_b-a_l}$$

$$-\frac{1}{3}\cdot\frac{(a_r-a_c)^2}{a_r+a_c-a_b-a_l}\Bigg]$$

$$=\mu(\{x_{i_0}\})\left[\frac{1}{3}\cdot\frac{a_r^2+a_ra_c+a_c^2-a_b^2-a_ba_l-a_l^2}{a_r+a_c-a_b-a_l}\right].$$

Let us consider two special cases: (1) when A is a triangular fuzzy number (that is, $a_b = a_c = a_0$), the above result becomes

$$(C)\int f d\mu = \mu(\{x_{i_0}\})\left[\frac{1}{3}\cdot\frac{a_r^2+a_ra_0-a_l^2-a_la_0}{a_r-a_l}\right] = \mu(\{x_{i_0}\})\left[\frac{a_r+a_0+a_l}{3}\right];$$

(2) when A is a rectangular fuzzy number (that is, $a_l = a_b$ and $a_c = a_r$), which is actually a crisp interval, we have

$$(C)\int f d\mu = \mu(\{x_{i_0}\})\left[\frac{1}{3}\cdot\frac{3(a_r^2-a_l^2)}{2(a_r-a_l)}\right] = \mu(\{x_{i_0}\})\left[\frac{a_r+a_l}{2}\right].$$

More specifically, if A collapses to a real number a (that is $a_l = a_b = a_c = a_r = a$), then

$$(C)\int f d\mu = \mu(\{x_{i_0}\}) \cdot a.$$

Now, we turn to the general case. Function f may have fuzzy values at every variable. Expressing f as $(\langle a_{1l}, a_{1b}, a_{1c}, a_{1r}\rangle, \langle a_{2l}, a_{2b}, a_{2c}, a_{2r}\rangle, \ldots, \langle a_{nl}, a_{nb}, a_{nc}, a_{nr}\rangle)$, for any given $\alpha \in R$, the α-level set of f is a fuzzy subset of X. Its degree of membership at x_i is expressed by the formula

$$m_{F_\alpha}(x_i) = \begin{cases} 1 & \text{when } \alpha \le a_{il} \\ 1 - \dfrac{(\alpha - a_{il})^2}{(a_{ir} + a_{ic} - a_{il} - a_{ib})(a_{ib} - a_{il})} & \text{when } \alpha \in (a_{il}, a_{ib}] \\ \dfrac{a_{ir} + a_{ic} - 2\alpha}{a_{ir} + a_{ic} - a_{il} - a_{ib}} & \text{when } \alpha \in (a_{ib}, a_{ic}] \\ \dfrac{(a_{ir} - \alpha)^2}{(a_{ir} + a_{ic} - a_{il} - a_{ib})(a_{ir} - a_{ic})} & \text{when } \alpha \in (a_{ic}, a_{ir}] \\ 0 & \text{when } \alpha > a_{ir} \end{cases}$$

for $i = 1, 2, \ldots, n$.

In this case, it is rather difficult to express $\mu(F_\alpha)$ in an explicit form involving only fundamental functions of α and to compute the precise value of $\int_0^\infty \mu(F_\alpha)d\alpha$. However, we can numerically calculate its approximate value through the following algorithm. In Step 4 of the algorithm, the given fuzzy-valued function is translated to be nonnegative; Step 7 is the stop controller; the membership function of F_α is calculated in Step 8; Step 9 is used to find the value of $\mu(F_\alpha)$ via the Choquet integral, and the Simpson method is used to calculate the value of the involved Riemann integral approximately.

Algorithm 14.1. Computing an approximate value of $\int_0^\infty \mu(F_\alpha)d\alpha$.

1. Input: n (the number of variables in X), K (the number of subintervals required in the Simpson method, with a default value $K = 100$), values of function f, $f(x_i) = \langle a_{il}, a_{ib}, a_{ic}, a_{ir}\rangle$ for $i = 1, 2, \ldots, n$, and values of a given signed general measure

$$\mu_j = \mu\Big(\bigcup_{i:\mathrm{frc}(\frac{j}{2^i})\in[\frac{1}{2},1)} \{x_i\}\Big)$$

for $j = 1, 2, \ldots, 2^n - 1$.
2. If $a_{il} \le a_{ib} \le a_{ic} \le a_{ir}$ for every $i = 1, 2, \ldots, n$, then go to Step 3; otherwise, return a message "data error: ..." to indicate where and what are the errors and after correcting the data, go to Step 1.
3. Find $a = \min_{1\le i\le n} a_{il}$, $b = \max_{1\le i\le n} a_{ir}$, and $\delta = \dfrac{b-a}{K}$.

4. Replace a_{il}, a_{ib}, a_{ic}, and a_{ir} with $a_{il} - a$, $a_{ib} - a$, $a_{ic} - a$, and $a_{ir} - a$, respectively, for $i = 1, 2, \ldots, n$.
5. Initiate $\alpha = 0$ and $S = \frac{\mu_{2^n-1}}{2}$.
6. $\alpha + \delta \to \alpha$.
7. If $\alpha > b - a$, then $\delta \cdot (S - \frac{\Delta S}{2}) + a \cdot \mu_{2^n-1} \to S$, output S as an approximate value of $(C)\int f d\mu$, and stop; otherwise, continue.
8. Find

$$c_i = \begin{cases} 1 & \text{when } \alpha \le a_{il} \\ 1 - \dfrac{(\alpha - a_{il})^2}{(a_{ir} + a_{ic} - a_{il} - a_{ib})(a_{ib} - a_{il})} & \text{when } \alpha \in (a_{il}, a_{ib}] \\ \dfrac{a_{ir} + a_{ic} - 2\alpha}{a_{ir} + a_{ic} - a_{il} - a_{ib}} & \text{when } \alpha \in (a_{ib}, a_{ic}] \\ \dfrac{(a_{ir} - \alpha)^2}{(a_{ir} + a_{ic} - a_{il} - a_{ib})(a_{ir} - a_{ic})} & \text{when } \alpha \in (a_{ic}, a_{ir}] \\ 0 & \text{when } \alpha > a_{ir} \end{cases}$$

for $i = 1, 2, \ldots, n$.

9. Viewing $h = (c_1, c_2, \ldots, c_n)$ as a function on X, calculate $\Delta S = (C)\int h d\mu$ by the fomula

$$\Delta S = \sum_{j=1}^{2^n-1} z_j \cdot \mu_j$$

where

$$z_j = \begin{cases} \min\limits_{i:\mathrm{frc}(\frac{j}{2^i})\in[\frac{1}{2},1)} c_i - \max\limits_{i:\mathrm{frc}(\frac{j}{2^i})\in[0,\frac{1}{2})} c_i, & \text{if it is} > 0 \\ 0, & \text{otherwise} \end{cases} \quad \text{for } j = 1, 2, \ldots, 2^n - 1.$$

10. $S + \Delta S \to S$ and go to Step 6.

Example 14.9. In Example 14.5 suppose that the evaluation of submitted papers is based on three criteria: originality, significance, and presentation. They are denoted by x_1, x_2, and x_3 respectively. The importance of each individual criterion and their joint importances are described by a signed general measure μ defined on $\mathbf{P}(X)$, where $X = \{x_1, x_2, x_3\}$. Also suppose that the values of μ are $\mu_1 = 0.2$, $\mu_2 = 0.3$, $\mu_3 = 0.8$, $\mu_4 = 0.1$, $\mu_5 = 0.4$ $\mu_6 = 0.4$, and $\mu_7 = 1$. Now, a paper is evaluated as *excellent* for originality, *fair* for significance, and *weak* for presentation by a reviewer. This reviewer's evaluation can be regarded as a fuzzy-valued function $f = (E, F, W)$ on $X = \{x_1, x_2, x_3\}$. Thus, a global evaluation for the quality of the paper is given by the Choquet integral of

f with respect to μ, $(C)\int fd\mu$. Using Algorithm 14.1, a rather precise approximate value of $\int fd\mu$ can be obtained:

$$(C)\int fd\mu \approx 2.92176 \text{ when } K = 100,$$

$$(C)\int fd\mu \approx 2.92222 \text{ when } K = 1000.$$

For another paper evaluated as *bad* for originality, *good* for significance, and *excellent* for presentation, denoted as $g = (B, G, E)$, we have

$$(C)\int g\, d\mu \approx 1.96618 \text{ when } k = 100,$$

$$(C)\int g\, d\mu \approx 1.96611 \text{ when } k = 1000.$$

This means that the paper represented by function f is more valuable than the one represented by function g for publishing in the journal.

Since the procedure of calculating the value of the Choquet integral with fuzzy integrand is repeated for a large number in multiregression or classification problems, we should reduce its running time as much as possible. For most real problems in decision-making, the precision of the relevant results reaching three or four decimal digits is sufficient. So, this example also suggests to use $K = 100$ as the default value of K.

Notes

14.1. The concept of a σ-algebra of fuzzy sets was introduced by Butnariu [1983], primarily for the study of cooperative games with fuzzy coalitions [Butnariu, 1985; Butnariu and Klement, 1993]; see also [Aubin, 1981], [Bronevich, 2005b], and [Branzei et al., 2005].

14.2. The concept of a real-valued Choquet integral with a fuzzy-valued integrand is discussed in [Wang et al., 2006b]. This integral can be employed as a defuzzification tool for fuzzy data.

Chapter 15
Applications of Generalized Measure Theory

15.1 General Remarks

It is undeniable that classical measure theory, based on additive measures and signed additive measures, and the associated Lebesgue theory of integration, is not only an important area of mathematics, but it has also played an important role in many application domains. Perhaps its most visible is its crucial role in probability theory, as rigorously formulated by Kolmogorov. Examples of other notable applications of classical measure theory are in the areas of classical geometry as well as fractal geometry, ergodic theory of dynamical systems, harmonic analysis, potential theory, calculus of variations, and mathematical economics (see Note 15.1).

Notwithstanding the many demonstrated applications of classical measure theory, it has increasingly been recognized that a broadening of this area's applicability is severely limited by the additivity requirement of classical measures. Requiring additivity in measuring a property on sets of some kind is basically the same as assuming that there is no interaction among sets with respect to the measured property. However, there are many problem areas involving properties measured on sets that do interact. Measuring such properties on a measurable space $(X, \mathbf{F})$ by a set function μ requires that μ be capable to capture, for each given pair of sets A and B in $\mathbf{F}$ such that $A \cap B = \emptyset$, any of the following three situations:

(a) $A \cup B \in \mathbf{F}$ implies that $\mu(A \cup B) > \mu(A) + \mu(B)$, which expresses a *positive interaction* (synergy, cooperation, coalition, enhancement, amplification) between A and B in terms of the measured property;
(b) $A \cup B \in \mathbf{F}$ implies that $\mu(A \cup B) < \mu(A) + \mu(B)$, which expresses a *negative interaction* (incompatibility, rivalry, inhibition, downgrading, condensation);
(c) $A \cup B \in \mathbf{F}$ implies that $\mu(A \cup B) = \mu(A) + \mu(B)$, which expresses the fact that there is *no interaction* between A and B in terms of the measured property

It is clear that classical measures can capture only situation (c), which means that they are applicable only to properties that are noninteractive. Properties that exhibit for each pair of sets in $\mathbf{F}$ whose union is also in $\mathbf{F}$ either a positive

Z. Wang, G.J. Klir, *Generalized Measure Theory*,
DOI: 10.1007/978-0-387-76852-6_15,

interaction or no interaction can be captured by superadditive measures, and those that exhibit either a negative interaction or no interaction can be captured by subadditive measures. Properties that exhibit positive interactions for some pairs of sets and negative interactions for some other pairs of sets and, possibly, no interactions for some additional pairs of sets, can be captured by monotone measures or, if the property violates monotonicity, by general measures,

Properties measured on sets that exhibit some positive or negative interactions are not rare. The following are some examples.

Let X be a set of criteria in multicriteria decision-making and let μ be a measure that is supposed to quantify the importance of any subset of criteria on the outcome of the decision-making process. Requiring that μ be additive would not be realistic in this case. Indeed, the importance of two or more criteria taken together on the decision may be higher (or, perhaps, lower) than the sum of their individual importances.

As another example, assume that X is a set of criteria by which we evaluate some feature (such as quality, performance, risk, etc.) associated with individual objects of some relevant class of objects. The overall evaluation clearly depends not only on the importance of each individual criterion, but also on the considered importance, in the context of each application, of various sets of criteria. Again, requiring that the measure of importance of the criteria be additive would be very restrictive and highly unrealistic in many application contexts, as it would totally ignore positive or negative interactions among the criteria with respect to their importance. Under such a requirement, the evaluation of each object, say the quality q_a of object a, would simply be expressed in terms of the weighted average, $q_a = \Sigma_{x \in X}\, \mu(x) c_a(x)$, where $\mu(x)$ denotes the measure of importance of criterion x (assumed to be additive) and $c_a(x)$ denotes the degree (measured or assessed in some specified way) to which object a satisfies criterion x. Clearly, the use of an appropriate nonadditive measure (determined in the application context) and the Choquet integral (or some other nonlinear integral) offers a more flexible and more realistic approach to this broad class of problems.

In the next example, consider a set of workers in a workshop, X, who are involved in manufacturing products of a specific type. Assume that the set X is partitioned into subsets (working groups) $G_1, G_2, \ldots, G_n$, and let $\mu(G_i)$ denote the number of products made by group G_i within a given unit of time. When the groups work separately, set function μ is clearly an additive measure. However, when some of the groups work together and their cooperation is efficient, the measure is required to be superadditive. If, on the other hand, their cooperation is inefficient the measure is required to be subadditive. If some groups that work together cooperate efficiently while others cooperate inefficiently, then a more general measure is required, usually a monotone measure, but a general measure may be required to capture some extreme situations.

There are many more examples of application contexts in which the restriction to additive measure is inhibitory and not realistic. Some of them are discussed in detail in the remaining sections of this chapter, while some others are only surveyed in Notes to this chapter. The chapter is organized as follows. In Section 15.2 we begin with an overview of the essential role of generalized measure theory in an area that is referred to in the literature as *generalized information theory* (GIT). The objective of GIT is to study the dual concepts of information-based uncertainty and uncertainty-based information in all their manifestations. One subarea of GIT, which we cover in somewhat greater detail in Section 15.3, consists of the various theories of imprecise probabilities, which deal with dual pairs of nonadditive measures of various types. The utility of imprecise probabilities is illustrated by simple examples in Section 15.4. In the remaining sections, we discuss applications of generalized measure theory in information fusion (Section 15.5), multiregression (Section 15.6), classification (Section 15.7), and other areas (Section 15.8).

Notes to this chapter are particularly important. In order to keep the size of this book modest, we are not able to cover the various applications of generalized measure theory in sufficient detail. However, we compensate for the lack of detail by providing the reader with ample references to relevant publications, which cover the missing details.

15.2 Generalized Information Theory: An Overview

The term "generalized information theory" (GIT) was introduced in the early 1990s as a research program for studying the interrelated concepts of information-based uncertainty and uncertainty-based information in all their conceivable manifestations [Klir, 1991]. In GIT, as in classical information theory, the primary concept is uncertainty, and information is defined in terms of uncertainty reduction. While uncertainty in classical information theory is formalized in terms of probability measures, uncertainty in GIT is formalized within an expanded framework. The expansion is two-dimensional. In one dimension classical measure theory, which is the framework for formalizing the concept of probability, is replaced with the much broader framework of generalized measure theory. In the other dimension the formalized language of classical set theory, which is employed in both classical measure theory and generalized measure theory, is replaced with the more expressive formalized language of fuzzy set theory. As we know, generalized measure theory is a very broad framework under which theories of various special types of measures are subsumed, including the theory of classical measures. Similarly, fuzzy set theory is a very broad framework under which various types of formalized languages are subsumed, including the one of classical set theory. By combining these two frameworks we obtain a comprehensive framework for investigating the concepts information-based uncertainty and uncertainty-based information.

The described framework allows us to recognize many distinct theories of information (or uncertainty), including the classical information theory. Each of these theories is characterized by employing measures of some particular type and a formalized language of some particular type for formalizing a particular type of uncertainty and the associated uncertainty-based information. In order for us to fully develop any of these theories the following issues must be adequately addressed at each of the following four levels:

1. Uncertainty functions u of the theory must be characterized via appropriate axioms. In classical information theory, functions u are probability measures; in GIT they are monotone measures or some special types of monotone measures, which, in addition may be fuzzified.
2. Calculus must be developed for dealing with functions u. In classical information theory, it is the well-known calculus of probability theory; in GIT it is a calculus for dealing with a given type of monotone measures or, possibly, with its fuzzified version.
3. A justifiable functional U must be found, which for each particular function u in the theory measures the amount of uncertainty associated with u. When a particular unit of measurement is chosen, functional U is required to be unique. A visible example of this functional is the well-known *Shannon entropy* in classical information theory. In GIT, functional U is usually an aggregate of several coexisting types of uncertainties captured by functions u, and it is desirable to disaggregate it in a justifiable way into components, each of which measures the amount of uncertainty of one particular type.
4. Methodological aspects of the theory must be properly developed for dealing with applications for which the theory is fit, utilizing properties of uncertainty functions u as well as the functional U and its components.

Clearly, the number of prospective uncertainty theories that emerge from the expanded framework of GIT grows very rapidly with the number of recognized types of monotone measures and the number of considered types of formalized languages. It turns out that the rapidly growing diversity of theories subsumed under GIT is balanced by their unity, which is manifested by their many common properties. The diversity of GIT offers an extensive inventory of distinct uncertainty theories, each characterized by specific assumptions. This allows us to choose, in any application context, a theory whose assumptions are in harmony with application of concern. The unity of GIT, on the other hand, allows us to work within GIT as a whole. That is, it allows us to move from one theory to another as needed.

Among the many uncertainty theories that are recognized within the expanded conceptual framework of GIT, only a few of them have been sufficiently developed so far. By and large, these are theories based on various types of monotone measures defined within the language of classical set theory. Fuzzifications of some of them have been explored, but only to some extent. One class of uncertainty theories, perhaps the most visible one at this time, consists of theories whose aim is to represent and deal with imprecise probabilities. It is this class of uncertainty theories that we cover in greater detail in Section 15.3.

15.3 Theories of Imprecise Probabilities

One important insight emanating from research in the area of GIT is that the tremendous diversity of uncertainty theories that emerge from the expanded framework is made tractable due to some key properties that are invariant across the whole spectrum or, at least, within some broad classes of uncertainty theories. One such class, which is the subject of this section, consists of theories that can be viewed as theories of imprecise probabilities.

The need for enlarging the framework of classical probability theory by allowing imprecisions in probabilities has been discussed quite extensively in the literature, and many arguments for imprecise probabilities have been put forward. The following are some of the most common of these arguments:

- Imprecision of probabilities is needed to reflect the *amount of information* on which they are based. The imprecision should decrease with the amount of statistical information.
- *Total ignorance* can be properly modeled by *vacuous probabilities*, which are maximally imprecise (i.e., each covers the whole range [0, 1]), but not by any precise probabilities.
- Imprecise probabilities are *easier to assess and elicit* than precise ones.
- We may be unable to assess probabilities precisely in practice, even if that is possible in principle, because we *lack the time or computational ability*.
- A precise probability model that is defined on some class of events determines only imprecise probabilities for *events outside the class*.
- When *several sources of information* (sensors, individuals of a group in a group decision) are combined, the extent to which they are inconsistent can be expressed by the imprecision of the combined model.

All theories of imprecise probabilities that are based on classical set theory share some common characteristics. One of them is that evidence within each theory is fully described by a *lower probability function* μ_* or, alternatively, by an *upper probability function* μ^*. These functions are always normalized monotone measures on some measurable space (X, $\mathbf{F}$). Moreover, they are superadditive and subadditive, respectively, and satisfy the inequalities

$$\sum_{x \in X} \mu_*(\{x\}) \leq 1, \sum_{x \in X} \mu^*(\{x\}) \geq 1.$$

In the various special theories of uncertainty, they possess additional special properties.

When evidence is expressed (at the most general level) in terms of an arbitrary convex polytope C of discrete probability distribution functions p (such a set is often referred to as a *credal set*) on a finite set X, functions μ_* and μ^* associated with C are determined for each $A \in \mathbf{F}$ by the formulas

$$\mu_*(A) = \inf_{p \in C} \sum_{x \in A} p(x), \mu^*(A) = \sup_{p \in C} \sum_{x \in A} p(x). \quad (15.1)$$

Since

$$\sum_{x \in A} p(x) + \sum_{x \notin A} p(x) = 1$$

for each $p \in C$ and each $A \in \mathbf{F}$, it follows that

$$\mu^*(A) = 1 - \mu_*(\bar{A}). \quad (15.2)$$

Due to this property, functions μ_* and μ_* are called *dual* (or *conjugate*). One of them is sufficient for capturing given evidence; the other one is uniquely determined by the duality equation. It is common to use the lower probability function to capture the evidence.

When X is a finite set, it is well known that any given lower probability function μ_* is uniquely represented by a set function m for which $m(\emptyset) = 0$ and $\Sigma_{A \in \mathbf{F}}\, m(A) = 1$. This function is called a *Möbius representation* of μ_* when it is obtained for all $A \in \mathbf{F}$ via the *Möbius transform*

$$m(A) = \sum_{B|B \subset A} (-1)^{|A-B|} \mu_*(B). \quad (15.3)$$

The inverse transform is defined for all $A \in \mathbf{F}$ by the formula

$$\mu_*(A) = \sum_{B|B \subset A} m(B). \quad (15.4)$$

It follows directly from the duality equation that

$$\mu^*(A) = \sum_{B|B \cap A \neq \emptyset} m(B). \quad (15.5)$$

for all $A \in \mathbf{F}$.

Example 15.1. Let $X = \{a, b, c, d\}$. An example of the three set functions connected via Eqs. (15.2)–(15.5) — m, μ_*, and μ^* — defined on the power set of X is shown in Table 15.1. Given any one of these functions, the other two are uniquely determined by the equations. When μ_* is given, we calculate m by Eq. (15.3). For example, $m(X) = \mu_*(X) - \mu_*(\{a,b,c\}) - \mu_*(\{a,b,d\}) - \mu_*(\{a,c,d\}) - \mu_*(\{b,c,d\}) + \mu_*(\{a,b\}) + \mu_*(\{a,c\}) + \mu_*(\{a,d\}) + \mu_*(\{b,c\}) + \mu_*(\{b,d\}) + \mu_*(\{c,d\}) - \mu_*(\{a\}) - \mu_*(\{b\}) - \mu_*(\{c\}) - \mu_*(\{d\}) = 3.2 - 3.4 = -0.2$. When m is given, μ_* is calculated by Eq. (15.4). For example, $\mu_*(\{a,b,c\}) = m(\{a,b,c\}) + m(\{a,b\}) + m(\{a,c\}) + m(\{b,c\}) + m(\{a\}) + m(\{b\}) + m(\{c\}) = 0.8$.

Table 15.1 Set functions in Example 15.1

A	$m(A)$	$\mu_*(A)$	$\mu^*(A)$
$\emptyset$	0.0	0.0	0.0
$\{a\}$	0.1	0.1	0.3
$\{b\}$	0.3	0.3	0.4
$\{c\}$	0.3	0.3	0.4
$\{d\}$	0.0	0.0	0.2
$\{a, b\}$	0.0	0.4	0.7
$\{a, c\}$	0.0	0.4	0.7
$\{a, d\}$	0.1	0.2	0.4
$\{b, c\}$	0.0	0.6	0.8
$\{b, d\}$	0.0	0.3	0.6
$\{c, d\}$	0.0	0.3	0.6
$\{a, b, c\}$	0.1	0.8	1.0
$\{a, b, d\}$	0.1	0.6	0.7
$\{a, c, d\}$	0.1	0.6	0.7
$\{b, c, d\}$	0.1	0.7	0.9
X	−0.2	1.0	1.0

Assume now that evidence is expressed in terms of a given lower probability function μ_* or, alternatively, in terms of a given upper probability function μ_*. Then, the set of probability distribution functions p that dominate function $\mu_*, C(\mu_*)$ is the same as the set of those that are dominated by function $\mu*, C(\mu*)$. This set, which is always closed and convex, can thus be characterized in either of the following ways:

$$\begin{aligned} C(\mu_*) &= \{p|\mu_*(A) \leq \Sigma_{x\in A}\, p(x) \text{ for all } A \in \mathbf{F}\}, \\ C(\mu^*) &= \{p|\mu^*(A) \geq \Sigma_{x\in A}\, p(x) \text{ for all } A \in \mathbf{F}\}. \end{aligned} \tag{15.6}$$

A well-defined category of theories of imprecise probabilities is based on Choquet capacities of various orders. The most general theory in this category is the theory based on *capacities of order 2*. Less general theories are then based on *capacities of higher orders*. The least general of all these theories is the one based on Choquet capacities of order ∞. This theory is usually referred to as *evidence theory* or *Dempster–Shafer theory* (DST). In this theory, lower and upper probabilities are referred to as *belief* and *plausibility* measures. These measures are introduced and examined in Section 4.5, where it is also shown that Sugeno λ-measures are special belief measures when $\lambda > 0$, and they are special plausibility measures when $\lambda < 0$. An important feature of DST is that the Möbius representation of evidence m (usually called a *basic probability assignment function* in this theory) is a nonnegative set function ($m(A) \in [0, 1]$).

Special plausibility measures, which are called *possibility measures* (or *consonant plausibility measures*) are introduced and discussed in Section 4.6. Their dual measures, which are special belief measures, are called *necessity measures* (or *consonant belied measures*). Theory that is based on possibility and necessity measures is usually called a *possibility theory*.

In another important theory of imprecise probabilities, which is computationally more efficient than DST, lower and upper probabilities μ_* and μ_* are determined for all sets ($A \in \mathbf{F}$) by intervals $[l(x), u(x)]$ of probabilities on singletons ($x \in X$). Clearly, $l(x) = P_*\{(x\}) \in [0,1]$ and $u(x) = P^*(\{x\}) \in [0,1]$. Each given tuple of probability intervals, $I = \langle [l(x), u(x)] | x \in X \rangle$, is associated with a closed convex set, $C(I)$, of probability distribution functions, p, defined as follows:

$$C(I) = \left\{ p(x) | x \in X, \, p(x) \in [l(x), u(x)], \sum_{x \in X} p(x) = 1 \right\}. \tag{15.7}$$

Sets defined in this way are clearly special credal sets. Their special feature is that they always form an $(n-1)$-dimensional polyhedron, where $n = |X|$, whose number c of extreme points is bounded by the inequalities

$$n \le c \le n(n-1).$$

Each probability distribution function contained in the set is expressed as a linear combination of these extreme points.

A given tuple I of probability intervals may be such that some combinations of values taken from the intervals do not correspond to any probability distribution function. This indicates that the intervals are unnecessarily broad. To avoid this deficiency the concept of "reachability" was introduced in the theory. A given tuple I is called *reachable* (or *feasible*) if and only if for each $x \in X$ and every value $v(x) \in [l(x), u(y)]$ there exists a probability distribution function p for which $p(x) = v(x)$. The reachability of any given tuple I can be easily checked: the tuple is reachable if and only if it passes the following tests:

(a) $\sum_{x \in X} l(x) + u(y) - l(y) \le 1$ for all $y \in X$;

(b) $\sum_{x \in X} u(x) + l(y) - u(y) \ge 1$ for all $y \in X$.

If I is not reachable it can be easily converted to the tuple $I' = \langle [l'(x), u'(y)] | x \in X \rangle$ of reachable intervals by the formulas

$$\begin{aligned} l'(x) &= \max\{l(x), 1 - \sum_{y \neq x} u(y)\}, \\ u'(\{x\}) &= \min\{u(x), 1 - \sum_{y \neq x} l(y)\} \end{aligned} \tag{15.8}$$

for all $x \in X$.

Given a reachable tuple I of probability intervals, the lower and upper probabilities are determined for each $A \in \mathbf{F}$ by the formulas

$$\mu_*(A) = \max\left\{\sum_{x\in A} l(x), 1 - \sum_{x\notin A} u(x)\right\}$$
$$\mu^*(A) = \min\left\{\sum_{x\in A} u(x), 1 - \sum_{x\notin A} l(x)\right\}. \tag{15.9}$$

It is known that the theory based on reachable probability intervals is not comparable with DST in terms of their generalities. However, they both are subsumed under a theory based on Choquet capacities of order 2.

Although Choquet capacities of order 2 do not capture all credal sets, they are quite general. Their significance is that they are computationally easier to handle than arbitrary credal sets. In particular, it is easier to compute $C(\mu_*)$ when μ_* is a Choquet capacity of order 2.

Let $X = \{x_1, x_2, \ldots, x_n\}$ and let $\sigma = (\sigma(x_1), \sigma(x_2), \ldots, \sigma(x_n))$ denote a permutation by which elements of X are reordered. Then, for any given Choquet capacity of order 2, $C(\mu_*)$ is determined by its extreme points, which are probability distributions p_σ computed as follows:

$$\begin{aligned}
p_\sigma(\sigma(x_1)) &= \mu_*(\{\sigma(x_1)\}),\\
p_\sigma(\sigma(x_2)) &= \mu_*(\{\sigma(x_1), \sigma(x_2)\}) - \mu_*(\{\sigma(x_1)\})\\
&\vdots\\
p_\sigma(\sigma(x_{n-1})) &= \mu_*(\{\sigma(x_1), \ldots, \sigma(x_{n-1})\}) - \mu_*(\{\sigma(x_1), \ldots, \sigma(x_{n-2})\}),\\
p_\sigma(\sigma(x_n)) &= \mu_*(\{\sigma(x_1), \ldots, \sigma(x_n)\}) - \mu_*(\{\sigma(x_1), \ldots, \sigma(x_{n-1})\}).
\end{aligned} \tag{15.10}$$

Each permutation defines an extreme point of $C(\mu_*)$, but different permutations can give rise to the same point. The set of distinct probability distributions p_σ is often called an *interaction representation of* μ_*.

Example 15.2. In order to illustrate the use of Eqs. (15.10), let us consider the lower probability functions ${}^1\mu_*$ and ${}^2\mu_*$ on the power set of $X = \{a,b,c\}$ that are specified in Fig. 15.1 (a) and (b), respectively, by their values shown in the diagram of the Boolean lattice of **P**(X). We can easily check that both ${}^1\mu_*$ and ${}^2\mu_*$ are 2-monotone measures and ${}^2\mu_*$ is even totally monotone (Choquet capacity of order ∞).

Hence, Eq. (15.10) can be used to determine in each case the extreme points of the convex set of probability distributions that dominate the lower probability function. Observe that each individual probability obtained by Eq. (15.10) for a particular permutation of elements in X is represented in the diagram by the sequence of differences of the lower probability function along a path in the diagram (defined by the permutation) from the node representing Ø to the one representing X. Probability distributions obtained

Fig. 15.1 Illustration to Example 15.2

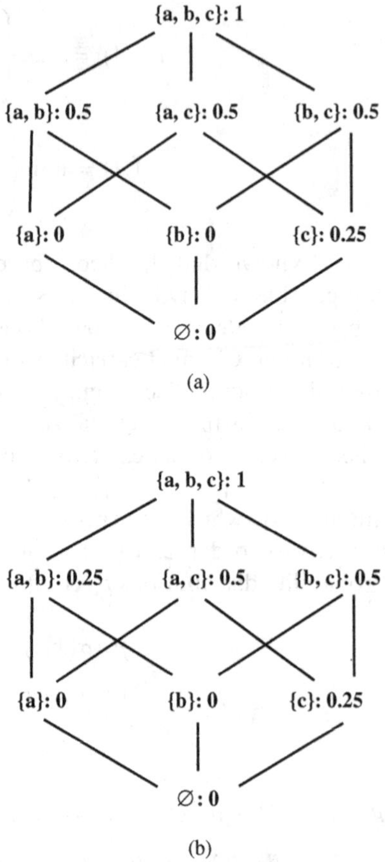

for functions ${}^1\mu_*$ and ${}^2\mu_*$ are shown for all permutations of elements of X in Table 15.2 (a) and (b), respectively.

For ${}^1\mu_*$ there are only four distinct probability distributions:

$$\begin{aligned}\mathbf{p_1} &= (0, 0.5, 0.5),\\ \mathbf{p_2} &= (0.5, 0, 0.5),\\ \mathbf{p_3} &= (0.25, 0.5, 0.25),\\ \mathbf{p_4} &= (0.5, 0.25, 0.25).\end{aligned}$$

For ${}^2\mu_*$ all obtained probabilities are distinct:

$$\begin{aligned}\mathbf{p_1} &= (0, 0.25, 0.75),\\ \mathbf{p_2} &= (0, 0.5, 0.5),\\ \mathbf{p_3} &= (0.25, 0, 0.75),\\ \mathbf{p_4} &= (0.5, 0, 0.5),\\ \mathbf{p_5} &= (0.25, 0.5, 0.25).\\ \mathbf{p_6} &= (0.5, 0.25, 0.25).\end{aligned}$$

This is clearly the maximum of possible number of extreme points ($3! = 6$).

Table 15.2 Interaction representation in Example 15.2

(a) Function ${}^1\mu_*$

$\sigma(a)$	$\sigma(b)$	$\sigma(c)$	$p_\sigma(\sigma(a))$	$p_\sigma(\sigma(b))$	$p_\sigma(\sigma(c))$
a	b	c	0.00	0.50	0.50
a	c	b	0.00	0.50	0.50
b	a	c	0.50	0.00	0.50
b	c	a	0.50	0.00	0.50
c	a	b	0.25	0.50	0.25
c	b	a	0.50	0.25	0.25

(b) Function ${}^2\mu_*$

$\sigma(a)$	$\sigma(b)$	$\sigma(c)$	$p_\sigma(\sigma(a))$	$p_\sigma(\sigma(b))$	$p_\sigma(\sigma(c))$
a	b	c	0.00	0.25	0.75
a	c	b	0.00	0.50	0.50
b	a	c	0.25	0.00	0.75
b	c	a	0.50	0.00	0.50
c	a	b	0.25	0.50	0.25
c	b	a	0.50	0.25	0.25

15.4 Classification of Pairs of Dual Measures

As is discussed earlier in this section, imprecise probabilities on a given measure space (X, **F**) can be explicitly defined by pairs of dual measures μ_* and μ^*, which define for each set $A \in \mathbf{F}$ the range $[\mu_*(A), \mu_*(A)]$ of acceptable probabilities. Due to the duality of these measures, the probability range can also be described by using only one of them, either as $[\mu_*(A), 1 - \mu_*(\overline{A})]$ or as $[1 - \mu^*(\overline{A}), \mu^*(A)]$.

While imprecise probabilities can always be represented by appropriate pairs of dual measures, it is by no means guaranteed that every given pair of dual measures, say pair (μ, ν), represents imprecise probabilities. In order to understand why some pairs of dual measures do not represent imprecise probabilities, we present in this section a fairly comprehensive classification of all possible pairs of dual measures defined on a special measurable space (X,**P**(X)), where X is assumed to be a finite set.

We begin with the most general class of dual pairs of measures, (μ, ν), for which it is only required that they qualify as general measures. For further reference we denote this class as GEN. Some, but not all, of the dual pairs of measures in this class are monotone measures. Measures that are not monotone clearly violate Eq. (15.1) and, therefore, they do not represent imprecise probabilities. This means that we need to restrict our consideration to the class of dual pair of monotone measures. We denote this class as MON.

Observe now that only those dual pairs (μ, ν) of monotone measures can represent imprecise probabilities that are ordered in the sense that either $\mu(A) \leq \nu(A)$ for all $A \in \mathbf{P}(X)$ *or* $\mu(A) \geq \nu(A)$ for all $A \in \mathbf{P}(X)$. The first pair in Table 15.3 is an example of a pair of dual monotone measures (defined on the

Table 15.3 Examples of dual pairs of measures on $(X, \mathbf{P}(X))$, where $X = \{a, b, c\}$

A	$\mu(A)$	$\nu(A)$	$\mu_*(A)$	$\mu_*(A)$	$\mu_*(A)$	$\mu_*(A)$
Ø	0.0	0.0	0.0	0.0	0.0	0.0
$\{a\}$	0.3	0.0	0.1	0.2	0.2	0.8
$\{b\}$	0.1	0.4	0.0	0.5	0.2	0.8
$\{c\}$	0.3	0.7	0.2	0.4	0.0	0.8
$\{a, b\}$	0.3	0.7	0.6	0.8	0.2	1.0
$\{a, c\}$	0.6	0.9	0.5	1.0	0.2	0.8
$\{b, c\}$	1.0	0.7	0.8	0.9	0.2	0.8
X	1.0	1.0	1.0	1.0	1.0	1.0

power set of $X = \{a,b,c\}$) that are not ordered. Clearly, neither of the measures qualifies as a lower probability function or an upper probability function. This means that we need to further restrict to the class of dual pairs of monotone measures that are ordered. We denote this class as ORD. In this class it is convenient to denote the dual pairs as (μ_*, μ^*), where it is understood by convention that $\mu_*(A) \leq \mu^*(A)$ for all $A \in \mathbf{P}(X)$.

Considering now a dual pair (μ_*, μ^*) of ordered monotone measures, it is still not guaranteed that the pair represents imprecise probabilities. If it does, then the measures μ_* and μ^* must be connected with a convex set of probability distribution functions via Eq. (15.1). Then, for any sets $A, B \in \mathbf{P}(X)$ such that $A \cap B = Ø$, we have

$$\mu_*(A \cup B) \geq \mu_*(A) + \mu_*(B),$$
$$\mu^*(A \cup B) \leq \mu^*(A) + \mu^*(B).$$

This means that μ_* must be a superadditive measure and μ^* must be a subadditive measure to qualify, respectively, as lower and upper probability functions. The second pair in Table 15.3 is an example of a dual pair of ordered monotone measures that violates this requirement. Observe that in this example, μ_* is superadditive, but μ^* is not subadditive. For example, $\mu^*(\{a,b\}) > \mu^*(\{a\}) + \mu^*(\{b\})$, which violates the subadditivity of μ^*. When μ_* is a superadditive measure and μ^* is a subadditive measure, the pair (μ_*, μ^*) is usually referred as an additivity-coherent pair of dual measures. We denote this class of pairs of dual measures as ACO.

In imprecise probabilities, the pair of dual measures represents bounds on probability measures. This requires that there exists a probability measure p for a given dual pair (μ_*, μ^*) of ordered monotone measures such that $\mu_*(A) \leq p(A) \leq \mu_*(A)$ for all $A \in \mathbf{P}(A)$. Any dual pair that satisfies this requirement is usually referred to as a probability limiting pair, and we denote this class as PRL. Dual pairs in this class are characterized by the following theorem.

Theorem 15.1. [Huber, 2004] *If* $(\mu_*, \mu^*) \in$ ORD*, then* $(\mu_*, \mu^*) \in$ PRL *iff* $\sum_{A \subset X} a_A\, \chi_A \leq 1$ *implies* $\sum_{A \subset X} a_A\, \mu_*(A) \leq 1$*, where, for all* $A \subset X$*,* χ_A *is the characteristic function of* A *and* a_A *is a nonnegative real number.*

Classes ACO and PRL are not comparable (neither is contained in the other one), as can be shown by the following two examples. One of them is the third pair of dual measures in Table 15.3. This pair is a probability limiting pair, as it represents bounds for a set of probability measures, one of which is defined by the probability distribution $p(\{a\}) = p(\{b\}) = 0.4, p(\{c\}) = 0.2$. However, the pair is not an additivity-coherent pair since μ_* is not superaddtive and μ^* is not subadditive. Hence, the pair is included in PRL, but not in ACO.

In the second example, adopted from [Lamata and Moral, 1989], $X = \{x_i \mid i = 1, 2, \ldots, 9\}$ and the following eight subsets of X are employed in defining a dual pair (μ_*, μ^*) of ordered monotone measures via μ_*:

$$A_1 = \{x_1, x_3, x_6, x_9\}, A_2 = \{x_1, x_4, x_7, x_8\},$$
$$A_3 = \{x_1, x_5, x_7, x_8\}, A_4 = \{x_2, x_3, x_7, x_8\},$$
$$A_5 = \{x_2, x_4, x_6, x_8\}, A_6 = \{x_1, x_2, x_6, x_9\},$$
$$A_7 = \{x_2, x_5, x_6, x_7\}, A_8 = \{x_3, x_4, x_5, x_9\}.$$

Now, μ_* is defined for all $A \in \mathbf{P}(X)$ by the formula

$$\mu_*(A) = \begin{cases} 1 & \text{when } A = X \\ 0.5 & \text{when } A \neq X \text{ when } A \supset A_k \quad \text{for some } k \in \{1,2,\ldots,8\} \\ 0 & \text{otherwise} \end{cases}$$

Then, (μ_*, μ^*) is clearly an additivity-coherent pair *ofdualmeasures.However,*

$$\sum_{k=1}^{7} 0.25\chi_{A_k} + 0.5\chi_8 = 1$$

and

$$\sum_{k=1}^{7} 0.25\mu_*(A_k) + 0.5\mu_*(A_8) = 1.125 > 1,$$

which violates the implication in Theorem 15.1. Hence, the pair (μ_*, μ^*) is in ACO, but not in PRL. It follows from the two examples that the classes ACO and PRL are not comparable.

In order to guarantee that a dual pair (μ_*, μ^*) of ordered monotone measures represents imprecise probabilities, it is essential to require that there exist a nonempty set C of probability measures such that μ_* and μ^* are obtained from C by Eq. (15.1). Dual pairs that satisfy this requirement are called "representable," and we denote the class of representable dual pairs by REP. Dual pairs of ordered monotone measures that are representable are characterized by the following theorem,

Theorem 15.2. [Wolf, 1977] *A dual pair of ordered monotone measures* (μ_*, μ^*) *on measurable space* $(X, \mathbf{P}(X))$, *where X is a finite set, is representable iff it satisfies the following property for all* $A \subset X$: *if* $\chi_A \le \sum_{B \subset X} a_B\, \chi_B - a$, *then* $\mu^*(A) \le \sum_{B \subset X} a_B\, \mu^*(B) - a$, *where* a_B *and a are nonnegative real numbers.*

It is interesting that the class REP of representable dual pairs is included in the intersection ACO ∩ PRL, but it is not equal to it. This is demonstrated by the following example.

Example 15.3. [Huber, 1981]. Consider the dual pair (μ_*, μ^*) of ordered monotone measures given in Table 15.4. It is easy to verify that μ_* is superadditive and μ^* is subadditive. This means that the pair belongs to class ACO. It can also be verified that there is a unique probability measure p such that $\mu_*(A) \le p(A) \le \mu_*(A)$. This probability measure is also shown in Table 15.4. This means that the dual pair is also probability-limiting, and it is thus belongs to class PRL. This means in turn that the considered dual pair belongs to ACO ∩ PRL. However, it is clearly not a representable pair.

It is well known that, in general, there are more than one closed and convex sets of probability measures that induce the same representable dual pair (μ_*, μ^*) via Eq. (15.1). The largest one of them is defined by Eq. (15.6), where clearly $C(\mu_*) = C(\mu^*)$.

The class REP of all dual pairs of representable measures provides us with the broadest framework for dealing with imprecise probabilities. Some of its subclasses are then bases for various special theories of imprecise probabilities. The following subclasses are introduced and examined at various places in this book:

Table 15.4 Example of a dual pair (μ_*, μ^*) of ordered monotone measures that is additivity-coherent and probability-limiting, but it is not representable

A	$\mu_*(A)$	$p(A)$	$\mu^*(A)$
Ø	0.0	0.00	0.0
$\{a\}$	0.0	0.25	0.5
$\{b\}$	0.0	0.25	0.5
$\{c\}$	0.0	0.25	0.5
$\{d\}$	0.0	0.25	0.5
$\{a, b\}$	0.5	0.50	0.5
$\{a, c\}$	0.5	0.50	0.5
$\{a, d\}$	0.5	0.50	0.5
$\{b, c\}$	0.5	0.50	0.5
$\{b, d\}$	0.5	0.50	0.5
$\{c, d\}$	0.5	0.50	0.5
$\{a, b, c\}$	0.5	0.75	1.0
$\{a, b, d\}$	0.5	0.75	1.0
$\{a, c, d\}$	0.5	0.75	1.0
$\{b, c, d\}$	0.5	0.75	1.0
X	1.0	1.00	1.0

- Dual pairs consisting of Choquet capacities and alternating Choquet capacities of order k, where $k \geq 2$ (Section 4.2). Among them, the most general are those of order 2, and we denote this class by CH2. The least general of them are those of order ∞, where the dual pairs are belief and plausibility measures of the Dempster–Shafer theory (Section 4.5); we denote this class by DST.
- A subclass of DST is the class of dual pairs consisting of necessity and possibility measures (Section 4.6). The class of these dual pairs is the base of the theory of graded possibilities, and we denote it by GPO.
- Another subclass of DST is the class of dual pairs of Sugeno λ-measures (Section 4.3), which we denote by LAM.
- A subclass of GPO is the class of dual pairs consisting of crisp necessity and possibility measures, whose values are restricted to the set {0,1}. We denote this class by CPO.
- A subclass of LAM is the class of dual pairs consisting of probability measures (characterized by $\lambda = 0$. Observe that these pairs are autodual in

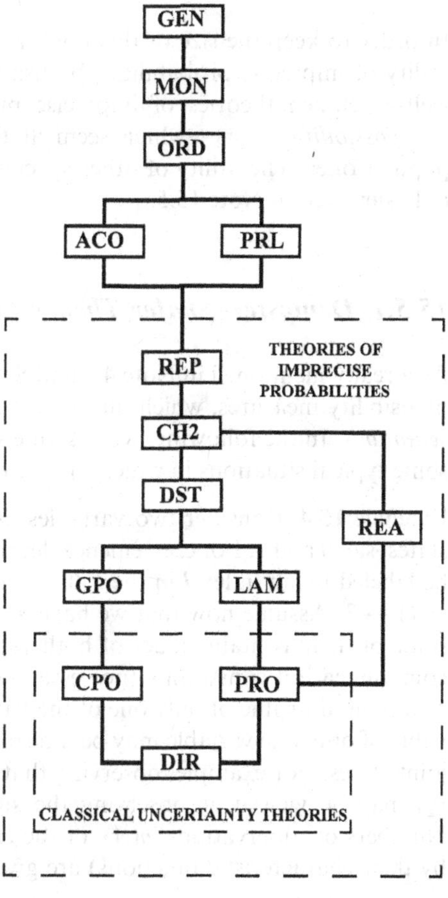

Fig. 15.2 Inclusion ordering of classes of pairs of dual measures introduced in Section 15.4

the sense that $\mu_* = \mu^*$, which means that they do not capture any imprecision. We denote this class by PRO.

- A subset of CPO as well as PRO is the class of autodual pairs of Dirac measures (Section 3.2, Example 3.3), which we denote by DIR. Clearly, DIR = CPO ∩ PRO.
- Another important class of representable pairs of dual measures consists of those that are determined by reachable intervals of probabilities on singletons (introduced earlier in Section 15.3). We denote this class by REA. It is well known that REA is a subset of CH2, but it is not comparable with the class DST.

The introduced classes of pairs of dual measures and their inclusion relationship are summarized in Fig. 15.2. Also indicated in the figure are those of these classes that represent the various theories of imprecise probabilities and those that are considered as classical theories of uncertainty.

15.5 Utility of Some Special Theories of Imprecise Probabilities

In order to keep the size of this book reasonable, we decided to illustrate the utility of imprecise probabilities by using examples formulated in two of the well-developed theories of imprecise probabilities—*Dempster-Shafer theory and possibility theory*—which seem at this time to be the most visible and popular ones. The utility of other special theories of imprecise probabilities is only surveyed in Note 15.5.

15.5.1 Dempster–Shafer Theory (DST)

As already mentioned in Note 4.5 and Section 15.3, DST is based on belief and plausibility measures, which, in turn, are based on *nonnegative Möbius representations*. In the following, we describe several simple examples that illustrate some typical situations in which the use of DST is fitting.

Example 15.4. Consider two variables, ν_1, ν_2, each of which has two possible states, say 0 and 1. For convenience, let the joint states of the variables, (ν_1, ν_2), be labeled by an index I in the following way: (0,0) → 0, (0,1) →1, (1,0) → 2, (1,1) → 3. Assume now that we have a record of 1000 observations, but only some of them contain values of both variables. Due to some measurement or communication constraints (not essential for our discussion), some observations contain value of only one of the variables. Observing in this example the value of only one variable may be interpreted in DST as observing a set of two joint states. For example, observing that $\nu_1 = 1$ (and not knowing the state of ν_2) may be viewed as observing the subset {2, 3} of the four joint states. Numbers of observations, $n(A)$, of the eight relevant sets of states A (defined by their characteristic functions) are given in section (a) of Table 15.5, which

Table 15.5 Belief and plausibility measures derived from incomplete data (Example 15.4)

					(a)				(b)			
$i =$	0	1	2	3	$n(A)$	$m(A)$	Bel(A)	Pl(A)	$n(A)$	$m(A)$	Bel(A)	Pl(A)
A:	1	1	0	0	212	0.212	0.373	0.839	5	0.005	0.812	0.835
	0	0	1	1	128	0.128	0.161	0.627	12	0.012	0.165	0.188
	1	0	1	0	315	0.315	0.446	0.786	8	0.008	0.588	0.605
	0	1	0	1	151	0.151	0.214	0.554	15	0.015	0.395	0.412
	1	0	0	0	106	0.106	0.106	0.633	555	0.555	0.555	0.568
	0	1	0	0	55	0.055	0.055	0.418	252	0.252	0.252	0.272
	0	0	1	0	25	0.025	0.025	0.468	25	0.025	0.025	0.045
	0	0	0	1	8	0.008	0.008	0.287	128	0.128	0.128	0.155

also contains values of the estimated basic assignment m based on frequency interpretation, as well as the corresponding degrees of belief Bel and plausibility Pl. Belief and plausibility degrees can readily be calculated for any of the eight remaining subsets of states. For example, Bel({1, 2, 3}) = 0.713 and Pl({1, 2, 3}) = 0.992.

Viewing values Bel(A) and Pl(A) as lower and upper probabilities, we obtain intervals [Bel(A), Pl(A)] of estimated probabilities, $p(A)$, of subsets A—for example, $p(\{1\}) \in [0.055, 0.418]$ and $p(\{0, 2\}) \in [0.446, 0.786]$.

Assume now that we have another record of 1000 observations of the same two variables, which is given in section (b) of Table 15.5. This record contains substantially more observations that are specific (i.e., they contain values of both variables). As a consequence, the intervals [Bel(A), Pl(A)] of estimated probabilities $p(A)$ are substantially smaller. For example, the value of $p(\{1\})$ is now in the interval [0.252, 0.272] and $p(\{0, 2\}) \in [0.588, 0.605]$.

Thus far, we calculated values $m(A)$ as relative frequencies of observations. These values describe the available data, and, consequently, the derived belief and plausibility measures are descriptive. If, however, we want to use values $m(A)$ as predictive estimators of states yet to be observed, we have to admit that we have no information regarding the future states. That is, we only know that any future observation is contained in the universal set X, which consists of the four states of the variables. If we want to estimate predictive belief and plausibility measures regarding only the next state, we should augment the frequencies given in Table 15.5 by the value $n(X) = 1$, which expresses our ignorance regarding the next state. Then, $m(X) = 0.000999$ and the other values $m(A)$ are only slightly adjusted (the number of considered observations is now 1001). Clearly, the adjustment in values $m(A)$ is very small and may be neglected for practical purposes. This situation is different, however, when we want to predict a large sequence of future states. For example, to predict the sequence of 1000 future observations (i.e., 2000 observations in total), $m(X) = 0.5$ and all values $m(A)$ in Table 15.5 must be divided by two. The resulting situation is shown in Table 15.6.

Table 15.6 Predictive belief and plausibility measures (Example 15.4)

					(a)				(b)			
$i =$	0	1	2	3	$n(A)$	$m(A)$	Bel(A)	Pl(A)	$n(A)$	$m(A)$	Bel(A)	Pl(A)
A:	1	1	0	0	212	0.1060	0.1865	0.9195	5	0.0025	0.4060	0.9175
	0	0	1	1	128	0.0640	0.0805	0.8135	12	0.0060	0.0825	0.5940
	1	0	1	0	315	0.1575	0.2230	0.8930	8	0.0040	0.2940	0.8025
	0	1	0	1	151	0.0755	0.1070	0.7770	15	0.0075	0.1975	0.7060
	1	0	0	0	106	0.0530	0.0530	0.8165	555	0.2775	0.2775	0.7840
	0	1	0	0	55	0.0275	0.0275	0.7090	252	0.1260	0.1260	0.6360
	0	0	1	0	25	0.0125	0.0125	0.7340	25	0.0125	0.0125	0.5225
	0	0	0	1	8	0.0040	0.0040	0.6435	128	0.0640	0.0640	0.5775
	1	1	1	1	1000	0.5000	1.0000	1.0000	1000	0.5000	1.0000	1.0000

Example 15.5. Dong and Wong [1986] describe an example in which a group of experts give their estimates of possible location of the epicenter of an earthquake. Suppose that 15 estimates are given as shown in Fig. 15.3. Observe that these estimates are both nonspecific and conflicting with each other. Using the evidence on hand, what is the likelihood that the epicenter is inside any particular area of interest (for example, a densely populated area A or B)? Each of the estimates has a weight of evidence 1/15, provided that we consider all reports as equally reliable and otherwise equivalent in their value. Then, degrees of belief and plausibility can be readily calculated: Bel(A) = 2/15 = 0.13, Pl(A) = 5/15 = 0.33; Bel(B) = 1/15 = 0.07, Pl(B) = 3/15 = 0.2. Hence, we obtain the following interval-valued estimates of probabilities $p(A)$ and $p(B)$ that the epicenter is in area A or B, respectively: $p(A) \in [0.13, 0.33]$, $p(B) \in [0.07, 0.2]$.

Example 15.6. Bogler [1987] describes how DST can be utilized in dealing with the following multiple sensor target identification in which intelligent reports are also employed as a source of information. It is assumed, based on an intelligent report, that there 100 possible target types. Let $X = \{x_1, x_2, \ldots, x_{100}\}$ denote

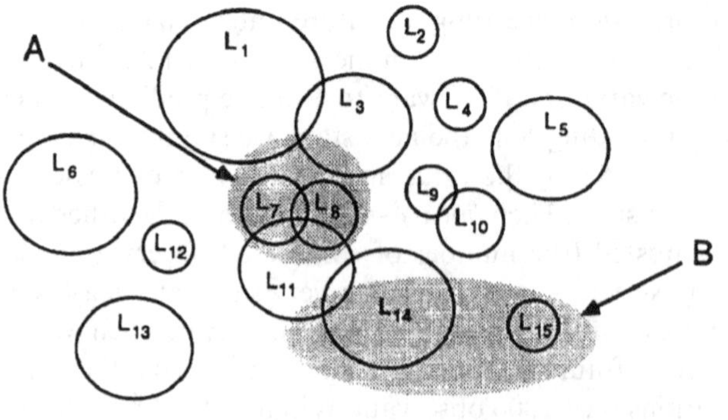

Fig. 15.3 Estimates of 15 experts regarding the location of the epicenter of an earthquake (Example 15.5)

the set of these target types. It is also known, from another intelligence report that only targets of type x_1 entered the relevant tactical area, but the reporting agent had access only to records pertaining to 40% of the targets entering the tactical area. Thus, according to this evidence, e_1, we have $m_1(\{x_1\}) = 0.4$ and $m_1(X) = 0.6$. We have also evidence from sensors that only targets of types $x_1, x_2, \ldots, x_{11}$ have been observed in the population of the incoming targets. According to this evidence, e_2, we have $m_2(\{x_1, x_2, \ldots, x_{11}\}) = 1$. Now, we need to combine evidence from the two sources. The standard way of combining evidence in Dempster–Shafer theory is expressed by the formula

$$m_{12}(A) = \frac{\sum_{B \cap C = A} m_1(B) \cdot m_1(C)}{1 - c}$$

for $A \neq \emptyset$, where

$$c = \sum_{B \cap C = \emptyset} m_1(B) \cdot m_2(C),$$

and $m_{12}(\emptyset) = 0$. This formula is called a *Dempster rule of combination*. This rule has been the subject of an ongoing debate in the literature. On the one hand, the uniqueness of the rule can be proven under the usual axioms of Dempster–Shafer theory and under the assumption that the two sources of information are independent of each other (observations made by one source do not constrain observations made by the other source) [Dubois and Prade, 1986]. On the other hand, the requirement that $m_{12}(\emptyset) = 0$, which leads to the normalization factor $1 - c$, is considered unnecessarily restrictive and leading, in some cases, to counterintuitive results. The latter position is, in fact, critical not only of the Dempster rule, but also of one of the axioms of Dempster–Shafer theory, the axiom that $m(\emptyset) = 0$. Allowing $m_{12}(\emptyset) > 0$, the value $m_{12}(\emptyset)$ may be interpreted as evidence pointing to a hypothesis that is outside the universal set under consideration. That is, when $m(\emptyset) = 0$ is required, it is implicitly assumed that all relevant hypotheses in a given context are included in the accepted universal set (closed-world position); on the other hand, when $m(\emptyset) > 0$ is allowed, it is recognized that our universal set may not be complete in a given context (open-world position). When $m_{12}(\emptyset) > 0$ and we want to take the closed-world position, the following *alternative rule of combination* is more appropriate:

$$m_{1,2}(A) = \begin{cases} \sum_{B \cap C = A} m_1(B).m_2(C) & \text{when } A \neq \emptyset \text{ and } A \neq X \\ m_1(X).m_2(X) + \sum_{B \cap C = \emptyset} m_1(B).m_2(C) & \text{when } A = X \\ 0 & \text{when } A = \emptyset \end{cases}$$

According to this alternative rule of combination, m_{12} is normalized by moving the value c to $m_{12}(X)$. This means that the conflict between the two sources of evidence is not hidden, but it is explicitly recognized as a contributor to our ignorance. The alternative rule is thus better justified on epistemological ground when $c > 0$.

Combining m_1 and m_2 in our example by the Dempster rule results in $m_{12}(\{x_1\}) = 0.4$ and $m_{12}(\{x_1, x_2, \ldots, x_{11}\}) = 0.6$. Since $c = 0$ in this case, the alternative rule gives the same result. From the two focal elements of m_{12}, we can calculate Bel_{12} and Pl_{12} for any subset of X and determine thus the lower and upper probability that targets of the types in the subset have entered the relevant tactical area.

Example 15.7. Let us discuss another example in the area of multiple sensor target identification described by Bogler [1987]. Assume that the universal set X is again the set of 100 possible target types, but only two targets are involved in this example, a fighter and a bomber, denoted by f and b, respectively. Evidence comes in this case from two sensors. A short-range sensor provides a support of 0.6 that the target is a fighter, while the radar-warning receiver gives a support of 0.95 that the detected target is a bomber. In this case, clearly, the two sources provide us with conflicting evident and $c = 0.57$. The outcome of the Dempster rule is: $m_{12}(\{f\}) = 0.07$, $m_{12}(\{b\}) = 0.88$, and $m_{12}(X)=0.05$. Due to the large conflict between the two sources of evidence, the outcome of the alternative combination rule is very different and considerably more reasonable: $m_{12}(\{f\}) = 0.03$, $m_{12}(\{b\}) = 0.38$, and $m_{12}(X) = 0.59$.

Example 15.8. Consider two packs of bridge cards. Assume that we distinguish the cards from each other only by their colors, red (r) and black (b). We know that all cards in Pack 1 are red and we know nothing about the properties of red and black cards in Pack 2. We also know that a card is drawn from Pack 1 with probability 0.8 and from Pack 2 with probability 0.2. A natural way of formalizing this knowledge is to express it in terms of the basic probabilistic assignment, $m(\{r\}) = 0.8$ and $m(\{r, b\}) = 0.2$, over the universal set $X = \{r, b\}$. This results in $\text{Bel}(\{r\}) = 0.8$, $\text{P1}(\{r\}) = 1$, $\text{Bel}(\{b\}) = 0$, and $\text{P1}(\{b\}) = 0.2$. Hence, the estimated ranges of probabilities of drawing a red card or a black card are, respectively, $p(r) \in [0.8, 1]$ and $p(b) \in [0, 0.2]$.

Observe that the information available in this case does not allow us to estimate the probabilities more specifically than by these intervals. When using probability theory, however, we would obtain fully specific estimates, $p(r) = 0.9$ and $p(b) = 0.1$, which claim more than the given information warrants.

Example 15.9. Let us modify a hydrological example proposed by Kong [1986]. Consider water permeation among three sites. We know that water can permeate, unless the permeation is for some reason blocked, only from site 1 to sites 2 and 3, and from site 2 to site 3. No information is available about the locations of water sources. They may be directly at some of the sites, but they may equally well be at other places we are not aware of, from which water permeates to some of the sites. We are primarily interested in whether the sites are dry or wet.

Let $s_i = 0$ or $s_i = 1$ denote that site i ($i = 1, 2, 3$) is dry or wet, respectively, and let p_{12}, p_{13}, p_{23} denote the probabilities that the water permeation between the respective sites is not blocked. The eight states of variables s_1, s_2, s_3, are labeled by an index i as shown in Table 15.7a. Assume now that three independent pieces of evidence are available, each of which provides us with the estimate of one of the probabilities p_{12}, p_{13}, p_{23}. Basic assignments expressing these pieces of evidence, which are given in Table 15.7b, are readily obtained. For example, when water permeation from site 1 to site 2 is not blocked, then water at site 1 implies water at site 2. This corresponds to the set of states {(0, 0), (0, 1), (1, 1)} × {0, 1}, and these, according to the evidence, are supported with probability p_{12}. Hence, we have

$$m_{12}(\{(0,0,0),(0,0,1),(0,1,0),(0,1,1),(1,1,0),(1,1,1)\}) = p_{12}.$$

When the percolation is blocked, there is no known relationship between the two sides and, hence, the probability $1 - p_{12}$ is allocated to the universal set $\{0, 1\}^3$; that is,

Table 15.7 Hydrological application (Example 15.9)

(a)

s_1	s_2	s_3	i
0	0	0	0
0	0	1	1
0	1	0	2
0	1	1	3
1	0	0	4
1	0	1	5
1	1	0	6
1	1	1	7

(b)

$i =$	0	1	2	3	4	5	6	7	m_{12}	m_{13}	m_{23}
A:	1	1	1	1	0	0	1	1	p_{12}	0	0
	1	1	1	1	0	1	0	1	0	p_{13}	0
	1	1	0	1	1	1	0	1	0	0	p_{23}
	1	1	1	1	1	1	1	1	$1-p_{12}$	$1-p_{13}$	$1-p_{23}$

(c)

$i =$	0	1	2	3	4	5	6	7	m_{123}	m_{123}	Bel_{123}
A:	1	1	0	1	0	0	0	1	$p_{12}p_{13}p_{23} + p_{12}p_{23}(1-p_{13})$	0.560	0.560
	1	1	1	1	0	0	0	1	$p_{12}p_{13}(1-p_{23})$	0.144	0.704
	1	1	0	1	0	1	0	1	$p_{13}p_{23}(1-p_{12})$	0.084	0.644
	1	1	1	1	0	0	1	1	$p_{12}(1-p_{13})(1-p_{23})$	0.096	0.800
	1	1	1	1	0	1	0	1	$p_{13}(1-p_{12})(1-p_{23})$	0.036	0.791
	1	1	0	1	1	1	0	1	$p_{23}(1-p_{12})(1-p_{13})$	0.056	0.700
	1	1	1	1	1	1	1	1	$(1-p_{12})(1-p_{13})(1-p_{23})$	0.024	1.000

$$m_{12}(\{0,1\}^3) = 1 - p_{12}.$$

Basic assignments m_{13} and m_{23} are determined similarly.

Using now the Dempster rule of combination twice, we obtain the combined body of evidence given in Table 15.7c in terms of all focal elements and expressions specifying respective values of the combined basic assignment m_{123}. Also shown in the table are numerical values of m_{123} for $p_{12} = 0.8$, $p_{13} = 0.6$, and $p_{23} = 0.7$, as well as values of the associated belief measure, Bel_{123}, for the focal elements.

15.5.2 Possibility Theory

A theory based upon both possibility measures and necessity measures is usually referred to as *possibility theory*. There are two dominant views about possibility theory. According to one view, possibility theory is obtained by restricting the Dempster–Shafer theory to bodies of evidence that are nested [Shafer, 1987]. According to the other view, it is formulated in terms of fuzzy sets [Zadeh, 1978].

Fuzzy sets and possibility measures are connected in the following way. Let F be a normal fuzzy set defined on X and let v be a variable that takes values in X. Then, the fuzzy proposition "v is F" induces a possibility profile function

$$r_{v,F} : X \to [0,1]$$

defined for all $x \in X$ by the equation

$$r_{v,F}(x) = \mu_F(x),$$

where $r_{v,F}(x)$ is viewed as the degree of possibility that the value of v is x. When F is a subnormal fuzzy set with height h_{F}, this equation must be replaced with the equation

$$r_{v,F}(x) = \mu_F(x) + 1 - h_F,$$

as is shown in [Klir, 1999]. Possibility measure, $\pi_{v,F}$, based upon $r_{v,F}$ is then defined for each $A \in \mathbf{P}(X)$ by the formula

$$\pi_{v,F}(A) = \sup_{x \in A} r_{v,F}(x).$$

Here, $\pi_{v,F}(A)$ is the (degree of) possibility that the value of v belongs to the crisp set A. The corresponding necessity measure, $\nu_{v,F}$, is then obtained by the equations $\nu_{v,F}(A) = 1 - \pi_{v,F}(\bar{A})$.

The two views of possibility theory are formally equivalent. Focal subsets in Dempster–Shafer theory correspond to distinct α-cuts of the associated fuzzy set. For each element of a focal subset, the value α of the corresponding α-cut is equal to the value of plausibility of that element. For a finite X the correspondence between a possibility profile on X and the associated basic probability assignment is introduced in the proof of Theorem 4.24.

The connection of possibility theory with fuzzy sets gives the former a very broad applicability. In general, possibility theory is applicable to problem areas in which the use of fuzzy sets is significant. These are particularly problem areas in which the role of natural language is essential. The use of fuzzy sets to capture vague concepts, for which natural languages are notorious, is of great advantage.

It is beyond the scope of this book to cover the many applications of possibility theory, which are surveyed in Note 15.5. Let us only illustrate the utility of extensions of possibility measures investigated in Chapter 5. In particular, let us describe how extensions of possibility measures can be utilized for constructing membership grade functions of fuzzy sets.

Example 15.10. The living region of elephants in a primeval forest may be viewed as a fuzzy subset, denoted by E, of the whole primeval forest, which is regarded as the universe of discourse and denoted by X.

Suppose we have received several reports, from which we can infer (for each set $A \in \mathbf{C}$, where $\mathbf{C}$ is a relevant class of crisp subsets of X), the possibility $\pi(A)$ that some elephants live in set (area) A. The sets in $\mathbf{C}$ may overlap with each other arbitrarily. Using this information we want to estimate the membership function μ_E of E.

Since the membership function of a fuzzy subset of X can be regarded as a generalized possibility profile on X, we can use the extension theory of generalized possibility measures (Chapter 5) to estimate the membership function as follows:

$$\mu_E : X \to [0, 1]$$

$$x \mapsto \inf_{A|x\in A\in \mathbf{C}} \pi(A).$$

This is the most optimistic estimate. When π is P-consistent on $\mathbf{C}$, we have

$$\sup_{x\in A} \mu_E(x) = \pi(A) \text{ for all } A \in \mathbf{C}.$$

In general, if π is not P-consistent on $\mathbf{C}$, then

$$\sup_{x\in A} \mu_E(x) \leq \pi(A) \text{ for all } A \in \mathbf{C}.$$

15.6 Information Fusion

Let $X = \{x_1, x_2, \ldots, x_n\}$ denote a set of information sources. Assume that in each experiment (or observation) we obtain numerical information from each source, x_i denoted by $f(x_i)$, $i = 1, 2, \ldots, n$. Function f is regarded as a real-valued function defined on X. That is, $f: X \to (-\infty, \infty)$. If we want to make a proper decision based on the information obtained by doing the experiment (or observation) once, an aggregation tool is needed to fuse the obtained numerical information from n information sources to a single real number. The aggregation tool is essentially a projection from the n-dimensional Euclidean space onto the one-dimensional Euclidean space. The most common and elementary aggregation tool is the *weighted average*. That is, the aggregation is expressed by

$$y = \sum_{i=1}^{n} w_i f(x_i) \tag{15.11}$$

where $w_i \geq 0$, $i = 1, 2, \ldots, n$, and $\sum_{i=1}^{n} w_i = 1$. If weights w_i are allowed to take negative values and the restriction $\sum_{i=1}^{n} w_i = 1$ is deleted, y obtained by Eq. (15.11) is called a *weighted sum*. The weighted average (or the weighted sum) can be regarded as the Lebesgue integral of function f with respect to a classical additive measure μ on $\mathbf{P}(X)$ determined by $\mu(x_i) = w_i$ for all $i = 1, 2, \ldots, n$, that is

$$y = \int f \, d\mu.$$

To use weighted average (or the weighted sum) for information fusion, a basic assumption is needed: there is no interaction among the contribution rates from various attributes towards the fusion result with the effect that the joint contribution from any set of attributes towards the fusion result is just the simple sum of the contributions from each individual attribute in this set. Such kind of interaction is totally different from the statistical correlation that describes the relation among the appearing values of function f at various attributes.

However, in most real-world problems, the above-mentioned interaction cannot be ignored. Direct tools for describing the interaction are monotone measures (or, in some cases, general measures, or signed general measures). In Example 11.3, set function μ is a monotone measure representing the individual and joint efficiencies of three workers. It is not additive. The nonadditivity of μ describes the interaction among the contribution rates of these three workers towards the total amount of manufactured toys. As is discussed in previous chapters, the linear Lebesgue integral fails when a nonadditive monotone measure is used. In this case, a nonlinear integral with respect to a monotone measure (or a signed general measure) should be applied as an aggregation tool in information fusion. Example 11.3 is, in fact, an example of the Choquet integral used for information fusion. Of course, in various situations of information fusion, different types of nonlinear integrals (with the linear Lebesgue integral as a special example), such as the upper and the lower integral and the Sugeno integral, can be chosen as the aggregation tool.

To illustrate the use of generalized measure theory in synthetic evaluation of objects, let an object be given that we want to evaluate and let $X = \{x_1, x_2, \ldots, x_n\}$ be the set of all quality factors regarding the objects of interest. For convenience, let X be called a *factor space* of the object.

Assume now that our goal is to get a numerical synthetic evaluation of the quality of the given object in terms of evaluations obtained for each individual quality factor. A classical and common way to solve this problem is to use the method of a weighted mean. This method is based on the assumption that the effects (weighted evaluations) of individual quality factors are independent of one another and, consequently, are additive. But, in most real problems these effects are interactive, as illustrated by an example given later. In such cases, we need a method that can describe the interaction involved. In the following we show how monotone measures and Sugeno integrals can be used for this purpose.

Assume that each subset E of the factor space X is associated with a real number $\mu(E)$ between 0 and 1 that indicates the importance of E. Such a real number should be the maximum possible score that the object can gain relying only on the quality factors in E. Obviously, the empty set Ø (the set which does not include any quality factor we are interested in) has the minimum importance 0, and the whole factor space X has the maximum importance 1. Moreover, if each factor in a factor set E belongs to another factor set F, then E is at most as important as F. Hence, the set function μ must satisfy the following conditions:

(1) $\mu(\emptyset) = 0$ and $\mu(X) = 1$;
(2) If $E \subset F \subset X$, then $\mu(E) = \mu(F)$.

Since the number of quality factors in which we are interested for the given object is always finite (that is, X is a finite space), the continuity of μ holds naturally. Therefore, μ is a normalized monotone measure on a measurable space $(X, \mathbf{P}(X))$. We call μ an importance measure on X. This measure is similar to the weights in the method of weighted mean. It is an aggregated summary of experts' opinions, which can be obtained by consultations or questionnaires. That is, the importance measure is regarded as a universally accepted criterion employed in the evaluation, which is given before the evaluation begins.

Example 15.11. Consider the problem of evaluating a dish of Chinese cuisine. Assume that the quality factors we consider are the taste, smell, and appearance (including, e.g., the color, shape and general arrangement of the dish). We denote these factors by T, S, and A, respectively; hence, $X = \{T, S, A\}$. Assume further that the following set function μ is employed as an importance measure:

$\mu(\{T\}) = 0.7$, $\mu(\{S\}) = 0.1$, $\mu(\{A\}) = 0, \mu(\{T, S\}) = 0.9, \mu(\{T, A\}) = 0.8$, $\mu(\{S, A\}) = 0.3, \mu(X) = 1$, and $\mu(\emptyset) = 0$. Observe that this importance measure, which is intuitively quite reasonable, is not additive. For instance, $\mu(\{T, S\}) \neq \mu(\{T\}) + \mu(\{S\})$ or $\mu(\{S, A\}) \neq \mu(\{S\}) + \mu(\{A\})$.

It is often convenient, especially when the number of quality factors is large, to use some special kinds of monotone measures, such as possibility measures, Sugeno measures, or belief measures, as the importance measures. The extension method of monotone measures can then be used to establish the importance measure involved.

Given a particular object to be evaluated, a factor space of the object, and an importance measure, the object is evaluated by an adjudicator for each individual quality factor $x_1, x_2, \ldots, x_n$, and we obtain scores $f(x_1), f(x_2), \ldots, f(x_n)$. Function f may be regarded as a measurable function defined on $(X, \mathbf{P}(X))$ such that $f(x_i) \in [0, 1]$ for each $x_i \in X$. Now, it is natural to use the Sugeno integral ($⨍ f\, d\mu$) of the scores $f(x_i)$ with respect to the importance measure μ to obtain a synthetic evaluation of the quality of the given object. Let the quality evaluation be denoted by q.

Example 15.12. Let us return to the problem of evaluating the Chinese cuisine. The quality factors and importance measure are given in Example 15.11. An expert is invited as an adjudicator to judge each quality factor of a particular dish, and he scores the quality factors as follows: $f(T) = 0.9$, $f(S) = 0.6$, $f(A) = 0.8$. The synthetic evaluation of the quality of this dish, q, is then calculated as follows:

$$\begin{aligned} q = ⨍ f d\mu &= [0.6 \wedge \mu(F_{0.6})] \vee [0.8 \wedge \mu(F_{0.8})] \vee [0.9 \wedge \mu(F_{0.9})] \\ &= [0.6 \wedge \mu(X)] \vee [0.8 \wedge \mu(\{T, A\})] \vee [0.9 \wedge \mu(\{T\})] \\ &= 0.6 \vee (0.8 \wedge 0.8) \vee (0.9 \wedge 0.7) \\ &= 0.8. \end{aligned}$$

Consider now a different dish with $f(T) = 1$, $f(S) = f(A) = 0$; then, we have

$$q = [1 \wedge \mu(\{T\})] \vee 0 = \mu(\{T\}) = 0.7.$$

Considering two additional dishes, one with $f(T) = f(S) = 1$, $f(A) = 0$, and the other with $f(T) = f(S) = f(A) = 1$, we obtain, respectively,

$$q = [1 \wedge \mu(\{T, S\})] \vee 0 = \mu(\{T, S\}) = 0.9.$$

and

$$q = [1 \wedge \mu(X)] = \mu(X) = 1.$$

These results confirm the requirement (stated earlier in this section) that $\mu(F)$ should be the maximum possible score that the object can gain relying only on the quality factors in F.

An evaluation undertaken by a single adjudicator is always influenced by his or her subjectivity. We can imagine, however, that each quality factor x_i of a given object has also its inherent quality index $g(x_i) \in [0, 1], i = 1, 2, \ldots, n$. That is, we assume the existence of an objective evaluation function g: $X \rightarrow [0, 1]$. The most ideal evaluation q_0 for the quality of the object is the Sugeno integral $q_0 = \fint g \, d\mu$ of this function g with respect to the importance measure μ, which we call the *objective synthetic evaluation*. Since the scores produced by each individual adjudicator are not fully consistent and involve some randomness (even if the same adjudicator judges the same quality factor of the same object at two different times, the scores are likely to be different), the score $f(x_i)$ is often not exactly equal to $g(x_i)$ and, consequently, the subjective evaluation q deviates from the objective evaluation q_0.

To reduce the influence of subjective biases of the individual adjudicators and get a more reasonable evaluation, we can use an arithmetic average of scores given by a number of adjudicators. Assume we invite several adjudicators (say, m adjudicators) to judge all quality factors (say, n factors) in X, and they give independently (without any discussion) scores $f_j(x_i)$, $j = 1, 2, \ldots, m$, for each quality factor x_i, $i = 1, 2, \ldots, n$. We can imagine that, for some fixed I, all scores for I_i given independently by an infinite number of adjudicators form a general population G_I with a mathematical expectation $g(x_i)$. Then, $\{f_1(x_i), f_2(x_i), \ldots, f_m(x_i)\}$ may be viewed as a simple random sample of this general population G_i . $f_1(x_i), f_2(x_i\}, \ldots, f_m(x_i)$ are independent random variables with the same distribution (and, therefore, the same mathematical expectation $g(x_i)$). By Kolmogorov's strong law of large numbers (see Halmos [1950]), we have

$$\lim_m \frac{1}{m} \sum_{j=1}^{m} f_j\,(x_i) = g(x_i)$$

with probability 1 for each $i = 1, 2, \ldots, n$. Since the number n of quality factors is finite, we have

$$\lim_m \frac{1}{m} \sum_{j=1}^{m} f_j\,(x_i) = g(x_i) \text{ for all i} = 1, 2, \ldots, n$$

with probability 1. Noting that the importance measure is finite, and using Theorem 9.7, we get

$$\lim_m \fint \frac{1}{m} \sum_{j=1}^{m} f_j \, d\mu = \fint g \, d\mu = q_0$$

with probability 1. This implies

$$\int \frac{1}{m} \sum_{j=1}^{m} f_j \, d\mu \to q_0$$

in probability. This means, in general, that the synthetic evaluation given by

$$q_m = \int \frac{1}{m} \sum_{j=1}^{m} f_j \, d\mu$$

is always very close to the objective synthetic evaluation q_0 provided that m is large enough. The greater the number of attending adjudicators, the closer to q_0 is the evaluation q_m. Let q_m be called an *approximate objective synthetic evaluation* of the given object.

Example 15.13. Consider the same object, quality factors, and importance measure as in Example 15.11. Assume that four experts, labeled as 1, 2, 3, 4, were invited as adjudicators to judge the factors T, S, and A. Their scores are given in Table 15.8.

First, we calculate the average of scores for each factor:

$$\frac{1}{4} \sum_{j=1}^{4} f_j (x_i) = \begin{cases} 0.75 & \text{if } i = 1 \\ 0.80 & \text{if } i = 2 \\ 0.70 & \text{if } i = 3. \end{cases}$$

Then, we have

$$q_m = \int \frac{1}{4} \sum_{j=1}^{4} f_j \, d\mu = 0.75.$$

This approximate objective synthetic evaluation, $q_m = 0.75$, is more reasonable than that obtained in Example 15.12.

Let us discuss now in more detail the motivation for using monotone measures in synthetic evaluations. Let $X = \{x_1, x_2, \ldots x_n\}$ denote again the factor space of an object that we want to evaluate. The method of weighted

Table 15.8 Scores given by four experts in Example 15.13

		i		
		1(T)	2(S)	3(A)
j	1	0.9	0.6	0.8
	2	0.7	0.8	0.8
	3	0.8	0.9	0.6
	4	0.6	0.9	0.6

mean requires that the weights corresponding to the individual factors be given before the evaluation is made. Let $w_1, w_2, \ldots, w_n$, where $0 \le w_i \le 1$, $i = 1, 2, \ldots, n$, and $w_1 + w_2 + \ldots + w_n = 1$, denote the weights. For any set of scores $\{f(x_i) \mid i = 1, 2, \ldots n\}$ given by an adjudicator, the method of weighted mean yields the evaluation

$$q_w = \sum_{i=1}^{n} w_i f(x_i).$$

As illustrated by the following example, this method is not always reasonable.

Example 15.14. We intend to evaluate three TV sets. For the sake of simplicity, we consider only two quality factors: "picture" and "sound." These are denoted by x_1 and x_2, respectively, and the corresponding weights are $w_1 = 0.7$ and $w_2 = 0.3$. Now, an adjudicator gives the following scores for each factor and each TV set:

TV Set No.	x_1 (picture)	x_2 (sound)
1	1	0
2	0	1
3	0.45	0.45

Using the method of weighted mean, we get these synthetic evaluations of the three TV sets:

$$q_{w1} = w_1 \times 1 + w_2 \times 0 = 0.7,$$

$$q_{w2} = w_1 \times 0 + w_2 \times 1 = 0.3,$$

$$q_{w3} = w_1 \times 0.45 + w_2 \times 0.45 = 0.45.$$

According to these results, the first TV set is the best. Such a result is hardly acceptable since it does not agree with our intuition: A TV set without any sound is not practical at all, even though it has an excellent picture. It is significant to realize that the cause of this counterintuitive result is not an improper choice of the weights. For example, if we chose $w_1 = 0.4$ and $w_2 = 0.6$, we would have obtained $q_{w1} = 0.4$, $q_{w2} = 0.6$, and $q_{w3} = 0.45$. Now, the second TV set is identified as the best one, which is also counterintuitive: A TV set with good sound but no picture is not a real TV set, but just a radio. We may conclude that, according to our intuition, the third TV set should be identified as the best one: among the three TV sets, only the third one is really practical, even though neither picture nor sound are perfect. Unfortunately, when using the method of weighted mean, no choice of the weights would lead to this expected result under the given scores.

The crux of this problem is that the method of weighted mean is based on an implicit assumption that the factors $x_1, x_2, \ldots, x_n$ are "independent" of one another. That is, their effects are viewed as additive. This, however, is not justifiable in this example, where the importance of the combination of picture and sound is much higher than the sum of importances associated with picture and sound alone. If we adopt a nonadditive set function (a monotone measure) to characterize the importances of the two factors and, relevantly, use the Sugeno integral as a synthetic evaluator of the quality of the three TV sets, a satisfactory result may be obtained. For instance, given the importance measure $\mu(\{x_1) = 0.3$, $\mu(\{x_2\}) = 0.1$, $\mu(X) = 1$, and $\mu(\emptyset) = 0$, and using the Sugeno integral, we obtain the following synthetic evaluations:

$$q_1 = ⨍ f_1 d\mu = (1 \wedge 0.3) \vee (0 \wedge 1) = 0.3,$$

$$q_2 = ⨍ f_2 d\mu = (1 \wedge 0.1) \vee (0 \wedge 1) = 0.1,$$

$$q_3 = ⨍ f_3 d\mu = 0.45 \wedge 1 = 0.45;$$

here, f_1, f_2, and f_3 characterize the scores given for the three TV sets: $f_1(x_1) = 1$, $f_1(x_2) = 0$, $f_2(x_2) = 0$, $f_2(x_2) = 1$, and $f_3(x_1) = f_3(x_2) = 0.45$. Hence, we get a reasonable conclusion—"the third TV set is the best"—which agrees with our intuition.

When using the same monotone measure but employing the Choquet integral instead of the Sugeno integral, we obtain

$$q_{c1} = (C)\int f_1 d\mu = \int \mu(F_\alpha^{(1)}) dm = 1 \times 0 + 0.3 \times 1 = 0.3,$$

$$q_{c2} = (C)\int f_2 d\mu = \int \mu(F_\alpha^{(2)}) dm = 1 \times 0 + 0.1 \times 1 = 0.1,$$

$$q_{c3} = (C)\int f_3 d\mu = \int \mu(F_\alpha^{(3)}) dm = 1 \times 0.45 + 0 \times 0.55 = 0.45,$$

where

$$F_\alpha^{(1)} = \begin{cases} X & \text{if } \alpha = 0 \\ \{x_1\} & \text{if } 0 < \alpha \le 1, \end{cases}$$

$$F_\alpha^{(2)} = \begin{cases} X & \text{if } \alpha = 0 \\ \{x_2\} & \text{if } 0 < \alpha \le 1, \end{cases}$$

$$F_\alpha^{(3)} = \begin{cases} X & \text{if } 0 \le \alpha \le 0.45 \\ \emptyset & \text{if } 0 .45 < \alpha \le 1. \end{cases}$$

Consequently, the result is again satisfactory: The third TV set is the best.

When using the pan-integral to evaluate the quality of the three TV sets, and choosing the common addition and multiplication as operations in the pan-integral, a similar result is obtained. Hence, the crux in this example is to choose a proper monotone importance measure.

15.7 Multiregression

Given a database involving some attributes, we often want to know how a specified objective attribute depends on other attributes. This is one of the most common problems in data mining, a problem usually referred to as *multiregression*. A traditional model of multiregression is the linear multiregression that has the form

$$y = a_1 x_1 + a_2 x_2 + \ldots + a_n x_n + N(a_0, \sigma^2),$$

where y is the objective attribute (dependent variable), $x_1, x_2, \ldots, x_n$ are n feature attributes (independent variables), and $N(a_0, \sigma^2)$ is a normally distributed random variable with mean a_0 and variance σ^2. This linear model can be viewed as the Lebesgue integral

$$y = \int f d\mu + N(a_0, \sigma^2)$$

where the integrand f represents the observations of feature attributes $x_1, x_2, \ldots, x_n$, and μ is an additive signed measure determined by $\mu(\{x_i\}) = a_i$, $i = 1, 2, \ldots, n$. Using this linear model makes sense under the assumption that there is no interaction among feature attributes towards the objective attribute such that the global contribution from feature attributes towards the objective attribute is just the simple sum of their individual contributions. However, the interaction cannot be ignored in many real-world problems, ones in which the linear model fails, as seen in the last section. Some nonlinear transformations, such as quadratic functions and splines, have been tried to deal with such problems, but their efficiency is limited. A proper idea for improving the classical linear model is to adopt the signed general measure to describe the above-mentioned interaction, as has been done for information fusion. Of course, in this case, the Lebesgue integral fails and then a nonlinear integral with respect to the signed general measure should be used as an appropriate aggregation tool. An initial nonlinear multiregression model may have the following form:

$$y = c + (\mathrm{C})\int f d\mu + N(0, \sigma^2),$$

where c is a constant, (C) $\int f\,d\mu$ is the Choquet integral of function f with respect to signed general measure μ, and $N(0, \sigma^2)$ is a normally distributed random perturbation with mean 0 and variance σ^2 . In this model, constant c as well as the values of μ are regression coefficients. This nonlinear multiregression problem is just a generalization of the method for constructing signed general measures based on data, which is discussed in Section 13.5. Once the data set

x_1	$x_2 \ldots$	x_n	y
f_{11}	$f_{12} \ldots$	f_{1n}	y_1
f_{21}	$f_{22} \ldots$	f_{2n}	y_2
$\vdots$			
f_{l1}	$f_{l2} \ldots$	f_{ln}	y_l

is available, the regression coefficients can be determined by minimizing the squared error

$$e^2 = \sum_{j=1}^{l} [y_j - c - (\mathrm{C})\int f_j\, d\mu]^2,$$

where function f_j is the j-th observation of attributes $x_1, x_2, \ldots, x_n$, i.e., $f_j(x_i) = f_{ji}$, $i = 1, 2, \ldots, n$, for $j = 1, 2, \ldots, l$. The algebraic least square method can be used for solving this minimization problem since the Choquet integral can be expressed as a linear function of unknown regression coefficients. Such a model is nonlinear with respect to the observation f since the Choquet integral is nonlinear. It is a real generalization of classical linear multiregression and can be regarded as the inverse problem of information fusion discussed in Section 15.6.

To deal with the data with some categorical attributes, a numericalization method is developed by using another relevant genetic algorithm. In this method, based on the given data the algorithm searches the optimal value for each state of all categorical attributes. Furthermore, an n-dimensional vector of weights, $w = (w_1, w_2, \ldots, w_n)$, is added on the observations f to balance the various scales of the feature attributes. Thus, the multiregression model becomes

$$y = c + (\mathrm{C})\int wf\, d\mu + N(0,\sigma^2)$$

and we should minimize

$$e^2 = \sum_{j=1}^{l} [y_j - c - (\mathrm{C})\int wf_j\, d\mu]^2$$

to get the optimal estimation for the regression coefficients. In such a model the unknown parameters (including w) are not in a linear form so that using only the least square method is not sufficient to obtain the optimal estimation for the unknown regression coefficients. In this case, we use a hybrid optimization method where the weights are determined through a genetic algorithm and the other regression coefficients are still determined via the least square method. To improve the multiregression model again, finally, an n-dimensional vector $a = (a_1, a_2, \ldots, a_n)$ is introduced and then w is replaced by n-dimensional vector $b = (b_1, b_2, \ldots, b_n)$ in the model. Then it becomes

$$y = c + (\mathrm{C})\int (a + bf)\ d\mu + N(0, \sigma^2),$$

where a represents the phase of an attribute when it interacts with the others. Vectors a and b should satisfy the following constraints:

$$a_i \geq 0 \text{ for } i = 1, 2, \ldots, n, \text{with} \min_{1 \leq i \leq n} a_i = 0;$$

$$-1 \leq b_i \leq 1 \text{ for } i = 1, 2, \ldots, n, \text{with} \max_{1 \leq i \leq n} |b_i| = 1.$$

In the final model, constant c, vectors a and b, and the values of μ are regression coefficients. Once the above-mentioned data are available, these regression coefficients can be determined by minimizing the squared error

$$e^2 = \sum_{j=1}^{l} [y_j - c - (\mathrm{C}) \int (a + bf_j) d\mu]^2.$$

The genetic algorithm is a global search method. Its advantage is that it ignores any local extremum of the objective function in the optimization problem since there is no risk of falling into a local extremum in the search process. However, the genetic algorithm is time-consuming and has a risk of being premature. To speed the running of the algorithm, some strategies, such as diversity and self-adaptivity, can be adopted in the algorithm.

While the Choquet integral has been found useful in dealing with the problem of multiregression, it is not the only one that is applicable in this problem area. For example, the upper and the lower integrals, or even the integrals described by rule r shown in Section 8.4, can be also used to form a nonlinear multiregression model.

15.8 Classification

Recall that in multiregression, as discussed in Section 15.7, some feature attributes are allowed to be categorical. In this section, we consider the case that the objective attribute is categorical. Thus, we deal here with the problem of classification, which, in turn, is essentially a pattern recognition problem. It is one of the common problems in data mining.

Let a data set

x_1	$x_2 \cdots$	x_n	y
f_{11}	$f_{12} \cdots$	f_{1n}	y_1
f_{21}	$f_{22} \cdots$	f_{2n}	y_2
$\vdots$			
f_{l1}	$f_{l2} \cdots$	f_{ln}	y_l

be available, where $x_1, x_2, \ldots, x_n$ are feature attributes and y is the classifying attribute. The range of feature attributes $x_1, x_2, \ldots, x_n$ is called the *feature space*. It is a subset of n-dimensional Euclidean space. Unlike in the multi-regression problem, y is now categorical and has only a finite number of possible states. Usually, attribute y has only a few possible states, denoted by $s_1, s_2, \ldots, s_k$. Set $S = \{s_1, s_2, \ldots, s_k\}$ is called the *state set* of attribute y. Each row $f_{j1}, f_{j2}, \ldots, f_{jn}$ in the data set is the feature of the j-th sample and y_j is the corresponding state that indicates a specified class. The aim of a *classification* problem is to find a classification model that divides the feature space into k disjoint pieces, each of which corresponds to a class based on the given data set, so that we may determine the corresponding class to which any new sample belongs by using the model. The classification model is usually called a *classifier*. Since any classification problem can be decomposed as $k - 1$ classification problems, each of which has only two possible states for classifying attribute y, we only consider 2-class classification problems in this section.

The simplest classification model is linear, that is, the two pieces of the feature space corresponding to two classes is divided by an $(n - 1)$-dimensional hyper-plane that can be expressed by a linear equation of n variable $x_1, x_2, \ldots, x_n$:

$$a_1x_1 + a_2x_2 + \cdots + a_nx_n = c,$$

where a_i, $i = 1, 2, \ldots, n$, and c are unknown parameters that we want to determine based on the given data. This $(n - 1)$-dimensional hyperplane is called the classifying *boundary*. Essentially, a linear classification model is just a linear projection $y = a_1x_1 + a_2x_2 + \cdots + a_nx_n$ from the n-dimensional feature space onto a one-dimensional real line, on which a point c is selected as the critical value for optimally separating the projections of the samples in two classes in the data set. Point c corresponds the classifying *boundary*. In fact, the projection of the classifying boundary onto that one-dimensional real line is just the critical point c. In most linear models the criterion of the optimization is to minimize the misclassification rate when some (all or a part of) samples in the data set are used as the training set. Sometimes, the optimization criterion can also be formed by a certain function of the distance from sample points to the boundary in the feature space. The values of the parameters a_i, $i = 1, 2, \ldots, n$, and c of the optimal classifying boundary can be calculated via an algebraic method precisely or be found via a numerical method approximately.

Similar to the multiregression problems, to use the above-mentioned linear classification model we need a basic assumption that the interaction among the feature attributes towards the classification can be ignored. However, in many real-world classification problems, the samples in the data are not linearly separable, that is, the optimal classifying boundary is not approximately linear, since the above-mentioned interaction cannot be ignored. In this case, similar to the multiregression, we should adopt a nonlinear integral, such as the Choquet integral with respect to a signed general measure μ, to express the classifying boundary, that is, the classifying boundary is identified by equation

$$(\mathrm{C})\int (a + bf)\, d\mu = c$$

where a, b, and f have the same meaning as in Section 15.7. The nonadditivity of signed general measure μ describes the interaction among the contribution rate from feature attributes towards the classification. Thus, when the Choquet integral is used, the classifying boundary is not an $(n-1)$-dimensional hyperplane generally, but an $(n-1)$-dimensional broken hyperplane. The parameters, a, b, and c, as well as signed general measure μ, can be optimally determined by the training samples in the given data set via a soft computing technique such as the genetic algorithm, approximately. Such a nonlinear classification model is a real generalization of the classical linear classification model.

After determining the parameters, a, b, c, and μ based on the training data, if a new individual f is obtained, we only need to calculate the value of $y(f) = (\mathrm{C})\int (a + bf) d\mu$. Then, we can classify f into one of the two classes according to whether $y(f) \geq c$.

15.9 Other Applications: An Overview

In this section, we briefly introduce some additional areas within which applications of generalized measure theory have been described in the literature. To cover these applications in detail would require that relevant knowledge regarding each application domain be introduced. However, such extensive coverage is beyond the scope of this book. Therefore, the use of generalized measure theory in each area is described in this section only in general, easy to understand terms. For readers interested in full coverage, each area is associated with a particular Note in which relevant literature is surveyed.

Let us begin with the area of decision-making, which is perhaps one of the most visible application areas of generalized measure theory. Although numerous applications of generalized measure theory has been recognized within this very large area, it seems useful to classify them into the following two broad categories:

1. Applications of the various types of imprecise probabilities to capture uncertainty regarding the states of the world. It was traditionally assumed that this uncertainty could be captured by a single probability measure, but it has increasingly been recognized this assumption is not realistic in the context of many decision-making problems, especially those emerging in recent years.
2. Applications of nonadditive measures to capture the importance of various subsets of given criteria in multicriteria decision-making, the importance of various coalitions in multiperson decision-making or, more generally, the importance of coalitions within the framework of cooperative game theory. The rather extensive literature in this area is surveyed in Note 15.7.

To our knowledge, there are several other areas in which nonadditive measures and nonlinear integrals have been found useful. They include economics, image processing and computer vision, and pattern recognition. Some references pertaining to these areas are given in Note 15.8.

Notes

15.1. The significance of classical measure theory to formalizing the concept of probability, which was established by Kolmogorov [1950], is well known (see also Note 1.6). There are many books that cover the connection of classical measure theory and probability theory. A few excellent representatives are the books by Billingsley [1986], Halmos [1950], Kingman and Taylor [1966], Parthasarathy [2005], and Pollard [2002]. Good overviews of the use of classical measure theory in classical as well as fractal geometries are the book by Morgan [1988] and Chapters 24 and 25 in Pap [2002a]. Literature on the role of classical measures in ergodic theory (a theory dealing with problems regarding long-term behaviors of dynamic systems that are formalized in terms of measure-preserving transformations on measure spaces of systems states) is fairly extensive. A few representative publications include Chapter 29 in Pap [2002a] and books by Aaronson [1997], Billingsley [1965], Halmos [1956], Petersen [1983], and Walters [1982]. The role of classical measure theory in economics and other social sciences is thoroughly discussed in the book by Faden [1977]. Among other application areas of classical measure theory are harmonic analysis [Pitt, 1985], potential theory [Helms, 1963; Du Plessis, 1970; Dellacherie and Meyer, 1978], and calculus of variations (Chapter 24 in [Pap, 2002a]).

15.2. Generalized information theory (GIT) emerged in the late 1980)s and was formally proposed as a long-term research program by Klir [1991]. The basic tenet of GIT, that the concept of uncertainty is broader than the concept of probability, has been debated in the literature since the late 1980s. An overview of the various published debates can be found in Klir [2001]. As a result of these debates, as well as convincing advances in GIT, limitations of classical probability theory to deal with uncertainty and uncertainty-based information have increasingly been recognized. A comprehensive coverage of results obtained by research within GIT prior to 2006 is the subject of a recent book [Klir, 2006].

15.3. The first thorough investigation of imprecise probabilities was carried out by Dempster [1967a,b]. Although his papers stimulated interest in imprecise probabilities, as manifested by the literature at that time, most early publications (in the 1970s and 1980s) in this area were oriented to special types of imprecise probabilities. Some notable exceptions were papers by Walley and Fine [1979] and Kyburg [1987]. Since the early 1990s, coinciding with the emergence of GIT, a greater emphasis can be observed in the literature on studying imprecise probabilities within broader frameworks. It is likely that this trend was influenced by the publication of an important book by Walley [1991]. Employing simple, but very fundamental, principles of *avoiding sure loss*, *coherence*, and *natural extension*, Walley presented in this book a highly general theory of imprecise probabilities and discussed successfully its importance from philosophical, mathematical, and practical points of view. Short versions of the material covered in this rather large book (706 pages) and some additional ideas are presented in [Walley, 1996, 2000]. Viewing imprecise probabilities from the standpoint of general lower and upper probability functions and the associated convex sets of probability measures, as is only briefly outlined in Section 15.3, shows the crucial role of generalized measure theory in formalizing imprecise probabilities. Further details, which are beyond the scope of this book, can be found in [Klir, 2006] as well as in the principal papers that contributed to the development of this broad view, including papers by Chateauneuf and Jaffray [1989], De Campos and Bolanos [1989], De Campos and Huete [1993], De Campos et al. [1990a,b], Grabisch [1997 a–c], Lamata and Moral [1989], and Miranda et al. [2003].

15.4. The classification of dual pairs of measures, which is summarized in Fig. 15.1, was inspired by [Lamata and Moral, 1989]. We do not cover the class of dual pairs of decomposable measures since this class is not contained in the class of dual pairs of ordered monotone measures and, therefore, it does not qualify, as a whole, for a theory of imprecise probabilities. While some of its subclasses do qualify, this area is still not sufficiently developed and it is beyond the scope of this book.

15.5. Among the various theories of imprecise probabilities, the most visible ones in terms of applications have been the Dempster–Shafer theory (DST), possibility theory, and the theory based on reachable probability intervals (RPI). Some representative publications that describe applications of these theories are: [Kong, 1986; Strat, 1990; Inagaki, 1991;

Caselton and Luo, 1992; Yager et al., 1994; Schubert, 1994; Kohlas and Monney, 1995; Resconi et al., 1998; Bell et al. 1998; Tanaka and Klir, 1999; Kriegler and Held, 2005; Helton et al., 2006] for DST; [Yager, 1982; Prade and Testemale, 1987; Tanaka and Hayashi, 1989; Wang and Li, 1990; De Cooman et al., 1995; Cai, 1996; Klir, 2002; Wolkenhauer, 1998; Delmotte, 2007] for possibility theory; and [Weichselberger and Pöhlman, 1990; Weichselberger, 2000; Pan and Klir, 1997] for RPI.

15.6. Applications of monotone measures and nonlinear integrals in data mining involving the problems of information fusion, multiregression, and classification are overviewed in [Wang et al., 2005]. Some additional references are: [Ishi and Sugeno, 1985], [Keller and Osborn, 1996], [Leung and Wang, 1998], [Wang et al., 1999a,b], [Liginlal et al., 2006], and [Näther and Wälder, 2007] for information fusion; [Xu et al., 2000, 2001b] and [Wang, 2002, 2003] for multiregression; and [Grabisch and Nicolas, 1994] and [Xu et al., 2001a, 2003] for classification.

15.7. Useful overview articles of applications of monotone measures in the area of decision-making were written by Grabisch [1995b, 1997b]. These applications are also covered well in [Grabisch et al., 1995]. A broad framework for decision-making is cooperative game theory, in which generalized measure theory plays a crucial role; the following are some representative references: [Aubin, 1981], [Shapley, 1953, 1971], [Aumann and Shapley, 1974], [Delbaen, 1974], [Owen, 1988], [Butnariu, 1985], [Butnariu and Klement, 1993], and [Branzei, et al., 2005]. For decision-making based on imprecise probabilities, see [Wolfenson and Fine, 1982], [Walley, 1991], [Yager and Kreinovich, 1999], and [Troffaes, 2007]. Further information can be found in [Grabisch et al., 2000, Part 2].

15.8. The following are references to applications of *generalized* measure theory in some other areas: [Billot, 1992] in economics; [Keller et al., 1986], [Tahani and Keller, 1990], [Keller et al., 2000], and [Hocaoglu and Gader, 2003] in image processing and computer vision; [Keller et al., 1994], [Grabisch, 1995a], [Gader et al., 1996], and [Stanley et al., 2001], in pattern recognition; and [Tanaka and Sugeno, 1991] in subjective evaluation of printed color images.

Exercises

15.1. Show that all pairs of lower and upper probabilities in Table 15.9 are based on reachable interval-valued probability distributions. Show also that ${}^{1}\mu_*$ and ${}^{2}\mu_*$ are 2-monotone.

15.2. For the lower probability functions defined in Fig. 15.1, ${}^{1}\mu_*$ and ${}^{2}\mu_*$, determine the dual upper probability functions, ${}^{1}\mu^*$ and ${}^{2}\mu^*$, and repeat Example 15.2 for these dual measures.

Table 15.9 Lower and upper probabilities in Exercise 15.1

A	${}^1\mu_*(A)$	${}^1\mu^*(A)$	${}^1m(A)$	${}^2\mu_*(A)$	${}^2\mu^*(A)$	${}^2m(A)$	${}^3\mu_*(A)$	${}^3\mu^*(A)$	${}^3m(A)$
Ø	0.0	0.0	0.0	0.00	0.00	0.00	0.00	0.00	0.00
$\{a\}$	0.0	0.5	0.0	0.00	0.50	0.00	0.00	0.50	0.00
$\{b\}$	0.0	0.5	0.0	0.00	0.50	0.00	0.00	0.50	0.00
$\{c\}$	0.0	0.5	0.0	0.25	0.50	0.25	0.25	0.75	0.25
$\{a,b\}$	0.5	1.0	0.5	0.50	0.75	0.50	0.25	0.75	0.25
$\{a,c\}$	0.5	1.0	0.5	0.50	1.00	0.25	0.50	1.00	0.25
$\{b,c\}$	0.5	1.0	0.5	0.50	1.00	0.25	0.50	1.00	0.25
X	1.0	1.0	−0.5	1.00	1.00	−0.25	1.00	1.00	0.00

15.3. Let $X = \{a, b, c, d\}$. Consider a convex set of probability distributions on X,

$$C = \{\mathbf{p} = (p(a), p(b), p(c), p(c))\},$$

that is represented by the convex hull of the following four extreme points:

$$\mathbf{p}_1 = (0.25, 0.25, 0.25, 0.25),$$
$$\mathbf{p}_2 = (0, 0, 0.5, 0.5),$$
$$\mathbf{p}_3 = (0, 0.5, 0, 0.5),$$
$$\mathbf{p}_4 = (0, 0, 0, 1).$$

Determine the lower and upper probability functions and the Möbius function associated with C.

15.4. Convert the two lower probability functions defined in Fig. 15.1 to their Möbius representations and determine whether the given functions are Choquet capacities of order 2 or some higher order.

15.5. Determine for the two lower probability functions defined in Table 15.3 whether they are Choquet capacities of order 2.

15.6. Determine values of the belief and plausibility measures, which are defined only for some subsets of the universal set in Table 15.5, for all the remaining subsets.

15.7. Is the lower probability function defined in Table 15.4a Choquet capacity of order 2? Is it a superadditive measure?

15.8. Let $X = \{a, b, c\}$. Consider the following probability intervals defined on X: $p(a) \in [0.3, 0.4]$, $p(b) \in [0.3, 0.5]$, and $p(c) \in [0.3, 0.5]$. If these intervals are not reachable, convert them to the corresponding reachable ones and calculate values of the associated lower and upper probability functions for all subsets of X.

15.9. What are some advantages and disadvantages of the uncertainty theory based on reachable probability intervals when compared with Dempster–Shafer theory?

15.10. Show that the Dempster rule of combination is associative so that the combined evidence does not depend on the order in which the sources are used.

15.11. Show that the alternative rule of combination in Dempster–Shafer theory is not associative so that the combined result depends on the order in which the sources are used. Can the alternative rule be generalized to more than two sources to be independent of the order in which the sources are used?

15.12. Repeat Example 15.9 by using the alternative rule of combination.

15.13. Repeat Example 15.14 for another TV set whose scores for x_1 and x_2 are 0.5 and 0.6, respectively.

Appendix A

Glossary of Key Concepts

1-alternating measure. The same concept as monotone measure, but which is subadditive.

1-monotone measure. The same concept as monotone measure, but which is superadditive.

λ-rule. A set function $\mu : \mathbf{C} \to [0, \infty]$ satisfies the λ-rule iff there exists

$$\lambda \in \left(-\frac{1}{\sup \mu}, \infty\right) \cup \{0\},$$

where $\sup \mu = \sup_{E \in \mathbf{C}} \mu(E)$, such that $\mu(E \cup F) = \mu(E) + \mu(F) + \lambda \cdot \mu(E) \cdot \mu(F)$ whenever $E, F, E \cup F \in \mathbf{C}$, and $E \cap F = \emptyset$.

σ-algebra. A σ-ring that contains X.

σ-ring. A nonempty class $\mathbf{F}$ such that $E - F \in \mathbf{F}$ for all $E, F \in \mathbf{F}$ and $\bigcup_{i=1}^{\infty} E_i \in \mathbf{F}$ for all $E_i \in \mathbf{F}$, $i = 1, 2, \ldots$.

Absolute Continuity. Given a pair of continuous monotone measures on $\mathbf{C}$, μ and ν, μ is said to be absolutely continuous with respect to ν iff for any $\varepsilon > 0$ there exists $\delta > 0$ such that $\mu(F) - \mu(E) < \varepsilon$ whenever $E \in \mathbf{C}$, $F \in \mathbf{C}$, $E \subset F$, and $\nu(F) - \nu(E) < \delta$.

Additivity. A set function $\mu : \mathbf{C} \to [0, \infty]$, where $\mathbf{C}$ is a nonempty class of X, is additive iff $\mu(E \cup F) = \mu(E) + \mu(F)$ whenever $E, F, E \cup F \in \mathbf{C}$ and $E \cap F = \emptyset$.

AI-class. A nonempty class $\mathbf{C}$ such that $\cap \mathbf{C}' \in \mathbf{C}$ implies $\cap \mathbf{C}' \in \mathbf{C}'$ for all $\mathbf{C}' \subset \mathbf{C}$.

Algebra. A nonempty class $\mathbf{R}$ such that $E \cup F \in \mathbf{R}$ and $\bar{E} \in \mathbf{R}$ for all $E, F \in \mathbf{R}$.

Alternating Choquet capacities of order k. For each particular integer $k \geq 2$, a monotone measure μ on a measurable space $(X, \mathbf{F})$ that satisfies the inequalities

$$\mu(\bigcap_{j=1}^{k} A_j) \leq \sum_{\substack{K \subset N_k \\ K \neq \emptyset}} (-1)^{|K|+1} \mu(\bigcup_{j \in K} A_j)$$

for all families of k sets in **F**.

Antisymmetric relation on *E*. A relation $R \subset E \times E$ such that aRb and bRa imply $a = b$ for each pair $a, b \in E$.

Atom. For any point $x \in X$, the atom of **C** at x, $A(x/\mathbf{C})$, is the set $\bigcap\{E|x \in E \in \mathbf{C}\}$.

AU-class. A nonempty class **C** such that $\bigcup \mathbf{C}' \in \mathbf{C}$ implies $\bigcup \mathbf{C}' \in \mathbf{C}'$ for all $\mathbf{C}' \subset \mathbf{C}$.

Autocontinuity. A set function μ: $\mathbf{F} \to [-\infty, \infty]$ is autocontinuous iff it is autocontinuous from both above and below.

Autocontinuity from above (or from below). A set function $\mu : \mathbf{F} \to [-\infty, \infty]$, where **F** denotes a σ-algebra of sets in $\mathbf{P}(X)$, is autocontinuous from above (or from below) iff $\lim_n \mu(E \cup F_n) = \mu(E)$ [or $\lim_n \mu(E - F_n) = \mu(E)$] whenever $E, F_n \in \mathbf{F}$, $\lim_n \mu(F_n) = 0$, and $E \cap F_n = \emptyset$ (or $F_n \subset E$, respectively), $n = 1, 2, \ldots$.

Basic probability assignment. A set function m: $\mathbf{P}(X) \to [0, 1]$ such that $m(\emptyset) = 0$ and $\Sigma_{E \in \mathbf{P}(X)} m(E) = 1$.

Belief measure. A set function Bel: $\mathbf{P}(X) \to [0, 1]$ such that $\text{Bel}(E) = \Sigma_{F \subset E} m(F)$ for each $E \in \mathbf{P}(X)$, where m is a basic probability assignment.

Borel field. The σ-algebra generated by the class of all bounded, left closed, and right open intervals of the real line.

Choquet capacity of order *k*. For each particular integer $k \geq 2$, a monotone measure μ on a measurable space $(X, \mathbf{F})$ that satisfies the inequalities

$$\mu(\bigcup_{j=1}^{k} A_j) \geq \sum_{\substack{K \subset N_k \\ K \neq \emptyset}} (-1)^{|K|+1} \mu(\bigcap_{j \in K} A_j)$$

for all families of k sets in **F**.

Choquet integral. Given a measurable space $(X, \mathbf{F})$, a monotone set function μ: $\mathbf{F} \to [0, \infty]$, a nonnegative finite measurable function f on $(X, \mathbf{F})$ and $A \in \mathbf{F}$, the Choquet integral of f with respect to μ on A is defined by

$$(C)\int_A f d\mu = \int_0^{\infty} \mu(A \cap F_\alpha) d\alpha,$$

where the integral on the right side is the Riemann integral and

$$F_\alpha = \{x|f(x) \geq \alpha\}, \alpha \epsilon [0, \infty).$$

Commutative isotonic semigroup. Given a binary operation $\oplus$ on $\bar{R}_+ = [0, \infty]$, the pair $(\bar{R}_+, \oplus)$ is called a commutative isotonic semigroup iff $\oplus$ is commutative, associative, and such that $a \leq b$ implies $a \oplus c \leq b \oplus c$ for every c, $a \oplus 0 = a$, the existence of $\lim_n a_n$ and $\lim_n b_n$ implies the existence of $\lim_n (a_n \oplus b_n)$, and

$$\lim_n (a_n \oplus b_n) = \lim_n a_n \oplus \lim_n b_n.$$

Commutative isotonic semiring. Given two binary operations, $\oplus$ and $\otimes$, on $\bar{R}_+ = [0, \infty]$, a triple $(\bar{R}_+, \oplus, \otimes)$ is called a commutative isotonic semiring iff $\oplus$ has the same meaning as in the commutative isotonic semigroup and $\otimes$ is commutative, associative, and distributive with respect to $\oplus$, and such that $a \leq b$ implies $a \otimes c \leq b \otimes c$ for every c, $a \neq 0$ and $b \neq 0$ iff $a \otimes b \neq 0$, there exists $I \in \bar{R}_+$ such that $I \otimes a = a$ for every $a \in \bar{R}_+$, and the existence of a finite $\lim_n a_n$ and a finite $\lim_n b_n$ implies the equality

$$\lim_n (a_n \otimes b_n) = \lim_n a_n \otimes \lim_n b_n.$$

Continuous monotone measure. A set function $\mu : \mathbf{C} \to [0, \infty]$ on space $(X, \mathbf{C})$ that satisfies the requirements of both lower and upper-semicontinuous monotone measures (vanishing at Ø, monotonicity, and semicontinuity from both below and above).

Decreasing sequence $\{E_n\}$. A set sequence $\{E_n\}$ for which $E_n \supset E_{n+1}$ for all $n = 1, 2, \ldots$.

Dual monotone measure. Given a normalized monotone measure μ on $(X, \mathbf{C})$, the dual monotone measure, ν, is defined for all $A \in \mathbf{C}$ by the equation

$$\nu(A) = 1 - \mu(\overline{A}).$$

Equivalence class. Given an equivalence relation $R \subset E \times E$ and some $x \in E$, the set $\{y|xRy\}$.

Equivalence relation on E. A relation $R \subset E \times E$ that is reflexive, symmetric, and transitive.

Fuzzy power set of X. The set of all fuzzy subsets of X.

General measure. A set function $\mu : \mathbf{C} \to [0, \infty]$ on $(X, \mathbf{C})$ for which $\mu(\text{Ø}) = 0$ when Ø $\in \mathbf{C}$.

Generalized possibility measure. Maxitive measure on $(X, \mathbf{C})$ such that there exists $E \in \mathbf{C}$ for which $\mu(E) < \infty$.

Hole. For any point $x \in X$, the hole of $\mathbf{C}$ at x, $H(x/\mathbf{C})$, is the set

$$\bigcup\{E|x \in \bar{E} \in \hat{\mathbf{C}}\},$$

where $\hat{\mathbf{C}} = \{\bar{E}|E \in \mathbf{C}\}$.

Increasing sequence $\{E_n\}$. A set sequence $\{E_n\}$ for which $E_n \subset E_{n+1}$ for all $n = 1, 2....$

Inferior limit of $\{E_n\}$. The set of all points of X that belong to E_n for all but a finite number of values of n.

Inverse relation. Given a relation $R \subset E \times F$, its inverse relation, R^{-1}, is the set $R^{-1} = \{(b,a)|(a,b) \in R\}$.

***k*-alternating measure ($k \geq 2$).** The same concept as alternating Choquet capacity of order k.

***k*-monotone measure ($k \geq 2$).** The same concept as Choquet capacity of order k.

Limit of $\{E_n\}$. The set representing both superior and inferior limits of $\{E_n\}$, provided that they are equal.

Lower-semicontinuous monotone measure. A function $\mu : \mathbf{C} \to [0, \infty]$ on space $(X, \mathbf{C})$ that satisfies the following three requirements:

1. $\mu(\emptyset) = 0$ when $\emptyset \in \mathbf{C}$ (vanishing at $\emptyset$);
2. For any $E, F \in \mathbf{C}$, $E \subset F$ implies $\mu(E) \leq \mu(F)$ (monotonicity);
3. For every increasing sequence $\{E_n\}$, $\bigcup_{n=1}^{\infty} E_n \in \mathbf{C}$ implies $\lim_n \mu(E_n) = \mu(\bigcup_{n=1}^{\infty} E_n))$ (continuity from below).

Maxitive measure. A set function $\mu : \mathbf{C} \to [0, \infty]$ on space $(X, \mathbf{C})$ such that $\mu(\bigcup_{t \in T} E_t) = \sup_{t \in T} \mu(E_t)$ for any subclass $\{E_t|t \in T\}$ of $\mathbf{C}$ whose union is in $\mathbf{C}$, where T is an arbitrary index set.

Measurable function. Given a measurable space $(X, \mathbf{F})$, a real-valued function f: $X \to (-\infty, \infty)$ such that $f^{-1}(B) = \{x|f(x) \in B \in \mathbf{F}\}$ for any Borel set $B \in \mathbf{B}$.

Measurable partition. Given a measurable space $(X, \mathbf{F})$, a partition $\{E_i\}$ of X such that $E_i \in \mathbf{F}$ for every i.

Measurable space. The pair $(X, \mathbf{F})$, where $\mathbf{F}$ is a σ-ring (or σ-algebra) on X. (Sets in $\mathbf{F}$ are called measurable sets.)

Monotone class. A nonempty class $\mathbf{M}$ for which $\lim_n E_n \in \mathbf{M}$ for every monotone sequence $\{E_n\} \subset \mathbf{M}$.

Monotone measure. A set function $\mu : \mathbf{C} \to [0, \infty]$ on $(X, \mathbf{C})$ that satisfies the following requirements:

1. $\mu(\emptyset) = 0$ when $\emptyset \in$ C;
2. $E \in$ C,$F \in$ C, and $E \subset F$ imply $\mu(E) \leq \mu(F)$.

Monotone measure space. A triple $(X, \mathbf{F}, \mu)$, where μ is a monotone measure on a measurable space $(X, \mathbf{F})$.

Monotone sequence $\{E_n\}$. A sequence that is either increasing or decreasing.

Necessity measure. A set function $\nu : \mathbf{P}(X) \to [0, 1]$ such that $\nu(E) = 1 - \pi(\bar{E})$ for each $E \in \mathbf{P}(X)$, where π denotes a possibility measure.

Normalized monotone measure. A monotone measure on $(X, \mathbf{C})$ for which $X \in \mathbf{C}$ and $\mu(X) = 1$.

Null-additivity. A set function $\mu : \mathbf{F} \to [-\infty, \infty]$, where $\mathbf{F}$ denotes a σ-algebra of sets in $\mathbf{P}(X)$, is null-additive iff $\mu(E \cup F) = \mu(E)$ whenever $E, F \in \mathbf{F}$, $E \cap F = \varnothing$, and $\mu(F) = 0$.

Pan-addition. A binary operation $\oplus$ on $\bar{R}_+ = [0, \infty]$ employed in a commutative isotonic semigroup.

Pan-characteristic function. Given a pan-space $(X, \mathbf{F}, \mu, \bar{R}_+, \oplus, \otimes)$ and $E \subset X$, a function χ_E: $X \to \{0, I\}$ such that

$$\chi_E(x) = \begin{cases} I & \text{when } x \in E \\ 0 & \text{otherwise,} \end{cases}$$

where I is the unit element of the commutative isotonic semiring $(\bar{R}_+, \oplus, \otimes)$.

Pan-integral. Given a measurable space $(X, \mathbf{F})$, a finite nonnegative measurable function f defined on $(X, \mathbf{F})$, and a set $A \in \mathbf{F}$, the pan-integral of f on A with respect to a monotone measure μ is defined by

$$\sup_{0 \le s \le f,\, s \in \mathbf{Q}} \bigoplus_{i=1}^{n} [a_i \otimes \mu(A \cap E_i)],$$

where $\mathbf{Q}$ denotes the set of all pan-simple measurable functions and $s(x) = \oplus_{i=1}^{n} [a_i \otimes \chi_{E_i}(x)] \in \mathbf{Q}$.

Pan-multiplication. A binary operation $\otimes$ on $\bar{R}_+ = [0, \infty]$ employed in a commutative isotonic semiring.

Pan-simple measurable function. Given a pan-space $(X, \mathbf{F}, \mu, \bar{R}_+, \oplus, \otimes)$, a function s on X defined by

$$s(x) = \bigoplus_{i=1}^{n} [a_i \otimes \chi_{E_i}(x)],$$

where $a_i \in \bar{R}_+$, $i = 1, 2, \ldots, n$, and $\{E_i | i = 1, 2, \ldots, n\}$ is a measurable partition of X.

Pan-space. The sixtuple $(X, \mathbf{F}, \mu, \bar{R}_+, \oplus, \otimes)$, where $(X, \mathbf{F}, \mu)$ is a monotone measure space and $(\bar{R}_+, \oplus, \otimes)$ is a commutative isotonic semiring.

Partial ordering on *E*. A relation $R \subset E \times E$ that is reflexive, antisymmetric, and transitive.

Partition of *X*. A disjoint class $\{E_1, E_2, \ldots, E_n\}$ of non-empty subsets of X such that $\bigcup_{i=1}^{n} E_i = X$.

Plausibility measure. A set function $\mathrm{Pl} : \mathbf{P}(X) \to [0, 1]$ such that $Pl(E) = \sum_{F \cap E \neq \emptyset} m(F)$ for each $E \in \mathbf{P}(X)$, where m is a basic probability assignment.

Plump class. A nonempty class $\mathbf{F}_p$ such that $\bigcup_t E_t \in \mathbf{F}_p$ and $\bigcap_t E_t \in \mathbf{F}_p$ for all $\{E_t | t \in T\} \subset \mathbf{F}_p$, where T is an arbitrary index set.

Poset (partially ordered set). The pair (E, R), where E is a set and R is a partial ordering on E.

Possibility measure. A generalized possibility measure defined on $\mathbf{P}(X)$ that is normalized.

Quasi-measure. A set function $\mu : \mathbf{C} \to [0, \infty]$ for which there exists a T-function θ such that $\theta \circ \mu$ is a classical (additive) measure on $\mathbf{C}$.

Quasi-probability. A quasi-measure that is normalized.

Quotient of *E* by *R*. The class of all equivalence classes of E induced by an equivalence relation R.

Power set. The class of all subsets of X.

Reflexive relation on *E*. A relation $R \subset E \times E$ such that aRa for each $a \in E$.

Relation from *E* to *F*. A subset of $E \times F$.

Relation on *E*. A subset of $E \times E$.

Ring. A nonempty class $\mathbf{R}$ such that $E \cup F \in \mathbf{R}$ and $E - F \in \mathbf{R}$ for all $E, F \in \mathbf{R}$.

***S*-compact space.** A measurable space $(X, \mathbf{C})$ such that for any sequence of sets in $\mathbf{C}$ there exists some subsequence that has a limit and this limit belongs to $\mathbf{C}$.

Semiring. A nonempty class $\mathbf{S}$ that satisfies the following two requirements:

1. For all $E, F \in \mathbf{S}$, $E \cap F \in \mathbf{S}$;
2. For all $E, F \in \mathbf{S}$ such that $E \subset F$, there exists a finite class $\{C_0, C_1, \ldots, C_n\}$ of sets in $\mathbf{S}$ such that $E = C_0 \subset C_1 \subset \ldots \subset C_n = F$ and $C_i - C_{i-1} \in \mathbf{S}$ for all $i = 1, 2, \ldots, n$.

Signed additive measure. An extended real-valued and countably additive set function μ on a measurable space $(X, \mathbf{C})$ that assumes at most one of the values $+\infty$ and $-\infty$, and for which $\mu(\emptyset) = 0$.

Signed general measure. An extended real-valued set function μ on a measurable space $(X, \mathbf{C})$ that assumes at most one of the values $+\infty$ and $-\infty$, and for which $\mu(\emptyset) = 0$.

Simple function. A function $s : X \to (-\infty, \infty)$ expressed in the form

$$\sum_{i=1}^{m} a_i \chi_{A_i},$$

where each a_i is a real constant, $A_i \in \mathbf{F}$, and χ_{A_i} is the characteristic function of A_i $(i = 1, 2, \ldots, m)$.

Space. The pair $(X, \mathbf{C})$, where $\mathbf{C}$ is a nonempty class of subsets of X.

***S*-precompact space**. A space $(X, \mathbf{C})$ such that for any sequence of sets in $\mathbf{C}$ there exists some subsequence that has a limit.

Subadditive measure. A set function $\mu : \mathbf{C} \to [0, \infty]$ on space $(X, \mathbf{C})$ such that $\mu(E) \le \mu(E_1) + \mu(E_2)$ whenever $E, E_1, E_2 \in \mathbf{C}$ and $E = E_1 \cup E_2$.

Sugeno integral. Given a monotone measure space $(X, \mathbf{F}, \mu)$ with $X \in \mathbf{F}$, a finite nonnegative measurable function f defined on $(X, \mathbf{F})$, and a set $A \in \mathbf{F}$, the Sugeno integral of f on A with respect to μ is defined by $\sup_{\alpha \in [0, \infty]} [\alpha \wedge \mu(A \cap F_\alpha)]$, where $F_\alpha = \{x | f(x) \ge \alpha\}$ and $\wedge$ denotes the minimum operator.

Sugeno measure. A normalized λ-measure defined on a σ-algebra.

Superadditive measure. A set function μ: $\mathbf{C} \to [0, \infty]$ on space $(X, \mathbf{C})$ such that $\mu(E) \ge \mu(E_1) + \mu(E_2)$ whenever $E, E_1, E_2 \in \mathbf{C}$, $E_1 \cap E_2 = \emptyset$, and $E = E_1 \cup E_2$.

Superior limit of $\{E_n\}$. The set of all points of X that belong to E_n for infinitely many values of n.

Symmetric relation on *E*. A relation $R \subset E \times E$ such that aRb implies bRa for each pair $a, b \in E$.

***T*-function**. A real function $\theta : [0, a] \to [0, \infty]$ that is continuous, strictly increasing, and such that $\theta(0) = 0$, $\theta^{-1}(\{\infty\}) = \emptyset$ when a is finite and $\theta^{-1}(\{\infty\}) = \{\infty\}$ otherwise.

Transitive relation on *E*. A relation $R \subset E \times E$ such that aRb and bRc imply aRc for any $a, b, c \in E$.

Uniform autocontinuity. A set function $\mu : \mathbf{F} \to [-\infty, \infty]$, where $\mathbf{F}$ is a σ-algebra of sets in $\mathbf{P}(X)$, is uniformly autocontinuous iff it is uniformly autocontinuous from both above and below.

Uniform autocontinuity from above (or **from below**). A function $\mu : \mathbf{F} \to [-\infty, \infty]$, where $\mathbf{F}$ denotes a σ-algebra of sets in $\mathbf{P}(X)$, is uniformly autocontinuous from above (or from below) iff for any $\varepsilon > 0$ there exists $\delta = \delta(\varepsilon) > 0$ such that $\mu(E) - \varepsilon \le \mu(E \cup F) \le \mu(E) + \varepsilon$ [or $\mu(E) - \varepsilon \le \mu(E - F) \le \mu(E) + \varepsilon$] whenever $E, F \in \mathbf{F}$, $|\mu(F)| \le \delta$ and $E \cap F = \emptyset$ (or $F \subset E$, respectively).

Upper-semicontinuous monotone measure. A set function μ: $\mathbf{C} \rightarrow [0, \infty]$ on space $(X, \mathbf{C})$ that satisfies the following three requirements:

1. $\mu(\emptyset) = 0$ when $\emptyset \in \mathbf{C}$ (vanishing at $\emptyset$);
2. For any $E, F \in \mathbf{C}$, $E \subset F$ implies $\mu(E) \leq \mu(F)$ (monotonicity);
3. For every decreasing sequence $\{E_n\} \subset \mathbf{C}$ such that $\mu(E_1) < \infty$, $\bigcap_{n=1}^{\infty} E_n \in \mathbf{C}$ implies $\lim_n \mu(E_n) = \mu(\bigcap_{n=1}^{\infty} E_n)$ (continuity from above).

Appendix B

Glossary of Symbols

ε, δ	Positive real numbers
Δ_E	Identity relation on set E
θ	T-function
θ^{-1}	The inverse function of θ
μ	Set function
v	Necessity measure (or general measure that is distinct from measure μ)
π	Possibility measure
$\pi(x)$	Proposition concerning x
χ_E	Characteristic function of E
Π	Product
Σ	Summation
$\emptyset$	Empty set
$\Rightarrow$	Implication
$\Leftarrow$	Inverse implication
$\Leftrightarrow$	Logical equivalence
$\leq$	Partial ordering
$\ll$	To be absolutely continuous to
$\equiv$	Identically equal
$\rightarrow$	Mapping into
$\mapsto$	Mapping to
$\oplus$	Pan-addition
$\otimes$	Pan-multiplication
$\vee$	Logical "or" or maximum operator
$\wedge$	Logical "and" or minimum operator
$\forall$	Universal quantifier "for all"
$\exists$	Existential quantifier "there exists at least one"
$⨍ f\,d\mu$	Sugeno integral of f on X with respect to μ
$⨍_A f\,d\mu$	Sugeno integral of f on A with respect to μ
$(p)\int_A f\,d\mu$	Pan-integral of f on A with respect to μ
$(C)\int_A f\,d\mu$	Choquet integral of f on A with respect to μ
$(L)\int_A f\,d\mu$	Lower integral of f on A with respect to μ

(continued)

(U) $\int_A f d\mu$	Upper integral of f on A with respect to μ
$(\underline{W}) \int_A f d\mu$	Widened-lower integral of f on A with respect to μ
$(\overline{W}) \int_A f d\mu$	Widened-upper integral of f on A with respect to μ
$\bigcup \mathbf{C}$	Union of sets in class **C**
$\bigcap \mathbf{C}$	Intersection of sets in class **C**
$[a, b]$	Closed interval of real numbers from a to b
$[a, b)$	Interval of real numbers closed at a and open at b
$(a, b]$	Interval of real numbers open at a and closed at b
(a, b)	Open interval of real numbers from a to b
a.e.	Almost everywhere
$f_n \nearrow f$	Limit of an increasing sequence $\{f_n\}$ is f
$f_n \searrow f$	Limit of a decreasing sequence $\{f_n\}$ is f
$f_n \xrightarrow{\text{a.e.}} f$	$\{f_n\}$ converges to f almost everywhere
$f_n \xrightarrow{\text{a.u.}} f$	$\{f_n\}$ converges to f almost uniformly
$f_n \xrightarrow{\text{p.a.e.}} f$	$\{f_n\}$ converges to f pseudo-almost everywhere
$f_n \xrightarrow{\text{p.a.u.}} f$	$\{f_n\}$ converges to f pseudo-almost uniformly
$f_n \xrightarrow{\text{p.}\mu} f$	$\{f_n\}$ converges to f pseudo-in μ (in measure)
$f_n \xrightarrow{\mu} f$	$\{f_n\}$ converges to f in μ (in measure)
$f \circ g$	Composition of functions f and g
f, g, h	Functions defined on X
g_λ	λ-measure
iff	If and only if
inf	Infimum
$\lim_n E_n$	Limit of $\{E_n\}$
$\overline{\lim_n} E_n$	Superior limit of $\{E_n\}$
$\underline{\lim_n} E_n$	Inferior limit of $\{E_n\}$
$\limsup_n E_n$	Superior limit of $\{E_n\}$
$\liminf_n E_n$	Inferior limit of $\{E_n\}$
$\lim_{\beta \to \alpha -}$	Limit as β approaches to α from the left
$\lim_{\beta \to \alpha +}$	Limit as β approaches to α from the right
m	Basic probability assignment
m_A	Membership function of fuzzy set A
max	Maximum
min	Minimum
$n!$	n factorial
$\binom{n}{i}$	Number of combinations of n things taken i at a time: $\frac{n!}{(n-i)!i!}$
p	Probability measure
p.a.e.	Pseudo-almost everywhere
proj	Projection of
s	Pan-simple measurable function
sup	Supremum
$[x]$	The equivalence class containing x
$\{x, y, \ldots\}$	Set of elements $x, y, \ldots$
$\{x \mid \pi(x)\}$	Set determined by proposition π
xRy	There is the relation R from x to y
$x\not{R}y$	There is no relation R from x to y
$x \in E$	x is a member of set E
$x \notin E$	x is not a member of set E

(continued)

$A(x/\mathbf{C})$	Atom of **C** at x
Bel	Belief measure
$\bar{E}$	Complement of set E
$\|E\|$	Number of elements in a finite set E
$\{E_n\}$	Sequence of sets: $\{E_1, E_2, \ldots\}$
$E_n \nearrow E$	Limit of a increasing sequence $\{E_n\}$ is E
$E_n \searrow E$	Limit of a decreasing sequence $\{E_n\}$ is E
$E_n \to E$	Limit of $\{E_n\}$ is E
$E, F, \ldots$	Subsets of X
$E \cup F$	Union of sets E and F
$E \cap F$	Intersection of sets E and F
$E - F$	Difference of sets E and F: $E \cap \bar{F}$
$E \Delta F$	Symmetric difference of sets E and F: $(E - F) \cup (F - E)$
$E \times F$	Product of sets E and F
$E \subset F$	E is a subset of F
$E = F$	Sets E and F are equal ($E \subset F$ and $F \subset E$)
E/R	The quotient of set E by an equivalence relation R
F_α	The α-level set of f
$F_{\alpha+}$	The strict α-level set of f
h	Probability distribution function
$H(x/\mathbf{C})$	Hole of **C** at x
Pl	Plausibility measure
$(P, \leq)$	Partially ordered set
R	The interval $(-\infty, \infty)$
$\bar{R}$	The interval $[-\infty, \infty]$
R_+	The interval $[0, \infty)$
$\bar{R}_+$	The interval $[0, \infty]$
R^n	n-dimensional Euclidean space
T	Arbitrary index set
X	Universe of discourse (universal set)
A[**C**]	The class of all atoms of **C**
B	Borel field
$\mathbf{B}^{(n)}$	Borel field on R^n
C	Class of subsets of X (a set of subsets of X)
E	Measurable partition of X
$\mathbf{E}_\pi(\mu)$	The set of all generalized possibility measure extensions of μ
$\mathbf{E}_\nu(\mu)$	The set of all necessity measure extensions of μ
F	σ - ring
$\tilde{\mathbf{F}}$	Fuzzy $\sigma-$ algebra
F(**C**)	σ - ring generated by **C**
$\mathbf{F}_\mathrm{p}$	Plump class
$\mathbf{F}_\mathrm{p}(\mathbf{C})$	Plump class generated by **C**
G	The class of all finite nonnegative measurable functions on (X, **F**)
H[**C**]	The class of all holes of **C**
M	Monotone class
M(**C**)	Monotone class generated by **C**
$\hat{\mathbf{P}}$	The set of all measurable partitions of X
P(X)	Power set of X
$\tilde{\mathbf{P}}(X)$	Fuzzy power set of X
Q	The class of all pan-simple measurable functions
R	Ring
R(**C**)	Ring generated by **C**

(continued)

$\mathbf{R}_\sigma$	The class of all sets each of which is expressed by the limit of an increasing sequence of sets in an algebra **R**
$\mathbf{R}_\delta$	The class of all sets each of which is expressed by the limit of a decreasing sequence of sets in an algebra **R**
S	Semiring

Bibliography

Aaronson, J. [1997]. *An Introduction to Infinite Ergodic Theory*. American Mathematical Society, Providence, RI.

Anger, B. [1971]. Approximation of capacities by measures. In: H. Baver (ed.), *Seminar on Political Theory II*. Springer-Verlag, Berlin and New York, pp. 152–170.

Anger, B. [1977]. Representation of capacities. *Mathematical Annals*, 259, pp. 245–258.

Arslanov, M.Z. and Ismail, E.E. [2004]. On the existence of possibility distribution function. *Fuzzy Sets and Systems*, 148 (2), pp. 279–290.

Aubin, J.P. [1981]. Cooperative fuzzy games. *Mathematics of Operations Research*, 6, pp. 1–13.

Aubin, J.P. and Frankowska, H. [1990]. *Set-Valued Analysis*. Birkhäuser, Boston.

Aumann, R.J. [1965]. Integrals of set-valued functions. *J. of Mathematical Analysis and Applications*, 12 (1), pp. 1–12.

Aumann, R.J. and Shapley, L.S. [1974]. *Values of Non-Atomic Games*. Princeton University Press, Princeton, New Jersey.

Ban, A.I. and Gal, S.G. [2002]. *Defects of Properties in Mathematics*. World Scientific, Singapore.

Banon, G. [1981]. Distinction between several subsets of fuzzy measures. *Fuzzy Sets and Systems*, 5 (3), pp. 291–305.

Basile, A. [1987]. Sequential compactness for sets of Sugeno fuzzy measures. *Fuzzy Sets and Systems*, 21(2), pp. 243–247.

Batle, N. and Trillas, E. [1979]. Entropy and fuzzy integral. *J. of Mathematical Analysis and Applications*, 69, pp. 469–474.

Baudrit, C., Couso, I. and Dubois, D. [2007]. Joint propagation of probability and possibility in risk analysis: Towards a formal framework. *Intern. J. of Approximate Reasoning*, 45(1), pp. 82-105.

Bauer, H. [2001]. *Measure and Integration Theory*. Walter de Gruyter, Berlin and New York.

Baumont, C. [1989]. Theory of possibility as a basis for analyzing business experience process. *Fuzzy Sets and Systems*, 31(1), pp. 1–12.

Bell, D.A., Guan, J.W. and Shapcott, C.M. [1998]. Using the Dempster–Shafer orthogonal sum for reasoning which involves space. *Kybernetes*, 27(5), pp. 511–526.

Bělohlávek, R. [2003]. Cutlike semantics for fuzzy logic and its applications. *Intern. J. of General Systems*, 32(4), pp. 305–319.

Benvenuti, P. and Mesiar, R. [2000]. Integrals with respect to a general fuzzy measure, In: M. Grabisch et al. (eds.), *Fuzzy Measures and Integrals*. Springer-Verlag, New York, pp. 203–232.

Benvenuti, P., Mesiar, R. and Vivona, D. [2002]. Monotone set functions-based integrals. In: E. Pap (eds), *Handbook of Measure Theory* (Chapter 33). Elsevier, Amsterdam, pp. 1329–1379.

Berberian, S.K. [1965]. *Measure and Integration*. Macmillan, New York.

Berres, M. [1988]. λ-additive measures on measure spaces. *Fuzzy Sets and Systems*, 27(2), pp. 159–169.

Biacino, L. [2007]. Fuzzy subsethood and belief functions of fuzzy events. *Fuzzy Sets and Systems*, 158(1), pp. 38–49.

Billingsley, P. [1965]. *Ergotic Theory and Information*. John Wiley, New York.

Billingsley, P. [1986]. *Probability and Measure* (Second Edition). John Wiley, New York.

Billot, A. [1992]. From fuzzy set theory to non-additive probabilities: How have economists reacted? *Fuzzy Sets and Systems*, 49(1), pp. 75–89.

Bingham, N.H. [2000]. Measure into probability: From Lebesgue to Kolmogorov. *Biometrika*, 87(1), pp. 145–156.

Black, P.K. [1997]. Geometric structure of lower probabilities. In: J. Goutsias, R.P.S. Mahler, and H.T. Nguen (eds.), *Random Sets*, Springer, New York, pp. 361–383.

Bogler, G. [1981]. Shafer–Dempster reasoning with applications to multisensor target identification systems. *IEEE Transactions on Systems, Man, and Cybernetics*, SMC-17(6), pp. 968–977.

Bolanos, M.J., de Campos Ibanez, L.M. and Munoz, G. [1989]. Convergence properties of the monotone expectation and its application to the extension of fuzzy measures. *Fuzzy Sets and Systems*, 33(2), pp. 201–212.

Borel, É. [1898]. *Lessons on a Theory of Functions*. Gauthier-Villars, Paris (in French).

Borgelt, C. and Kruse, R. [2002]. *Graphical Models: Methods for Data Analysis and Mining*. John Wiley, New York.

Bouchon, B. and Yager, R.R., eds. [1987]. *Uncertainty in Knowledge-Based Systems*. Springer-Verlag, New York.

Bouchon-Meunier, B., Mesiar, R. and Ralescu, D. A. [2004]. Linear non-additive set-functions. *Intern. J. of General Systems*, 33(1), pp. 89–98.

Branzei, R., Dimitrov, D. and Tijs, S. [2005]. *Models in Cooperative Game Theory*. Springer, Berlin and Heidelberg.

Bronevich, A. G. [2005a]. An investigation of ideals in the set of fuzzy measures. *Fuzzy Sets and Systems*, 152(2), pp. 271–288.

Bronevich, A. G. [2005b]. On the closure of families of fuzzy measures under eventwise aggregation. *Fuzzy Sets and Systems*, 153(1), pp. 45–70.

Bronevich, A. G. [2007]. Necessary and sufficient consensus conditions for the eventwise aggregation of lower probabilities. *Fuzzy Sets and Systems*, 158(8), pp. 881–894.

Bronevich, A.G. and Karkishchenko, A.N. [2002a]. The structure of fuzzy measure families induced by upper and lower probabilities. In: C. Bertoluzza et al. (eds.), *Statistical Modeling, Analysis, and Management of Fuzzy Data*. Physica-Verlag, Heidelberg and New York, pp. 160–172.

Bronevich, A.G. and Karkishchenko, A.N. [2002b]. Statistical classes and fuzzy set theoretical classification of probability distributions. In: C. Bertoluzza et al. (eds.), *Statistical Modeling, Analysis, and Management of Fuzzy Data*, Physica-Verlag, Heidelberg and New York, pp. 173–195.

Brüning, M. and Denneberg, D. [2002]. Max-min σ-additive representation of monotone measures. *Statistical Papers*, 34, pp. 23–35.

Burk, F. [1998]. *Lebesgue Measure and Integration: An Introduction*. Wiley-Interscience, New York.

Butnariu, D. [1983]. Additive fuzzy measures and integrals. *J. of Mathematical Analysis and Applications*, 93, pp. 436–452.

Butnariu, D. [1985]. Non-atomic fuzzy measures and games. *Fuzzy Sets and Systems*, 17(1), pp. 39–52.

Butnariu, D. and Klement, E.P. [1993]. *Triangular Norm-Based Measures and Games with Fuzzy Coalitions*. Kluwer, Dordrecht and Boston.

Cai, K.Y. [1996]. *Introduction to Fuzzy Reliability*. Kluwer, Boston and Dordrecht.

Cano, A. and Moral, S. [2000]. Algorithms for imprecise probabilities. In: J. Kohlas and S. Moral (eds.), *Algorithms for Uncertainty and Defeasible Reasoning*. Kluwer, Dordrecht and Boston, pp. 369–420.

Caratheodory, C. [1963]. *Algebraic Theory of Measure and Integration*. Chelsea, New York (first published in German in 1956).

Caselton, W.F. and Luo, W. [1992]. Decision making with imprecise probabilities: Dempster–Shafer theory and applications. *Water Resource Research*, 28(12), pp. 3071–3083.

Chae, S.B. [1995]. *Lebesgue Integration* (Second Edition). Springer-Verlag, New York.

Chambers, R.G. and Melkonyan, T. [2007]. Degree of imprecision: Geometric and algoritmic approaches. *Intern. J. of Approximate Reasoning*, 45(1), pp. 106–122.

Chateauneuf, A. [1988]. Uncertainty aversion and risk aversion in modes with nonadditive probabilities. In: B.R. Munier (eds), *Risk, Decision and Rationality*, D. Reidel, Dordrecht, pp. 615–627.

Chateauneuf, A. [1991]. On the use of capacities in modeling uncertainty aversion and risk aversion. *J. of Mathematical Economics*, 20(4), pp. 343–369.

Chateauneuf, A. [1995]. Elsberg paradox intuition and Choquet expected utility. In: Coletti et al., (eds.), *Mathematical Models for Handling Partial Knowledge in Artificial Intelligence*, Plenum Press, New York, pp. 1–20.

Chateauneuf, A. [1996]. Decomposable capacities, distorted probabilities, and concave capacities. *Mathematical Social Sciences*, 31, pp. 19–37.

Chateauneuf, A. and Jaffray, J.Y. [1989]. Some characterizations of lower probabilities and other monotone capacities through the use of Möbius inversion. *Mathematical Social Sciences*, 17, pp. 263–283.

Chateauneuf, A. and Vergnaud, J.-C. [2000]. Ambiguity reduction through new statistical data. *Intern. J. of Approximate Reasoning*, 24(2–3), pp. 283–299.

Chellas, B.F. [1980]. *Modal Logic: An Introduction*. Cambridge University Press, Cambridge, UK and New York.

Chen, T.Y., Wang, J.C. and Tzeng, G.H. [2000]. Identification of general fuzzy measures by genetic algorithms based on partial information. *IEEE Trans. on Systems, Man, and Cybernetics* (Part B: *Cybernetics*), 30(4), pp. 517–528.

Chiang, J.H. [1999]. Choquet fuzzy integral-based hierarchical networks for decision analysis. *IEEE Trans. on Fuzzy Systems*, 7(1), pp. 63–71.

Chiang, J.H. [2000]. Aggregating membership values by a Choquet fuzzy integral-based operator. *Fuzzy Sets and Systems*, 114(3), pp. 367–375.

Choquet, G. [1953–54]. Theory of capacities. *Annales de l'Institut Fourier*, 5, pp. 131–295. (Also *Technical Note*, No. 1, Department of Mathematics, University of Kansas, Lawrence, KS, May 1954.)

Choquet, G. [1969]. *Lectures on Analysis* (3 volumes). W. A. Benjamin, Reading, MA.

Coletti, G. and Scozzafava, R. [2001]. From conditional events to conditional measures: A new axiomatic approach. *Annals of Mathematics and Artificial Intelligence*, 32(1–4), pp. 373–392.

Coletti, G. and Scozzafava, R. [2002]. *Probabilistic Logic in a Coherent Setting*. Kluwer, Boston.

Colyvan, M. [2004]. The philosophical significance of Cox's theorem. *Intern. J. of Approximate Reasoning*, 37(1), pp. 71–85.

Constantinescu, C. and Weber, K. [1985]. *Integration Theory*, Vol. 1: *Measure and Integration*. Wiley-Interscience, New York.

Couso, S. Moral, P. and Walley, P. [2000]. A survey of concepts of independence for imprecise probabilities. *Risk Decision and Policy*, 5, pp. 165–181.

Cox, R.T. [1946]. Probability, frequency, and reasonable expectation. *American J. of Physics*, 14(1), pp. 1–13.

Cox, R.T. [1961] *The Algebra of Probable Inference*. John Hopkins Press, Baltimore.

De Campos, L.M. and Bolaños, M.J. [1989]. Representation of fuzzy measures through probabilities. *Fuzzy Sets and Systems*, 31(1), pp. 23–36.

De Campos, L.M. and Bolaños, M.J. [1992]. Characterization and comparison of Sugeno and Choquet integrals. *Fuzzy Sets and Systems*, 52(1), pp. 61–67.

De Campos, L.M. and Huete, J.F. [1993]. Independence concepts in upper and lower probabilities. In: B. Bouchon-Meunier, L. Velverde, and R.R. Yager (eds.), *Uncertainty in Intelligent Systems*, North-Holland, Amsterdam, pp. 85–96.

De Campos, L.M. and Huete, J.F. [1999]. Independence concepts in possibility theory. *Fuzzy Sets and Systems, Part I*, 103(1), pp. 127–152, *Part II*, 103(3), pp. 487–505.

De Campos, L. M. and Moral, S. [1995]. Independence concepts for convex sets of probabilities. In: P. Bernard and S. Hanks (eds.), *Proc. Eleventh Conf. on Uncertainty in Artificial Intelligence*, Morgan Kaufmann, San Mateo, CA, pp. 108–115.

De Campos, L.M., Huete, J.F. and Moral, S. [1994]. Probability intervals: A tool for uncertain reasoning. *Intern. J. of Uncertainty, Fuzziness, and Knowledge-Based Systems*, 2(2), pp. 167–196.

De Campos, L.M., Lamata, M.T. and Moral, S. [1990a]. Distances between fuzzy measures through associated probabilities: Some applications. *Fuzzy Sets and Systems*, 35(1), pp. 57–68.

De Campos, L.M., Lamata, M. T. and Moral, S. [1990b]. The concept of conditional fuzzy measure. *Intern. J. of Intelligent Systems*, 5(3), pp. 237–246.

De Campos, L.M., Lamata, M.T. and Moral, S. [1991]. A unified approach to define fuzzy integrals. *Fuzzy Sets and Systems*, 39(1), pp. 75–90.

De Cooman, G. [1997]. Possibility theory – I, II, III. *Intern. J. of General Systems*, 25(4), pp. 291–371.

De Cooman, G. [2005]. A behavioural model for vague probability assessments. *Fuzzy Sets and Systems*, 154(3), pp. 305–358.

De Cooman, G. and Aeyels, D. [1999]. Supremum-preserving upper probabilities. *Information Sciences*, 118, pp. 173–212.

De Cooman, G., Ruan, D. and Kerre, E.E., eds. [1995]. *Foundations and Applications of Possibility Theory*. World Scientific, Singapore.

Delbaen, F. [1974]. Convex games and extreme points. *J. of Mathematical Analysis and Applications*, 45, pp. 210–233.

Delgado, M. and Moral, S. [1989]. Upper and lower fuzzy measures. *Fuzzy Sets and Systems*, 33(2), pp. 191–200.

Dellacherie, C. and Meyer, P.A. [1978]. *Probabilities and Potential*. North-Holland, Amsterdam.

Delmotte, F. [2001]. Comparison of the performances of decision aimed algorithms with Bayesian and beliefs basis. *Intern. J. of Intelligent Systems*, 16(8), pp. 963–981.

Delmotte, F. [2007]. Detections of defective sources in the setting of possibility theory. *Fuzzy Sets and Systems*, 158(5), pp. 555–571.

Dempster, A.P. [1967a]. Upper and lower probabilities induced by multi-valued mapping. *Annals of Mathematical Statistics*, 38(2), pp. 325–339.

Dempster, A.P. [1967b]. Upper and lower probability inferences based on a sample from a finite univariate population. *Biometrika*, 54, pp. 515–528.

Dempster, A.P. [1968a]. A generalization of Bayesian inference. *J. of the Royal Statistical Society*, Ser. B, 30, pp. 205–247.

Dempster, A.P. [1968b]. Upper and lower probabilities generated by a random closed interval. *Annals of Mathematical Statistics*, 39, pp. 957–966.

Denneberg, D. [1994a]. *Non-Additive Measure and Integral*. Kluwer, Boston.

Denneberg, D. [1994b]. Conditioning (updating) non-additive measures. *Annals of Operations Research*, 52(1), pp. 21–42.

Denneberg, D. [1997]. Representation of the Choquet integral with the σ-additive Möbius transform. *Fuzzy Sets and Systems*, 92(2), pp. 139–156.

Denneberg, D. [2000a]. Non-additive measure and integral, basic concepts and their role for applications. In: M. Grabisch et al. (eds.), *Fuzzy Measures and Integrals*. Physica-Verlag, Heidelberg and New York, pp. 42–69.

Denneberg, D. [2000b]. Totally monotone core and products of monotone measures. *Intern. J. of Approximate Reasoning*, 24(2–3), pp. 273–281.

Denneberg, D. [2002]. Conditional expectation for monotone measures, the discrete case. *J. of Mathematical Economics*, 37, pp. 105–121.

Denneberg, D. and Grabisch, M. [1999]. Interaction transform of set functions over a finite set. *Information Sciences*, 121, pp. 149–170.

Denneberg, D. and Grabisch, M. [2004]. Measure and integral with purely ordinal scales. *Journal of Mathematical Psychology*, 48(1), pp. 15–26.

Dong, W. and Wong, F.S. [1986]. From uncertainty to approximate reasoning. Part 1: Conceptual models and engineering interpretations. *Civil Engineering Systems*, 3, pp. 143–202.

Doob, J. L. [1994]. The development of rigor in mathematical probability. In: J.P. Pier (eds), *Development of Mathematics 1900–1950*. Birkhäuser, Basel.

Dubois, D., Nguyen, H.T. and Prade, H. [2000]. Possibility theory, probability theory, and fuzzy sets: Misunderstandings, bridges, and gaps. In: D. Dubois and H. Prade (eds.), *Fundamentals of Fuzzy Sets*, Kluwer, Boston, pp. 343–438.

Dubois, D. and Prade, H. [1980]. *Fuzzy Sets and Systems: Theory and Applications*. Academic Press, New York.

Dubois, D. and Prade, H. [1982]. A class of fuzzy measures based on triangular norms. *Intern. J. of General Systems*, 8(1), pp. 43–61.

Dubois, D. and Prade, H. [1985]. Evidence measures based on fuzzy information. *Automatica*, 21, pp. 547–562.

Dubois, D. and Prade, H. [1986a]. A set-theoretic view of belief function. *Intern. J. of General Systems*, 12(3), pp. 193–226.

Dubois, D. and Prade, H. [1986b]. On the unicity of Dempster rule of combination. *Intern. J. of Intelligent Systems*, 1(2), pp. 133–142.

Dubois, D. and Prade, H. [1986c]. Fuzzy sets and statistical data. *European J. of Operations Research*, 25(3), pp. 345–356.

Dubois, D. and Prade, H. [1988]. *Possibility Theory*. Plenum Press, New York (translated from the French original published in 1985).

Dubois, D. and Prade, H. [1990a]. Consonant approximations of belief functions. *Intern. J. of Approximate Reasoning*, 4,(5–6), pp. 419–449.

Dubois D. and Prade, H. [1990b]. Aggregation of possibility measures. In: J. Kacprzyk and M. Fedrizzi (eds.), *Multiperson decision making using fuzzy sets and possibility theory*. Dordrecht: Kluwer, pp. 55–63.

Dubois, D. and Prade, H. [1992]. When upper probabilities are possibility measures. *Fuzzy Sets and Systems*, 49(1), pp. 65–74.

Dubois, D. and Prade, H. [1998]. Possibility theory: Qualitative and quantitative aspects. In: P. Smets (eds), *Quantified Representation of Uncertainty and Imprecision*. Kluwer, Boston, pp. 199–226.

Dubois, D. and Prade, H., eds. [2000]. *Fundamentals of Fuzzy Sets*. Kluwer, Boston.

Dubois, D., Fodor, J.C., Prade, H. and Roubens, M. [1996]. Aggregation of decomposable measures with application to utility theory. *Theory and Decision*, 41, pp. 59–95.

Du Plessis, N. [1970]. *An Introduction to Potential Theory*. Oliver & Boyd, Edinburgh.

Faden, A.M. [1977]. *Economics of Space and Time: The Measure-Theoretic Foundations of Social Science*. Iowa State University Press, Ames, Iowa.

Fine, T.L. [1973]. *Theories of Probability: An Examination of Foundations*. Academic Press, New York.

Fujimoto, K. and Murofushi, T. [2007]. Some relations among values, interactions, and decomposibility of non-additive measures. *Intern. J. of Uncertainty, Fuzziness, and Knowledge-based Systems*, 15(2), pp. 175–191.

Gader, P.D., Mohamed, M. and Keller, J.M. [1996]. Dynamic programming based hand-written word recognition using Choquet fuzzy integral as the match function. *J. of Electronic Imaging*, 51(1), pp. 15–25.

Garmendia, L. [2005]. The evolution of the concept of fuzzy measure. In: D. Ruan et al. (eds.), *Intelligent Data Mining: Techniques and Applications*. Springer, Berlin, pp. 185–200.

Gertler, J.J. and Anderson, K. C. [1992]. An evidential reasoning extension to qualitative model-based failure diagnosis. *IEEE Transactions on Systems, Man, and Cybernetics*, 22, pp. 275–289.

Ghirardato, P. [1997]. On independence for non-additive measures, with a Fubini theorem. *J. of Economic Theory*, 73(2), pp. 261–291.

Gilboa, I. [1987]. Expected utility with purely subjective non-additive probabilities. *J. of Mathematical Economics*, 16, pp. 65–88.

Gilboa, I. and Schmeidler, D. [1995]. Canonical representation of set functions. *Mathematics of Operations Research*, 20(1), pp. 197–212.

Goguen, J.A. [1967]. L-fuzzy sets. *Journal of Mathematical Analysis and Applications*, 18(1), pp. 145–174.

Good, I.J. [1962]. The measure of a non-measurable set. In: E. Nagel, P. Suppes, and A. Tarski (eds.), *Logic, Methodology, and Philosophy of Science*. Stanford University Press, Stanford, CA, pp. 319–329.

Goutsias, J., Mahler, R.P.S. and Nguyen, H.T., eds. [1997]. *Random Sets: Theory and Applications*. Springer-Verlag, New York.

Grabisch, M. [1995a]. A new algorithm for identifying fuzzy measures and its application to pattern recognition. *Proc. FUZZ-IEEE/IFES'95*, Yokohama, Japan, pp.145–150.

Grabisch, M. [1995b]. Fuzzy integral in multicriteria decision making. *Fuzzy Sets and Systems*, 69(3), pp. 279–298.

Grabisch, M. [1997a]. k-order additive discrete fuzzy measures and their representation. *Fuzzy Sets and Systems*, 92(2), pp. 167–189.

Grabisch, M. [1997b]. Alternative representations of discrete fuzzy measures for decision making. *Intern. J. of Uncertainty, Fuzziness, and Knowledge-Based Systems*, 5(5), pp. 587–607.

Grabisch, M. [1997c]. Fuzzy measures and integrals: A survey of applications and recent issues. In: D. Dubois, H. Prade, and R.R. Yager (eds.), *Fuzzy Information Engineering*. John Wiley, New York, pp. 507–529.

Grabisch, M. [1997d]. On the representation of k-decomposable measures. *Proc. IFSA'97*, Academia, Prague, pp. 478–483.

Grabisch, M. [2000]. The interaction and Möbius representations of fuzzy measures on finite spaces, k-additive measures: A survey. In: M. Grabisch et al. (eds.), *Fuzzy Measures and Integrals: Theory and Applications*. Springer-Verlag, New York, pp. 70–93.

Grabisch, M. [2006]. Capacities and games on lattices: A survey of results. *Intern. J. of Uncertainty, Fuzziness, and Knowledge-Based Systems*, 14(4), pp. 371–392.

Grabisch, M. and Labreuche, C. [2005]. Bi-capacities: I. Definition, Möbius transform and integration, pp. 211–236; II. Choquet integral, pp. 237–259. *Fuzzy Sets and Systems*, 151(2).

Grabisch, M., Murofushi, T., and Sugeno, M., eds. [1992]. Fuzzy measure of fuzzy events defined by fuzzy integrals. *Fuzzy Sets and Systems*, 50(3), pp. 293–313.

Grabisch, M., Murofushi, T. and Sugeno, M., eds. [2000]. *Fuzzy Measures and Integrals: Theory and Applications*. Springer-Verlag, New York.

Grabisch, M. and Nicolas, J.M. [1994]. Classification by fuzzy integral: Performance and tests. *Fuzzy Sets and Systems*, 65(2/3), pp. 255–271.

Grabisch, M., Nguyen, T. and Walker, E.A. [1995]. *Fundamentals of Uncertainty Calculi with Applications to Fuzzy Inference*. Kluwer, Dordrecht and Boston.

Guan, J.W. and Bell, D.A. [1991–92]. *Evidence Theory and Its Applications*: Vol. 1 (1991), Vol. 2 (1992). North-Holland, New York.

Guth, M.A.S. [1988]. Uncertainty analysis of rule-based expert systems with Dempster–Shafer mass assignment. *Intern. J. of Intelligent Systems*, 3(2), pp. 123–139.

Hacking, I. [1975]. *The Emergence of Probability*. Cambridge University Press, Cambridge.

Hájek, P. [1998]. *Metamathematics of Fuzzy Logic*. Kluwer, Boston.

Halmos, P.R. [1950]. *Measure Theory*. Van Nostrand, New York.

Halmos, P.R. [1956]. *Lectures on Ergodic Theory*. Chelsea, New York.

Halpern, J.Y. [1999]. A counterexample to theorems of Cox and Fine. *J. of Artificial Intelligence Research*, 10, pp. 67–85.

Harmanec, D., Klir, G.J. and Resconi, G. [1994]. On modal logic interpretation of Dempster–Shafer theory of evidence. *Intern. J. of Intelligent Systems*, 9(10), pp. 941–951.

Harmanec, D., Klir, G.J. and Wang, Z. [1996]. Modal logic interpretation of Dempster–Shafer theory: An infinite case. *Intern. J. of Approximate Reasoning*, 14(2–3), pp. 81–93.

Hartley, R.V.L. [1928]. Transmission of information. *The Bell System Technical J.*, 7(3), pp. 535–563.

Hawkins, T. [1975]. *Lebesgue's Theory of Integration: Its Origins and Development*. Chelsea, New York.

Helms, L.L. [1963]. *Introduction to Potential Theory*. John Wiley, New York.

Helton, J.C. and Oberkampf, W. L. [2004]. Special Issue on Alternative Representations of Epistemic Uncertainty. *Reliability Eng. & System Safety*, 85(1–3), pp. 1–369.

Helton, J.C. et al. [2006]. A sampling-based computational strategy for the representation of epistemic uncertainty in model predictions with evidence theory. *Sandia Report SAND 2006-5557*, Sandia National Laboratory, Albuquerque, NM.

Higashi, M. and Klir, G.J. [1982]. On measures of fuzziness and fuzzy complements. *Intern. J. of General Systems*, 8(3), pp. 169–180.

Hocaoglu, A.K. and Gader, P.D. [2003]. An interpretation of discrete Choquet integrals in morphological image processing. *Proc. of FUZZ-IEEE '03*, pp. 1291–1295.

Höhle, U. [1982]. A general theory of fuzzy plausibility measures. *J. of Mathematical Analysis and Applications*, 127, pp. 346–364.

Höhle, U. [1984]. Fuzzy probability measures. In: H.J. Zimmermann, L.A. Zadeh, and B.R. Gaines (eds.), *Fuzzy Sets and Decision Analysis,* North-Holland, New York, pp. 83–96.

Hua, W. [1988]. The properties of some non-additive measures. *Fuzzy Sets and Systems*, 27(3), pp. 373–377.

Huber, P.J. [1972]. Robust statistics: A review. *Annals of Mathematical Statistics*, 43, pp. 1041–1067.

Huber, P.J. [1973]. The use of Choquet capacities in statistics. *Bulletin of the International Statistical Institute*, 45(4), pp. 181–188.

Huber, P.J. [1981]. *Robust Statistics*. John Wiley, New York (reprinted in 2004).

Huber, P.J. and Strassen, V. [1973]. Minimax tests and the Neyman-Pearson lemma for capacities. *Annals of Statistics*, 1, pp. 251–263, 2, pp. 223–224.

Hughes, G.E. and Cresswell, M.J. [1996]. *A New Introduction to Modal Logic*, Routledge, London and New York.

Ichihashi, H., Tanaka, H., and Asai, K. [1988]. Fuzzy integrals based on pseudo-additions and multiplications. *J. of Mathematical Analysis and Applications*, 130, pp. 354–364.

Inagaki, T. [1991]. Interdependence between safety-control policy and multiple-sensor schemes via Dempster–Shafer theory. *IEEE Trans. on Reliability*, 40(2), pp. 182–188.

Ishi, K. and Sugeno, M. [1985]. A model of human evaluation process using fuzzy measure. *Intern. J. of Man-Machine Studies*, 22, pp. 19–38.

Jang, L.C. and Kwon, J.S. [2000]. On the representation of Choquet integrals of set-valued functions and null sets. *Fuzzy Sets and Systems*, 112(2), pp. 233–239.

Jech, T. [2003]. *Set Theory*, Springer, Berlin.

Jiang, Q. and Wang, Z. [1995]. Property (p.g.p.) of fuzzy measures and convergence in measure. *Intern. J. of Fuzzy Mathematics*, 3(3), pp. 699–710.

Jiang, Q., Klir, G.J. and Wang, Z. [1996]. Null-additive fuzzy measures on S-compact spaces. *International Journal of General Systems*, 25(3), pp. 219–228.

Jiang, Q., Suzuki, H., Wang, Z. and Klir, G.J. [1998]. Exhaustivity and absolute continuity of fuzzy measures. *Fuzzy Sets and Systems*, 96(2), pp. 231–238.

Jiang, Q., Wang, S., Ziou, D., Wang, Z. and Klir, G.J. [2000]. Pseudometric generating property and autocontinuity of fuzzy measures. *Fuzzy Sets and Systems*, 112(2), pp. 207–216.

Kadane, J.B. and Wasserman, L. [1996]. Symmetric, coherent, Choquet capacities. *Annals of Statistics*, 24(3), pp. 1250–1264.

Keller, J., Qiu, H. and Tahani, H. [1986]. Fuzzy integral and image segmentation. *Proc. North American Fuzzy Information Processing Soc.*, New Orleans, pp. 324–338.

Keller, J.M. and Osborn, J. [1996]. Training the fuzzy integral. *Intern. J. of Approximate Reasoning*, 15(1), pp. 1–24.

Keller, J.M., Gader, P.D. and Hocaoglu, A.K. [2000]. Fuzzy integrals in image processing and recognition. In: M. Grabisch et al. (eds.), *Fuzzy Measures and Integrals*, Springer, Berlin, pp. 435–466.

Keller, J.M. et al. [1994]. Advances in fuzzy integration for pattern recognition. *Fuzzy Sets and Systems*, 65(2/3), pp. 273–283.

Kendall, D.G. [1973]. *Foundations of a Theory of Random Sets in Stochastic Geometry*. John Wiley, New York.

Kendall, D.G. [1974]. Foundations of a theory of random sets. In: E.F. Harding, and D.G. Kendall (eds.), *Stochastic Geometry*, John Wiley, New York, pp. 322–376.

Kingman, J.F.C. and Taylor, S.T. [1966]. *Introduction to Measure and Probability*. Cambridge University Press, New York.

Klement, E.P. and Ralescu, D. [1983]. Nonlinearity of the fuzzy integral. *Fuzzy Sets and Systems*, 11(3), pp. 309–315.

Klement, E.P. and Weber, S. [1991]. Generalized measures. *Fuzzy Sets and Systems*, 40(2), pp. 375–394.

Klement, E.P. and Weber, S. [1999]. Fundamentals of generalized measure theory. In: U. Höhle and S.E. Rodabaugh (eds.), *Mathematics of Fuzzy Sets*. Kluwer, Boston and Dordrecht, pp. 633–651.

Klement, E.P., Mesiar, R. and Pap, E. [2000]. *Triangular Norms*. Kluwer, Dordrecht, Boston, London.

Klir, G.J. [1991]. Generalized information theory. *Fuzzy Sets and Systems*, 40(1), pp. 127–142.

Klir, G.J. [1994]. Multivalued logics versus modal logics: Alternative frameworks for uncertainty modeling. In: P.P. Wang, (ed.) *Advances in Fuzzy Theory and Technology*; Vol. II, Bookwrights Press, Durham, NC, pp. 3–47.

Klir, G.J. [1997]. Fuzzy arithmetic with requisite constraints. *Fuzzy Sets and Systems*, 91(2), pp. 147–161.

Klir, G.J. [1999]. On fuzzy-set interpretation of possibility theory. *Fuzzy Sets and Systems*, 108(3), pp. 263–273.

Klir, G.J. [2001]. Foundations of fuzzy set theory and fuzzy logic: A historical overview. *Intern. J. of General Systems*, 30(2), pp. 91–132.

Klir, G.J. [2002]. Uncertainty in economics: The heritage of G.L.S. Shackle. *Fuzzy Economic Review*, VII(2), pp. 3–21.

Klir, G.J. [2006]. *Uncertainty and Information: Foundations of Generalized Information Theory*. Wiley-Interscience, Hoboken, New Jersey.

Klir, G.J. and Harmanec, D. [1994]. On modal logic interpretation of possibility theory. *Intern. J. Uncertainty, Fuzziness, and Knowledge-Based Systems*, 2(2), pp. 237–245.

Klir, G.J. and Pan, Y. [1998]. Constrained fuzzy arithmetic: Basic questions and some answers. *Soft Computing*, 2(2), pp. 100–108.

Klir, G.J. and Yuan, B. [1995]. *Fuzzy Sets and Fuzzy Logic: Theory and Applications*. Prentice Hall, Upper Saddle River, New Jersey.

Klir, G.J. and Yuan, B., eds. [1996]. *Fuzzy Sets, Fuzzy Logic, and Fuzzy Systems: Selected Papers by Lofti A. Zadeh*. World Scientific, Singapore.

Klir, G.J., Yuan, B. and Swan-Stone, J.F. [1995]. Constructing fuzzy measures from given data. *Proc. Sixth IFSA World Congress*, San Paulo, Brazil, pp. 61–64.

Klir, G.J., Wang, Z. and Harmanec, D. [1997]. Constructing fuzzy measures in expert systems. *Fuzzy Sets and Systems*, 92(2), pp. 251–264.

Klir, G.J., Wang, Z. and Wang, W. [1996]. Constructing fuzzy measures by transformations. *J. of Fuzzy Mathematics*, 4(1), pp. 207–215.

Kohlas, J. and Monney, P.A. [1995]. *A Mathematical Theory of Hints: An Approach to the Dempster–Shafer Theory of Evidence*. Springer, Berlin.

Kolmogorov, A.N. [1950]. *Foundations of the Theory of Probability*. Chelsea, New York (first published in German in 1933).

Kong, C.T.A. [1986]. Multivariate belief functions and graphical models. *Research Report S-107*, Harvard University, Cambridge, Massachusetts.

Koshevoy, G.A. [1998]. Distributive lattices and products of capacities. *J. of Mathematical Analysis and Applications*, 219, pp. 427–441.

Kramosil, I. [2001]. *Probabilistic Analysis of Belief Functions*. Kluwer Academic/Plenum Publishers, New York.

Kramosil, I. [2006]. Continuity and completeness of lattice-valued possibilistic measures. *Intern. J. of General Systems*, 35(5), pp. 555–574.

Krätschmer, V. [2003a]. Coherent lower previsions and Choquet integrals. *Fuzzy Sets and Systems*, 138(3), pp. 469–484.

Krätschmer, V. [2003b]. When fuzzy measures are upper envelopes of probability measures. *Fuzzy Sets and Systems*, 138(3), pp. 455–468.

Kriegler, E. and Held, H. [2005]. Utilizing belief functions for the estimation of future climate change. *Intern. J. of Approximate Reasoning*, 39(2–3), pp. 185–209.

Kruse, R. [1980]. *Zur Konstruktion von unscharfen, λ-additiven Massen*. Ph.D. Dissertation, Technischen Universität Carolo-Wilhelmina zu Braunschweig.

Kruse, R. [1982a]. A note on λ-additive fuzzy measures. *Fuzzy Sets and Systems*, 8(2), pp. 219–222.

Kruse, R. [1982b]. On the construction of fuzzy measures. *Fuzzy Sets and Systems*, 8(3), pp. 23–327.

Kruse, R. [1983]. Fuzzy integrals and conditional fuzzy measures. *Fuzzy Sets and Systems*, 10(3), pp. 309–313.

Kruse, R., Gebhardt, J. and Klawonn, F. [1994]. *Foundations of Fuzzy Systems*. John Wiley, Chichester, UK.

Kyburg, H.E. [1987]. Bayesian and non-Bayesian evidential updating. *Artificial Intelligence*, 31, pp. 271–293.

Kyburg, H.E. and Pittarelli, M. [1996]. Set-based Bayesianism. *IEEE Trans. on Systems, Man, and Cybernetics*, Part A, 26(3), pp. 324–339.

Lamata, M.T. and Moral, S. [1989]. Classification of fuzzy measures. *Fuzzy Sets and Systems*, 33(2), pp. 243–253.

Lebesgue, H. [1966]. *Measure and the Integral*. Holden-Day, San Francisco.

Leszczynski, K., Penczek, P. and Grochulski, W. [1985]. Sugeno's fuzzy measure and fuzzy clustering. *Fuzzy Sets and Systems*, 15(2), pp. 147–158.

Leung, K.S. and Wang, Z. [1998]. A new nonlinear integral used for information fusion. *Proc. of FUZZ-IEEE '98*, Anchorage, pp. 802–807.

Leung, K.S., Wong, M.L., Lam, W., Wang, Z. and Xu, K. [2002]. Learning nonlinear multi-regression networks based on evolutionary computation. *IEEE Trans. on Systems, Man, and Cybernetics*, 32(5), pp. 630–644.

Li, J. and Yasuda, M. [2005]. On Egoroff's theorems on finite monotone non-additive measure space. *Fuzzy Sets and Systems*, 153(1), pp. 71–78.

Li, J., Yasuda, M., Jiang, Q., Suzuki, H., Wang, Z. and Klir, G.J. [1997]. Convergence of sequence of measurable functions on fuzzy measure spaces. *Fuzzy Sets and Systems*, 87(3), pp. 317–323.

Li, S. [1987]. *S*-measure and its extension. *Journal of Hebei University*, 4, pp. 82–89.

Liau, C.J. and Lin, B. IP. [1993]. Proof methods for reasoning about possibility and necessity. *Intern. J. of Approximate Reasoning*, 9(4), pp. 327–364.

Lie, Z.Q. et al. [2001]. Dynamic image sequence analysis using fuzzy measures. *IEEE Trans. on Systems, Man, and Cybernetics*, Part B: Cybernetics, 31(4), pp. 557–572.

Liginlal, D., Ram, S., and Duckstein, L. [2006]. Fuzzy measure theoretical approach to screening product innovations. *IEEE Trans. of Systems, Man, and Cybernetics,* Part A, pp. 577–591.

Marinacci, M. [1999]. Limit laws for non-additive probabilities and their frequentist interpretation. *J. of Economics Theory*, 84, pp. 145–195.

Matheron, G. [1975]. *Random Sets and Integral Geometry*. John Wiley, New York.

Mesiar, R. [1995]. Choquet-like integrals. *J. of Mathematical Analysis and Applications*, 194, pp. 477–488.

Mesiar, R. and Rybárik, J. [1995]. Pan-operations structure. *Fuzzy Sets and Systems*, 74(3), pp. 365–369.

Mesiar, R. and Šipoš, J. [1994]. A theory of fuzzy measures: Integration and its additivity. *Intern. J. of General Systems*, 23(1), pp. 49–57.

Miranda, E., Couso, I. and Gil, P. [2003]. Extreme points of credal sets generated by 2-alternating capacities. *Intern. J. of Approximate Reasoning*, 33(1), pp. 95–115.

Miranda, E. and Grabisch, M. [2004]. p-symmetric bi-capacities. *Kybernetika*, 40(4), pp. 421–440.

Mohamed, M.A. and Xiao, W. [2003]. Q-measures: An efficient extension of the Sugeno λ-measures. *IEEE Trans. on Fuzzy Systems*, 11(3), pp. 419–426.

Molchanov, I. [2005]. *Theory of Random Sets*. Springer, New York.

Monney, P.A. and Chan, M. [2007]. Modelling dependence in Dempster–Shafer theory. *Intern. J. of Uncertainty, Fuzziness, and Knowledge-Based Systems*, 15(1), pp. 93–114.

Morgan, F. [1988]. *Geometric Measure Theory: A Beginner's Guide*. Academic Press, London.

Murofushi, T. [2003a]. A note on upper and lower Sugeno integrals. *Fuzzy Sets and Systems*, 138(3), pp. 551–558.

Murofushi, T. [2003b]. Duality and ordinality in fuzzy measure theory. *Fuzzy Sets and Systems*, 138(3), pp. 523–535.

Murofushi, T. and Sugeno, M. [1989]. An interpretation of fuzzy measure and the Choquet integral as an integral with respect to a fuzzy measure. *Fuzzy Sets and Systems*, 29(2), pp. 201–227.

Murofushi, T. and Sugeno, M. [1991a]. A theory of fuzzy measures: Representations, the Choquet integral, and null set. *J. of Mathematical Analysis and Applications*, 159, pp. 532–549.

Murofushi, T. and Sugeno, M. [1991b]. Fuzzy *t*-conorm integral with respect to fuzzy measures: Generalization of Sugeno integral and Choquet integral. *Fuzzy Sets and Systems*, 42(1), pp. 57–71.

Murofushi, T. and Sugeno, M. [1993]. Some quantities represented by the Choquet integral. *Fuzzy Sets and Systems*, 56(2), pp. 229–235.

Murofushi, T. and Sugeno, M. [2000]. Fuzzy measures and fuzzy integrals. In: M. Grabisch et al. (eds.), *Fuzzy Measures and Integrals: Theory and Applications*, Springer-Verlag, New York, pp.3–41.

Murofushi, T., Sugeno, M. and Machida, M. [1994]. Non-monotonic fuzzy measures and the Choquet integral. *Fuzzy Sets and Systems*, 64(1), pp. 73–86.

Narukawa, Y. and Murofushi, T. [2006]. Representation of Choquet integral — interpreter and Möbius transform. *Intern. J. of Uncertainty, Fuzziness, and Knowledge-Based Systems*, 14(5), pp. 579–589.

Narukawa, Y. and Torra, V. [2005]. Graphical interpretation of the twofold integral and its generalization. *Intern. J. of Uncertainty, Fuzziness, and Knowledge-Based Systems*, 13(4), pp. 415–424.

Narukawa, Y., Murofushi, T. and Sugeno, M. [2003]. Space of fuzzy measures and convergence. *Fuzzy Sets and Systems*, 138(3), pp. 497–506.

Näther, W. and Wälder, K. [2007]. Applying fuzzy measures for considering interaction effects in root dispersal models. *Fuzzy Sets and Systems*, 158(5), pp. 572–582.

Navara, M. [2005]. Triangular norms and measures of fuzzy sets. In: E.P. Klement and R. Mesiar (eds.), *Logical Algebraic, Analytic, and Probabilistic Aspects of Triangular Norms*. Elsevier, Amsterdam.

Nguyen, H.T. [1978]. On random sets and belief functions. *J. of Mathematical Analysis and Applications*, 65, pp. 531–542.

Nguyen, H.T. [2006]. *An Introduction to Random Sets*. Chapman & Hall / CRC, Boca Raton, FL.

Nguyen, H.T. and Walker, E.A. [1997]. *A First Course in Fuzzy Logic*. CRC Press, Boca Raton, FL.

Onisawa, T., Sugeno, M., Nishiwaki, Y., Kawai, H. and Harima, Y. [1986]. Fuzzy measure analysis of public attitude towards the use of nuclear energy. *Fuzzy Sets and Systems*, 20(3), pp. 259–289.

Owen, G. [1988]. Multilinear extensions of games. In: A.E. Roth (eds), *The Shapely Value: Essays in Honor of Lloyd S. Shapely*. Cambridge University Press, Cambridge, UK, pp. 139–151.

Pan, Y. and Klir, G.J. [1997]. Bayesian inference based on interval-valued prior distributions and likelihoods. *Intern. J. of Intelligent and Fuzzy Systems*, 5(3), pp. 193–203.

Pan, Y. and Yuan, B. [1997]. Bayesian inference of fuzzy probabilities. *Intern. J. of General Systems*, 26(1–2), pp. 73–90.

Pap, E. [1990]. Lebesgue and Saks $\wedge$-decompositions of decomposable measures. *Fuzzy Sets and Systems*, 38(3), pp. 345–353.

Pap, E. [1995]. *Null-Additive Set Functions*. Kluwer, Boston.

Pap, E. [1997a]. Decomposable measures and nonlinear equations. *Fuzzy Sets and Systems*, 92(2), pp. 205–221.

Pap, E. [1997b]. Pseudo-analysis as a mathematical base for soft computing. *Soft Computing*, 1(2), pp. 61–68.

Pap, E. [1999]. Applications of decomposable measures. In: U. Höhle and S.E. Rodabaugh (eds.), *Mathematics of Fuzzy Sets*. Kluwer, Boston and Dordrecht, pp. 675–700.

Pap, E., ed. [2002a]. *Handbook of Measure Theory* (2 volumes). Elsevier, Amsterdam.

Pap, E. [2002b]. Pseudo-additive measures and their applications. In: E. Pap (eds), *Handbook of Measure Theory*, Elsevier, Amsterdam, pp. 1403–1468.

Parthasarathy, K.R. [2005]. *Introduction to Probability and Measure*. Hindustan Book Agency, New Delhi, India.

Pedrycz, W. and Gomide, F. [1998]. *An Introduction to Fuzzy Sets: Analysis and Design*. MIT Press, Cambridge, MA.

Petersen, J. [1983]. *Ergodic Theory*. Cambridge University Press, Cambridge and New York.

Pitt, H.R. [1985]. *Measure and Integration in Use*. Clarendon Press, Oxford.

Pollard, D. [2002]. *A User's Guide to Measure Theoretic Probability*. Cambridge University Press, Cambridge, UK.

Prade, H. and Testemale, C. [1987]. Applications of possibility and necessity measures to documentary information retrieval. In: B. Bouchon and R.R. Yager (eds.), *Uncertainty in Knowledge-Based Systems*, Springer-Verlag, Berlin and Heidelberg, pp. 265–274.

Puri, M.L. and Ralescu, D. [1982]. A possibility measure is not a fuzzy measure. *Fuzzy Sets and Systems*, 7(3), pp. 311–313.

Qiao, Z. [1989]. On the extension of possibility measures. *Fuzzy Sets and Systems*, 32(3), pp. 315–320.

Qiao, Z. [1990]. On fuzzy measure and fuzzy integral on fuzzy set. *Fuzzy Sets and Systems*, 37(1), pp. 77–92.

Qiao, Z. [1991]. Fuzzy integrals on L-fuzzy sets. *Fuzzy Sets and Systems*, 38(1), pp. 61–79.

Ralescu, D. [1986]. Radon-Nikodym theorem for fuzzy set-valued measures. In: A. Jones et al. (eds.), *Fuzzy Set Theory and Applications*, D. Reidel, Dordrecht, pp. 39–50.

Ralescu, D. and Adams, G. [1980]. The fuzzy integral. *J. of Mathematical Analysis and Applications*, 75(2), pp. 562–570.

Reche, F. and Salmerón, A. [2000]. Operational approach to general fuzzy measures. *Intern. J. of Uncertainty, Fuzziness, and Knowledge-Based Systems*, 8(3), pp. 369–382.

Resconi, G., Klir, G.J. and Pessa, A. [1998]. Conceptual foundations of quantum mechanics: The role of evidence theory, quantum sets, and modal logic. *Intern. J. of Modern Physics*, Sec. C., 10(1), pp. 29–62.

Resconi, G., Klir, G.J. and St. Clair, U. [1992]. Hierarchical uncertainty metatheory based upon modal logic. *Intern. J. of General Systems*, 21(1), pp. 23–50.

Resconi, G., Klir, G.J., St. Clair, U. and Harmanec, D. [1993]. On the integration of uncertainty theories. *Intern. J. of Uncertainty, Fuzziness, and Knowledge-Based Systems*, 1(1), pp. 1–18.

Román-Flores, H. and Chalco-Cano, Y. [2007]. Sugeno integral and geometric inequalities. *Intern. J. of Uncertainty, Fuzziness, and Knowledge-Based Systems*, 15(1), pp. 1–11.

Ruspini, E.H., Bonissone, P.P. and Pedrycz, W., eds. [1998]. *Handbook of Fuzzy Computation*. Institute of Physics Publication, Bristol (UK) and Philadelphia.

Saks, S. [1937]. *Theory of the Integral* (Second Revised Edition). Hafner, New York.

Schmeidler, D. [1972]. Cores of exact games. *J. of Mathematical Analysis and Applications*, 40, pp. 214–225.

Schmeidler, D. [1986]. Integral representation without additivity. *Proc. American Mathematical Society*, 97, pp. 255–261.

Schmeidler, D. [1989]. Subjective probability and expected utility without additivity. *Econometrica*, 57(3), pp. 571–587.

Schubert, J. [1994]. *Cluster-Based Specification Techniques in Dempster–Shafer Theory for an Evidential Intelligence Analysis of Multiple Target Tracks*. Royal Institute of Technology, Stockholm.

Sgaro, A. [1997]. Bodies of evidence versus simple interval probabilities. *Intern. J. of Uncertainty, Fuzziness, and Knowledge-Based Systems*, 5(2), pp. 199–209.

Shackle, G.L.S. [1949]. *Expectation in Economics*. Cambridge University Press, Cambridge, UK.

Shackle, G.L.S. [1955]. *Uncertainty in Economics and Other Reflections*. Cambridge University Press, Cambridge, UK.

Shackle, G.L.S. [1961]. *Decision, Order and Time in Human Affairs*. Cambridge University Press, Cambridge, UK.

Shafer, G. [1976]. *A Mathematical Theory of Evidence*. Princeton University Press, Princeton, New Jersey.

Shafer, G. [1978]. Non-additive probabilities in the works of Bernoulli and Lambert. *Archive for History of Exact Sciences*, 19, pp. 309–370.

Shafer, G. [1979]. The allocation of probability. *The Annals of Probability*, 7(5), pp. 827–839.

Shafer, G. [1981]. Constructive probability. *Synthese*, 48, pp. 1–60.

Shafer, G. [1982]. Belief functions and parametric models. *J. of Royal Statistical Society*, B44, pp. 322–352.

Shafer, G. [1987]. Belief function and possibility measures. In: J. Bezdek (eds), *Analysis of Fuzzy Information*, CRC Press, Boca Raton, FL, pp. 51–83.

Shafer, G. [1990]. Perspectives on the theory and practice of belief functions. *Intern. J. of Approximate Reasoning*, 4(5), pp. 323–362.

Shafer, G. and Vovk, V. [2006]. The sources of Kolmogorov's Grundbegriffe. *Statistical Science*, 21(1), pp. 70–98.

Shapley, L. S. [1953]. A value for n-person games. In: H.W. Kuhnand and A.W. Tucker (eds.), *Contributions to the Theory of Games*, Princeton University Press, Princeton, pp. 307–317.

Shapley, L. S. [1971]. Core of convex games. *Intern. J. of Game Theory*, 1(1), pp. 11–26.
Sims, J. R. and Wang, Z. [1990]. Fuzzy measures and fuzzy integrals: An overview. *Intern. J. of General Systems*, 17(2–3), pp. 157–189.
Šipoš, J. [1979a]. Integral with respect to a pre-measure. *Mathematica Slovaca*, 29, pp. 141–155.
Šipoš, J. [1979b]. Non-linear integrals. *Mathematica Slovaca*, 29, pp. 257–270.
Skulj, D. [2005]. Products of capacities on separate spaces through additive measures. *Fuzzy Sets and Systems*, 152(2), pp. 289–301.
Smets, P. [1981]. The degree of belief in a fuzzy event. *Information Sciences*, 25, pp. 1–19.
Smets, P. [1988]. Belief functions. In: P. Smets et al. (eds.), *Non-Standard Logics in Automated Reasoning*. Academic Press, San Diego, pp. 253–286.
Smets, P. [1990]. The combination of evidence in the transferable belief model. *IEEE Trans. on Pattern Analysis and Machine Intelligence*, 12, pp. 447–458.
Smets, P. [1992]. Resolving misunderstanding about belief functions. *Intern. J. of Approximate Reasoning*, 6(3), pp. 321–344.
Smets, P. [1998]. The transferable belief model for quantified belief representation. In: Gabbay and Smets (eds.), *Handbook of Defeasible Reasoning and Uncertainty Management Systems*, Vol. 1, Kluwer, Boston, pp. 267–301.
Smets, P. [2002]. The application of the matrix calculus to belief functions. *Intern. J. of Approximate Reasoning*, 31(1–2), pp. 1–30.
Smets, P. and Kennes, R. [1994]. The transferable belief model. *Artificial Intelligence*, 66, pp. 191–234.
Soria-Frisch, A. [2006]. Unsupervised construction of fuzzy measures through self-organizing feature maps and its application in color image segmentation. *Intern. J. of Approximate Reasoning*, 41(1), pp. 23–42.
Squillante, M. and Ventre, A. G. S. [1989]. Representations of the fuzzy integral. *Fuzzy Sets and Systems*, 29(2), pp. 165–170.
Stanley, R. J., Keller, J. M., Caldwell, C. W., and Gader, P. D. [2001]. Abnormal cell detection using Choquet integral. *Proc. of IFSA/NAFIPS '01Conference*, pp. 1134–1139.
Strat, T.M. [1990]. Decision analysis using belief functions. *Intern. J. of Approximate Reasoning*, 4(5), pp. 391–417.
Strat, T.M. and Lowrance, J. D. [1989]. Explaining evidential analysis. *Intern. J. of Approximate Reasoning*, 3(4), pp. 299–353.
Struk, P. [2006]. Extremal fuzzy integrals. *Soft Computing*, 10(6), pp. 502–505.
Suárez, F. and Gill, P. [1986]. Two families of fuzzy integrals. *Fuzzy Sets and Systems*, 18(1), pp. 67–81.
Sugeno, M. [1974]. *Theory of Fuzzy Integrals and its Applications*. Ph.D. dissertation, Tokyo Institute of Technology.
Sugeno, M. [1977]. Fuzzy measures and fuzzy integrals: A survey. In: M.M. Gupta, G.N. Saridis, and B.R. Gaines (eds.), *Fuzzy Automata and Decision Processes*, North-Holland, New York, pp. 89–102.
Sugeno, M. and Murofushi, T. [1987]. Pseudo-additive measures and integrals. *J. of Mathematical Analysis and Applications*, 122, pp. 197–222.
Sun, Q. [1992]. On the pseudo-autocontinuity of fuzzy measures. *Fuzzy Sets and Systems*, 45(1), pp. 59–68.
Sun, Q. and Wang, Z. [1988]. On the autocontinuity of fuzzy measures. In: R. Trappl (eds), *Cybernetics and Systems '88*, Kluwer, Boston, pp. 717–721.
Suzuki, H. [1988]. On fuzzy measures defined by fuzzy integrals. *J. of Mathematical Analysis and Applications*, 132, pp. 87–101.
Suzuki, H. [1991]. Atoms of fuzzy measures and fuzzy integrals. *Fuzzy Sets and Systems*, 41(3), pp. 329–342.
Tahani, H. and Keller, J.M. [1990]. Information fusion in computer vision using the fuzzy integral. *IEEE Trans. on Systems, Man and Cybernetics*, 20, pp. 733–741.

Tanaka, H. and Hayashi, I. [1989]. Possibilistic linear regression analysis for fuzzy data. *European J. of Operations Research*, 40, pp. 389–396.

Tanaka, H. and Sugeno, M. [1991]. A study of subjective evaluation of printed color images. *Intern. J. of Approximate Reasoning*, 5(3), pp. 213–222.

Tanaka, H., Sugihara, K. and Maeda, Y. [2004]. Non-additive measures by interval probability functions. *Information Sciences*, 164, pp. 209–227.

Tanaka, K. and Klir, G.J. [1999]. A design condition for incorporating human judgment into monitoring systems. *Reliability Engineering and System Safety*, 65, pp. 251–258.

Temple, G. [1971]. *The Structure of Lebesgue Integration Theory*. Oxford University Press, London.

Terán, P. [2007]. Probabilistic foundations for measurement modeling with fuzzy random variables. *Fuzzy Sets and Systems*, 158(9), pp. 973–986.

Torra, V. and Narukawa, Y. [2006]. The interpretation of fuzzy integrals and their application to fuzzy systems. *Intern. J. of Approximate Reasoning*, 41(1), pp. 43–58.

Troffaes, M.C.M. [2007]. Decisionmaking under uncertainty using imprecise probabilities. *Intern. J. of Approximate Reasoning*, 45(1), pp. 17–29.

Tsiporkova, E., Boeva, V. and De Baets, B. [1999]. Evidence measures induced by Kripke's accessibility relations. *Intern. J. of Uncertainty, Fuzziness, and Knowledge-Based Systems*, 7(6), pp. 589–613.

Vicig, P. [2000]. Epistemic independence for imprecise probabilities. *Intern. J. of Approximate Reasoning*, 24(2–3), pp. 235–250.

Viertl, R. [1987]. Is it necessary to develop a fuzzy Bayesian inference? In: R. Viertl (eds), *Probability and Bayesian Statistics*, Plenum Press, New York, pp. 471–475.

Viertl, R. [1996]. *Statistical Methods for Non-Precise Data*. CRC Press, Boca Raton, Florida.

Vitali, G. [1997]. On the definition of integral of functions of one variable. *Rivista di matematica per le scienze economiche e sociali*, 20(2), pp. 159–168. [Originally published in Italian in 1925.]

Wagner, C.G. [1989]. Consensus for belief functions and related uncertainty measures. *Theory and Decision*, 26, pp. 295–304.

Wakker, P. [1990]. A behavioral foundation for fuzzy measures. *Fuzzy Sets and Systems*, 37(3), pp. 327–350.

Walley, P. [1987]. Belief function representations of statistical evience. *Annals of Statistics*, 15, pp. 1439–1456.

Walley, P. [1991]. *Statistical Reasoning with Imprecise Probabilities*. Chapman and Hall, London.

Walley, P. [1996]. Measures of uncertainty in expert systems. *Artificial Intelligence*, 83, pp. 1–58.

Walley, P. [2000]. Towards a unified theory of imprecise probability. *Intern. J. of Approximate Reasoning*, 24(2–3), pp. 125–148.

Walley, P. and De Cooman, G. [1999]. Coherence of rules for defining conditional probability. *Intern. J. of Approximate Reasoning*, 21(1), pp. 63–107.

Walley, P. and Fine, T.L. [1979]. Varieties of model (classificatory) and comparative probability. *Synthese*, 41, pp. 321–374.

Wallner, A. [2007]. Extreme points of coherent probabilities in finite spaces. *Intern. J. of Approximate Reasoning*, 44(3), pp. 339–357.

Walters, P. [1982]. *An Introduction to Ergodic Theory*. Springer-Verlag, New York.

Wang, J.C. and Chen, T.Y. [2005]. Experimental analysis of λ-fuzzy measure identification by evolutionary algorithms. *Intern. J. of Fuzzy Systems*, 7(1), pp. 1–10.

Wang, J. and Wang, Z. [1997]. Using neural networks to determine Sugeno measures by statistics. *Neural Networks*, 10(1), pp. 183–195.

Wang, R. and Ha, M. [2006]. On Choquet integrals of fuzzy-valued functions. *Journal of Fuzzy Mathematics*, 14(1), pp. 89–102.

Wang, R., Wang, L. and Ha, M. [2006]. Choquet integrals on L-fuzzy sets. *J. of Fuzzy Mathematics*, 14(1), pp. 151–163.

Wang, W. Klir, G.J. and Wang, Z. [1996]. Constructing fuzzy measures by rational transformations. *J. of Fuzzy Mathematics*, 4(3), pp. 665–675.

Wang, W., Wang, Z. and Klir, G.J. [1998b]. Genetic algorithms for determining fuzzy measures from data. *J. of Intelligent and Fuzzy Systems*, 6(2), pp. 171–183.

Wang, Z. [1981]. Une class de mesures floues—les quasi-mesures. *BUSEFAL*, 6, pp. 28–37.

Wang, Z. [1984]. The autocontinuity of set function and the fuzzy integral. *J. of Mathematical Analysis and Applications*, 99, pp. 195–218.

Wang, Z. [1985a]. Asymptotic structural characteristics of fuzzy measure and their applications. *Fuzzy Sets and Systems*, 16(3), pp. 277–290.

Wang, Z. [1985b]. Semi-lattice structure of all extensions of possibility measure and consonant belief function. In: D. Feng and X. Liu (eds.), *Fuzzy Mathematics in Earthquake Researches*, Seismological Press, Beijing, pp. 332–336.

Wang, Z. [1986]. Semi-lattice isomorphism of the extensions of possibility measure and the solutions of fuzzy relation equation. In: R. Trappl (ed.) *Cybernetics and Systems '86*, Kluwer, Boston, pp. 581–583.

Wang, Z. [1987]. Some recent advances on the possibility measure theory. In: B. Bouchon and R.R. Yager (eds.), *Uncertainty and Knowledge-Based Systems*, Springer-Verlag, New York, pp. 173–175.

Wang, Z. [1990a]. Absolute continuity and extension of fuzzy measures. *Fuzzy Sets and Systems*, 36(3), pp. 395–399.

Wang, Z. [1990b]. Structural characteristics of fuzzy measure on *S*-compact spaces. *Intern. J. of General Systems*, 17(4), pp. 309–316.

Wang, Z. [1992]. On the null-additivity and the autocontinuity of fuzzy measure. *Fuzzy Sets and Systems*, 45(2), pp. 223–226.

Wang, Z. [1996]. Constructing nonadditive set functions in systems. *Journal of Hebei University*, 16(3), pp. 44–47.

Wang, Z. [1997]. Convergence theorems for sequences of Choquet integrals. Intern . *J. of General Systems*, 26(1–2), pp. 133–143.

Wang, Z. [2002]. A new model of nonlinear multiregression by projection pursuit based on generalized Choquet integrals. *Proc. of FUZZ-IEEE '02*, pp. 1240–1244.

Wang, Z. [2003]. A new genetic algorithm for nonlinear multiregression based on generalized Choquet integrals. *Proc. of FUZZ-IEEE '03*, pp. 819–821.

Wang, Z. and Klir, G.J. [1992]. *Fuzzy Measure Theory*, Plenum Press, New York.

Wang, Z. and Klir, G.J. [1997a]. Choquet integrals and natural extensions of lower probabilities. *Intern. J. of Approximate Reasoning*, 16(2), pp. 137–147.

Wang, Z. and Klir, G.J. [1997b]. PFB-integrals and PFA-integrals with respect to monotone set functions. *Intern. J. of Uncertainty, Fuzziness, and Knowledge-Based Systems*, 5(2), pp. 163–175.

Wang, Z. and Klir, G.J. [2007]. Coordination uncertainty of belief measures in infomation fusion, *Proc.12th IFSA World Congress*, Cancun, Mexico.

Wang, Z. and Leung, K.S. [2006]. Uncertainty carried by fuzzy measures in aggregation. *Proc. IPMU '2006*, pp. 105–112.

Wang, Z. and Li, S.M. [1990]. Fuzzy linear regression analysis of fuzzy valued variables. *Fuzzy Sets and Systems*, 36(1), pp. 125–136.

Wang, Z. and Qiao, Z. [1990]. Transformation theorems for fuzzy integrals on fuzzy sets. *Fuzzy Sets and Systems*, 34(3), 355–364.

Wang, Z. and Wang, J. [1996]. Using genetic algorithms for λ-measure fitting and extension. *Proc. FUZZ-IEEE '96*, New Orleans, pp. 1871–1874.

Wang, Z. and Xu, K. [1998]. A brief discussion of a new type of nonlinear integrals with respect to nonadditive set functions. In: *Proc. of 1998 Conf. of the Chinese Fuzzy Mathematics and Fuzzy Systems Association*, pp. 95–103.

Wang, Z. et al. [1995a]. The preservation of structural characteristics of monotone set functions defined by fuzzy integral. *J. of Fuzzy Mathematics*, 3(1), pp. 229–240.

Wang, Z. et al. [1995b]. Expressing fuzzy measure by a model of modal logic: A discrete case. In: Z. Bien and K.C. Min (eds.), *Fuzzy Logic and Its Applications to Engineering, Information Sciences, and Intelligent Systems*, Kluwer, Boston, pp. 3–13.

Wang, Z. et al. [1996a]. Fuzzy measures defined by fuzzy integral and their absolute continuity. *J. of Mathematical Analysis and Applications*, 203(1), pp. 150–165.

Wang, Z. et al. [1996b]. Monotone set functions defined by Choquet integral. *Fuzzy Sets and Systems*, 81(2), pp. 241–250.

Wang, Z. et al. [1996c]. Pan-integrals with respect to imprecise probabilities. *Intern. J. of General Systems*, 25(3), pp. 229–243.

Wang, Z. et al. [1998a]. Neural networks used for determining belief measures and plausibility measures. *Intelligent Automation and Soft Computing*, 4(4), pp. 313–324.

Wang, Z. et al. [1999a]. A genetic algorithm for determining nonadditive set functions in information fusion. *Fuzzy Sets and Systems*, 102(3), pp. 463–469.

Wang, Z. et al. [1999b]. Using genetic algorithms to determine nonnegative monotone set functions for information fusion in environments with random perturbation. *Intern. J. of Intelligent Sytems*, 14(10), pp. 949–962.

Wang, Z. et al. [2000a]. A new type of nonlinear integrals and the computational algorithm. *Fuzzy Sets and Systems*, 112(2), pp. 223–231.

Wang, Z. et al. [2000b]. Nonlinear nonnegative multiregression based on Choquet integrals. *Intern. J. of Approximate Reasoning*, 25(2), pp. 71–87.

Wang, Z. et al. [2000c]. Determining nonnegative monotone set functions based on Sugeno's integral: An application of genetic algorithms. *Fuzzy Sets and Systems*, 112(1), pp. 155–164.

Wang, Z. et al. [2003]. Interdeterminate integrals with respect to nonadditive measures. *Fuzzy Sets and Systems*, 138(3), pp. 485–495.

Wang, Z. et al. [2005]. Applying fuzzy measures and nonlinear integrals in data mining. *Fuzzy Sets and Systems*, 156(3), pp. 371–380.

Wang, Z. et al. [2006a]. Integration on finite sets. *Intern. J. of Intelligent Systems*, 21(10), pp. 1073–1092.

Wang, Z. et al. [2006b]. Real-valued Choquet integrals with fuzzy-valued integrand. *Fuzzy Sets and Systems*, 157(2), pp. 256–269.

Wang, Z. et al. [2008a]. Lower integrals and upper integrals with respect to nonadditive set functions. *Fuzzy Sets and Systems* 159(6), pp. 646–660.

Wang, Z. et al. [2008b]. The Choquet integral with respect to fuzzy-valued signed efficiency measures. *Proc. of WCCI 2008*. pp. 2143–2148.

Wasserman, L.A. and Kadane, J. [1990]. Bayes' theorem for Choquet capacities. *Annals of Statistics*, 18(3), pp. 1328–1339.

Wasserman, L.A. and Kadane, J. [1992]. Symmetric upper probabilities. *Annals of Statistics*, 20(4), pp. 1720–1736.

Weber, S. [1984]. $\perp$-decomposable measures and integrals for Archimedean *t*-conorms. *J. of Mathematical Analysis and Applications*, 101, pp. 114–138.

Weber, S. [1986]. Two integrals and some modified versions – critical remarks. *Fuzzy Sets and Systems*, 20(1), pp. 97–105.

Weber, S. [1991]. Conditional measures and their applications to fuzzy sets. *Fuzzy Sets and Systems*, 42(1), pp. 73–85.

Weichselberger, K. [2000]. The theory of interval-probability as a unifying concept for uncertainty. *Intern. J. of Approximate Reasoning*, 24(2–3), pp. 149–170.

Weichselberger, K. and Pöhlmann, S. [1990]. *A Methodology for Uncertainty in Knowledge-Based Systems*, Springer-Verlag, New York.

Weir, A.J. [1973]. *Lebesgue Integration and Measure*. Cambridge University Press, New York.

Wheeden, R.L. and Zygmund, A. [1977]. *Measure and Integral: An Introduction to Real Analysis*, Marcel Dekker, New York.

Wierzchon, S.T. [1982]. On fuzzy measure and fuzzy integral. In: M.M. Gupta and E. Sanchez (eds.), *Fuzzy Information and Decision Processes*, North-Holland, New York, pp. 79–86.

Wierzchon, S.T. [1983]. An algorithm for identification of fuzzy measure. *Fuzzy Sets and Systems*, 9(1), pp. 69–78.

Wilson, N. [2000]. Algorithms for Dempster–Shafer theory. In: J. Kohlas and S. Moral (eds.), *Algorithms for Uncertainty and Defeasible Reasoning*, Kluwer, Dordrecht, and Boston, pp. 421–475.

Wolf, R.G. [1977]. *Obere und untere Wahrsteinlichkeiten*. Ph.D. Dissertation, Eidgen, Technische Hochschule, Zurich.

Wolfenson, M. and Fine, T.L. [1982]. Bayes-like decision making with upper and lower probabilities. *J. of American Statistical Association.*, 77, pp. 80–88.

Wolkenhauer, O. [1998]. *Possibility Theory with Applications to Data Analysis*. Research Studies Press, Taunton, UK.

Wong, S.K.M., Wang, L.S. and Yao, Y.Y. [1995]. On modeling uncertainty with interval structures. *Computational Intelligence*, 11(2), pp. 406–426.

Wong, S.K.M., Yao; Y.Y. and Bollmann, P. [1992]. Characterization of comparative belief structures. *Intern. J. of Man–Machine Studies*, 37, pp. 123–133.

Wu, C. and Ma, M. [1989]. Some properties of fuzzy integrable function space $L^1(\mu)$. *Fuzzy Sets and Systems*, 31(3), pp. 397–400.

Wu, C. and Traore, M. [2003]. An extension of Sugeno integral, *Fuzzy Sets and Systems*, 138(3), pp. 537–550.

Xu, K., Wang, Z. and Ke, Y. [2000]. A fast algorithm for Choquet-integral-based nonlinear multiregression used in data mining. *J. of Fuzzy Mathematics*, 8(1), pp. 195–201.

Xu, K., Wang, Z., Heng, P.A. and Leung, K.S. [2001a]. Using generalized Choquet integrals in projection pursuit based classification. *Proc. IFSA / NAFIPS'01*, pp. 506–511.

Xu, K., Wang, Z., Heng, P.A. and Leung, K.S. [2003]. Classification by nonlinear integral projections. *IEEE Trans. on Fuzzy Systems*, 11(2), pp. 187–201.

Xu, K., Wang, Z., Wong, M.L., and Leung, K.S. [2001b]. Discover dependency pattern among attributes by using a new type of nonlinear multiregression. *Intern. J. of Intelligent Systems*, 16(8), pp. 949–962.

Yager, R.R., ed. [1982]. Fuzzy Set and Possibility Theory. Pergamon Press, Oxford, UK.

Yager, R.R., Fedrizzi, M. and Kacprzyk, J., eds. [1994]. *Advances in the Dempster–Shafer Theory of Evidence*. John Wiley, New York.

Yager, R.R. and Kreinovich, V. [1999]. Decision making under interval probabilities. *Intern. J. of Approximate Reasoning*, 22(3), pp. 195–215.

Yager, R.R., Ovchinnikov, S., Tong, R.M., and Nguyen, H.T., eds. [1987]. *Fuzzy Sets and Applications: Selected Papers by L. A. Zadeh*. Wiley-Interscience, New York.

Yan, B. and Keller, J.M. [1991]. Conditional fuzzy measures and image segmentation. *Proc. NAFIPS'91*, Columbia, Missouri, pp. 32–36.

Yang, Q. [1985]. The pan-integral on the fuzzy measure space. *Fuzzy Mathematics*, 3, pp. 107–114 (in Chinese).

Yang, Q. and Song, R. [1985]. A further discussion on the pan-integral. *Fuzzy Mathematics*, 4, pp. 27–36 (in Chinese).

Yang, R., Wang, Z., Heng, P.A. and Leung, K.S. [2005]. Fuzzy numbers and fuzzification of the Choquet integral. *Fuzzy Sets and Systems*, 153(1), pp. 95–113.

Yang, R., Wang, Z., Heng, P.A. and Leung, K.S. [2007]. Classification of heterogeneous fuzzy data by Choquet integral with fuzzy-valued integrand. *IEEE Trans. on Fuzzy Systems*, in production.

Yang, R. Wang, Z., Heng, P.A. and Leung, K. S. [2008]. Fuzzified Choquet integral with fuzzy-valued integrand and its application on temperature prediction, *IEEE T. SMCB* 38(2), pp. 367–380.

Yen, J. [1989]. GERTIS: A Dempster–Shafer approach to diagnosing hierarchical hypotheses. *ACM Communications*, 32, pp. 573–585.

Yen, J. [1990]. Generalising the Dempster–Shafer theory to fuzzy sets. *IEEE Trans. on Systems, Man, and Cybernetics*, 20, pp. 559–570.

Yuan, B. and Klir, G.J. [1996]. Constructing fuzzy measures: A new method and its application to cluster analysis. *Proc. NAFIPS '96*, Berkeley, CA, pp. 567–571.

Yue, S., Li, P. and Yin, Z. [2005]. Parameter estimation for Choquet fuzzy integral based on Takagi-Sugeno fuzzy model. *Information Fusion*, 6, pp. 175–182.

Zadeh, L.A. [1965]. Fuzzy sets. *Information and Control*, 8, pp. 338–353.

Zadeh, L.A. [1968]. Probability measures of fuzzy events. *J. of Mathematical Analysis and Applications*, 23, pp. 421–427.

Zadeh, L.A. [1978]. Fuzzy sets as a basis for a theory of possibility. *Fuzzy Sets and Systems*, 1(1), pp. 3–28.

Zadeh, L.A. [1979]. On the validity of Dempster's rule of combination of evidence. *Memo UCB/ERL No. 97/24*, University of California, Berkeley.

Zadeh, L.A. [1981]. Possibility theory and soft data analysis. In: L. Cobb and R.M. Thrall (eds.), *Mathematical Frontiers of the Social and Policy Sciences*, Westview Press, Boulder, Colorado, pp. 69–129.

Zadeh, L.A. [1986]. A simple view of the Dempster rule. *The Artificial Intelligence Magazine*, VII(2), pp. 85–90.

Zadeh, L.A. [2002]. Toward a perception-based theory of probabilistic reasoning with imprecise probabilities. *J. of Statistical Planning and Inference*, 105, pp. 233–264.

Zadeh, L.A. [2005]. Toward a generalized theory of uncertainty (GTU) – an outline. *Information Sciences*, 172(1–2), pp. 1–40.

Zhang, D. and Guo, C. [1996]. Integrals of set-valued functions for ⊥– decomposable measures. *Fuzzy Sets and Systems*, 78(3), pp. 341–346.

Zhang, G. [1992a]. On fuzzy number-valued fuzzy measures defined by fuzzy number-valued integral—I. *Fuzzy Sets and Systems*, 45(2), pp. 227–237.

Zhang, G. [1992b]. On fuzzy number-valued fuzzy measures defined by fuzzy number-valued integral—II. *Fuzzy Sets and Systems*, 48(2), pp. 257–265.

Zhang, W.X. et al. [1990]. Set-valued measure and fuzzy set-valued measure. *Fuzzy Sets and Systems*, 36(1), pp. 181–188.

Zimmermann, H.J. [1996]. *Fuzzy Set Theory – and Its Applications* (Third Edition). Kluwer, Boston.

Subject Index

Name Index